REGRESSION ANALYSIS:
Statistical Modeling of
a Response Variable

REGRESSION ANALYSIS:
Statistical Modeling of a Response Variable

Rudolf J. Freund ■ **William J. Wilson**

Department of Statistics
Texas A & M University

Department of Mathematics and Statistics
University of North Florida

ACADEMIC PRESS

San Diego • London • Boston
New York • Sydney • Tokyo • Toronto

ACADEMIC PRESS
525 B Street, Suite 1900, San Diego, CA 92101-4495, USA
1300 Boylston Street, Chestnut Hill, MA 02167, USA
http://www.apnet.com

ACADEMIC PRESS LIMITED
24–28 Oval Road, London NW1 7DX, UK
http://www.hbuk.co.uk/ap/

Library of Congress Cataloging-in-Publication Data
Freund, Rudolf Jakob, 1927–
      Regression analysis : statistical modeling of a response variable
   / Rudolf J. Freund, William J. Wilson.
         p.   cm.
      Includes bibliographical references and index.
      ISBN 0-12-267475-8 (alk. paper)
      1. Regression analysis.   I. Wilson, William J., 1940–
   II. Title.
   QA278.2.F698   1998   97-48544
   519.5 36—dc21   CIP

Printed in the United States of America

97 98 99 00 01 02 IC 9 8 7 6 5 4 3 2 1

TABLE OF CONTENTS

2   Simple Linear Regression: Linear Regression with One Independent Variable 35

3   Multiple Regression 75

Part II Problems and Remedies

4   Problems with Observations

5   Multicollinearity   181

8   Intrinsically Linear and Nonlinear Models 303

9   The General Linear Model 339

PREFACE

The purpose of this book is to provide the tools necessary for using the modeling approach for the intelligent statistical analysis of a response variable. Although there is strong emphasis on regression analysis, there is coverage of other linear models such as the analysis of variance, the analysis of covariance, and the analysis of a binary response variable, as well as an introduction to nonlinear regression.

The common theme is that we have observed sample or experimental data on a response variable and want to perform a statistical analysis to explain the behavior of that variable. The analysis is based on the proposition that the behavior of the variable can be explained by a model that (usually) takes the form of an algebraic equation that involves other variables that describe experimental conditions, parameters that describe how these conditions affect the response variable, and a catchall expression, called the error, which simply says that the model does not completely explain the behavior of the response variable. The statistical analysis includes the estimation of the parameters, inferences (hypothesis tests and confidence intervals), and assessing the nature (magnitude) of the error. In addition there must be investigations of things that may have gone wrong: errors in the data, wrong choice of model, and other violations of assumptions underlying the inference procedures.

Data for such analyses can arise from experiments, sample surveys, observations of processes (operational data) or aggregated or secondary data. In all cases, but especially when operational and secondary data are used, the statistical analysis requires more than plugging numbers into formulas or running data sets through a computer program. Typically, an analysis will consist of an often poorly defined sequence of steps, which include problem definition, model formulation, data screening, selecting appropriate computer programs, proper interpretation of computer output, diagnosing results for data anomalies and model inadequacies, and interpreting the results and making recommendations within the framework of the purpose for which the data were collected.

Note that none of these steps includes plugging numbers into formulas. That is because this aspect of statistical analysis is performed by computers. Therefore, all presentations assume that computations will have been performed by computers, and we can therefore concentrate on all the other aspects of a proper analysis. This means that there will not be many formulas, and those that are presented are used to indicate how the computer performs the analysis and occasionally to show the rationale for some analysis procedures.

Because of the interdependence of the various topics in this book, the sequencing of materials is somewhat difficult. We have chosen the following sequencing:

(1) A review of prerequisites. After a brief introduction and review of terminology, the basic statistical methods are reviewed in the context of the linear model. These methods include one- and two-sample analysis of means and the analysis of variance.

(2) A thorough review of simple linear regression. This section is largely formula based, since the formulas are simple, have practical interpretations, and provide principles that have implications for multiple regression.

(3) A thorough coverage of multiple regression, assuming that the model is correct and there are no data anomalies. This section also includes formulas and uses matrices, for which a brief introduction is provided in the appendix. However, the greater emphasis is placed on model formulation, interpretation of results with special emphasis on the derivation and interpretation of *partial* coefficients, inferences on parameters using full and reduced or restricted models, and the relationships among the various statistics describing the fit of the model.

(4) Methods for identifying what can go wrong with either data or the model. This section shows how to diagnose potential problems, and what remedial methods may help. We begin with row diagnostics (outliers and problems with the assumptions on the error) and continue with column diagnostics (multicollinearity). Emphasis here is on both descriptive and inferential tools and includes warnings on the use of standard inferential methods in exploratory analyses. Although there is thorough coverage of variable selection procedures, considerable emphasis is given to on alternatives to variable selection as a remedy to multicollinearity. Also included is discussion of the interplay between row and column problems.

(5) Presentation of models that do not describe straight lines. This includes models that can be analyzed by adaptations of linear models such as polynomial models, log linear models, dichotomous dependent and independent variables, and the "general linear model" used for unbalanced data and the analysis of covariance. The analysis of strictly nonlinear models is also presented.

As noted, there is considerable interplay among the methodologies covered in the different sections. Therefore, it is only fair to warn the reader: The material in this book is not completely useful unless (almost) all of it has been covered!

Examples

Good examples are, of course, of utmost importance in a book on regression. Such examples should do the following:

Be understandable by students from all disciplines
Have a reasonable number of variables and observations
Have some interesting features

The examples in this book are largely "real", and thus usually have some interesting features. A large fraction do have agricultural origins, not only because they are more readily available, but also because many of these are more easily understood by a wide variety of audiences. Also, in order to be understandable and interesting, data may have been modified, abbreviated, or redefined. Occasionally, example data may be artificially generated. In general, most examples and exercises are unsuitable for manual calculation. We assume that, especially in courses designed for special audiences, additional examples will be supplied by the instructor or by students in the form of term projects.

In order to maintain consistency, most examples are illustrated with output from the SAS System, although a few examples of other output are provided for comparison and to make the point that most computer output gives almost identical information. Computer output is occasionally abbreviated to save space and avoid confusion. However, this book is intended to be usable with any computer package since all discussion of computer usage is generic and software-specific instruction is left to the instructor.

Exercises

Exercises are a very important part of learning about statistical methods. However, because of the computer, the purpose of exercises has been drastically altered. No longer do students need to plug numbers into formulas and insure numerical accuracy, and when that has been achieved, go to the next exercise. Instead, because numerical accuracy is essentially guaranteed, the emphasis now is on choosing the appropriate computer programs and subsequently using these programs to obtain the desired results. Also important is to properly interpret the results of these analyses to determine if additional analyses are needed. Finally, students now have the opportunity to study the results and consequently discuss the usefulness of the results. Because students' performance on exercises is related to proper usage and interpretation, it will probably take students rather long to do exercises, especially the rather open-ended ones in Chapter 4 and beyond.

Because proper use of computer programs is not a trivial aspect of an exercise, we strongly urge that instructors formally require students to do the examples, and

for that reason we have included the example data sets in the data diskette. Not only will this give students more confidence when they embark on the exercises but their conclusions may not match the ones we present!

We have included a modest set of exercises. Many exercises, especially in the later chapters, often have no universally correct answer: hence, the choice of methods and associated computer programs is of prime importance. For this reason, we chose to give only limited or sometimes no guidance as to the appropriate analysis. Finally, we expect that both the instructor and students will supply exercises that are challenging and of interest to the variety of students that are usually found in such a course.

Prerequisites for this course are two semesters of statistical methods, although one concentrated course may be adequate. Calculus is not required: however, if this tool is available, the estimation using calculus are available in Appendix C. A minimal exposure to matrices is provided in Appendix B.

Data Sets

Virtually all data sets for both examples and exercises are available on the data diskette. A README file provides the nomenclature for the files. SAS code is provided for the simulation examples, but will most likely give different answers due to the random number generating process.

Acknowledgments

First of all, we give our employers, The Department of Statistics of Texas A&M University, and the University of North Florida, without whose cooperation and encouragement the book could never have been completed. We also owe a debt of gratitude to the following reviewers whose comments have made this a much more readable work.

Professor Patricia Buchanan
Department of Statistics
Pennsylvania State University
University Park, PA 16802

Professor Robert Gould
Department of Statistics
University of California
  at Los Angeles
Los Angeles, CA 90024

Jack Reeves
Statistics Department
University of Georgia
Athens, GA 330605

Stephen Garren
Division of Statistics
University of Virginia
Charlottesville, VA 22903

Professor E. D. McCune
Department of Statistics
Stephen F. Austin University
Nacogdoches, TX 75962

Dr. Arvind K. Shah
Department of Mathematics
  and Statistics
University of South Alabama
Mobile, AL 36688

Professor James Schott
Statistics Department
University of Central Florida
Orlando, FL 32806

We also express our appreciation for the encouragement and guidance provided by Charles B. Glaser, Associate Editorial Director, Robert Ross, Acquisitions Editor, and Vanessa Gerhard, Production Editor at Academic Press, whose expertise have made this a much more readable book.

We acknowledge SAS Institute whose software (The SAS System) is used to illustrate computer output for virtually all examples. The SAS System was also used to produce the tables of the Normal, t, χ^2, and F distributions.

Finally we owe an undying gratitude to our wives, Marge and Marilyn, who have encouraged our continuing this project despite the often encountered frustrations.

AN OVERVIEW

This book is divided into three parts:

Part I, consisting of the first three chapters, starts with a review of elementary statistical methods recast as applications of linear models and continues with the methodology of simple linear and multiple regression analyses. All presentations include the methods of statistical inference necessary to evaluate the models.

Part II, consisting of Chapters 4 through 6, contains comprehensive discussions of the many practical problems most often encountered in regression analyses and presents some suggested remedies.

Part III, consisting of Chapters 7 through 10, contains presentations of additional uses of the regression model, including polynomial models, models using transformations of both dependent and independent variables, strictly non-linear models, and models with a categorical response variable. This section contains a chapter entitled "General Linear Model" that provides a unified approach to regression, analysis of variance, and analysis of covariance.

PART I

THE BASICS

The use of mathematical models to solve problems in the physical and biological sciences dates back to the first development of the scientific principle of discovery. The use of a theoretical model to explain natural phenomena has present-day applications in virtually all disciplines, including business, economics, engineering, the physical sciences, and the social, health, and biological sciences. Successful use of these models requires understanding of the theoretical underpinnings of the phenomena, the mathematical or statistical characteristics of the model, and the practical problems that may be encountered when using these models in real-life situations.

There are basically two approaches to using mathematical models to explain natural phenomena. The first attempts to use complex models to completely explain a phenomenon. In this case, models can result that defy solution. Even in many of the very simple cases, solutions can be obtained only through sophisticated mathematics. A model that completely explains the action of a response to a natural phenomenon is often called a deterministic model. A deterministic model, when it can be solved, yields an exact solution. The second approach to using models to solve problems involves using a simple model to obtain a solution that approximates the exact solution. This model is referred to as a statistical model, or often, a stochastic model. The statistical model usually has a simple solution that can be evaluated using probability distributions. That is, solutions to a statistical model are most useful when presented as a confidence interval or when the solutions can be supported by the results of a hypothesis test. It is this second approach that defines the discipline of statistics and is therefore the approach used in this book.

A statistical model contains two parts: (1) a deterministic or functional relationship among the variables, and (2) a stochastic or statistical part. The deterministic part may be simple or complex, and it is often the result of applications of mathematics to the underlying principles of the phenomenon. The model is expressed as a function, usually algebraic in nature, and parameters that specify the nature of the function. For example, the relationship between the circumference of a circle and its radius is an example of a deterministic relationship. The model $C = br$, where C = circumference, $b = 2\pi$, and r = radius, will give the exact circumference of a circle for a given radius. Written in this form, b is the parameter of the model, and in introductory geometry classes, an exercise might be conducted to determine a value of the parameter by measuring the radius and the circumference of a circle and solving for b.

On the other hand, if each student in the class were asked to draw a circle freehand, this deterministic model would not adequately describe the relationship between the radius and circumference of the figures drawn by students because the deterministic relationship assumes a perfect circle. The deviations from the deterministic model displayed by each student's figure would make up the statistical part of the model. The statistical portion of the model is usually considered to be of a random nature and is often referred to as the random error component of the model. We can explain the relationship between the circumference and the radius of the figures drawn by the students as $C = 2\pi r + \epsilon$, where ϵ is the statistical part of the model. It is easy to see that $\epsilon = C - 2\pi r$, the difference between the circumference of the hand-drawn figure and a perfect circle of the same radius. We would expect the value of this difference to vary from student to student, and we could even make some reasonable assumptions as to the distribution of this difference.

This is the basic idea for the use of statistical models to solve problems. We first hypothesize about the functional portion of the model. For example, the first part of this book deals strictly with linear models. Once the form of the function is identified, we then specify what parameters of this function need to be estimated. For example, in a simple linear relationship between two variables x and y (written in slope–intercept form, this would be $y = ax + b$), we need two parameters, a and b, to uniquely define the line. If the line represents a process that is truly linear, then a deterministic model would be appropriate. In this case, we would only need two points (a sample of size 2) to determine the values of the slope and the y-intercept. If the line is only an approximation to the process, or if a stochastic model is appropriate, we would write it in the form: $y = ax + b + \epsilon$. In this case, we would need a larger sample and would have to use the estimation procedures used in Chapter 2 to estimate a and b.

The random error component of a model is usually assumed to behave according to some probability distribution, usually the normal distribution. In fact, the standard assumption for most statistical models is that the error component is normal with mean zero and a constant variance. With this assumption it can be seen that the deterministic portion of the model is in fact the expected value of the response variable. For example, in the student circle example the expected value of the circumference of the figures would be $2\pi r$.

All of the models considered in Part I of this book are called linear models. This definition really means that the models are linear in the model parameters. It turns out that the most frequently used statistical methods involving a quantitative response variable are special cases of a linear model. This includes the one- and two-sample t tests, the analysis of variance, and simple linear regression. Because these topics are presumed to be prerequisite knowledge for those reading this book, they will be reviewed very briefly as they are normally presented, and then recast as linear models followed by the statistical analysis suggested by that model.

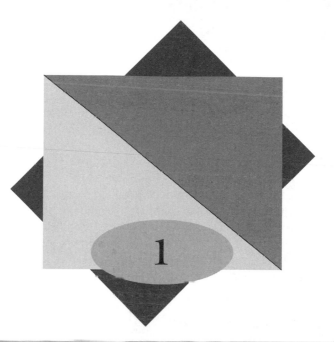

The Analysis of Means

1.1 Introduction

In this chapter we review the statistical methods for inferences on means using samples from one, two, and several populations. These methods are initially reviewed as they are presented in most basic textbooks, that is, using the principles of sampling distributions. Then these methods are recast as analyses of a linear model, using the concept of a linear model for making inferences. These methods also use sampling distributions, but in a different manner.

The purpose of this order of presentation is to introduce the linear-model approach for performing statistical analyses for situations where concepts are already familiar and formulas are easy to understand. Since these topics have been covered in prerequisite materials, there is no discussion of applications, and numerical examples are presented only to show the mechanics of the methods.

1.2 Sampling Distributions

In the usual approach to statistical inference, one or more parameters are identified that will characterize or describe a population. Then a sample of observations is taken from that population, one or more sample statistics are computed from the resulting data, and the statistics are used to make inferences on the unknown population parameter(s). There are several methods of obtaining appropriate statistics, called point estimators of the parameter. The standard methods of statistical analysis of data obtained as a random sample use the method of maximum likelihood (see Appendix C) to obtain estimates, called **sample statistics,** of the unknown parameters, and the sampling distributions associated with these estimates are used to make inferences on the parameters.

A **sampling distribution** describes the distribution of all possible values of a sample statistic. The concept of a sampling distribution is based on the proposition that a statistic computed from a random sample is a random variable whose distribution has a known relationship to the population from which the sample is drawn. We review here the sampling distributions we will use in this book.

Sampling Distribution of the Mean

Assume that a random sample of size n is drawn from a normal population with mean μ and standard deviation σ. Then the sample mean, $\bar{y}$, is a normally distributed random variable with mean μ and variance σ^2/n. The standard deviation, $\sigma/\sqrt{n}$, is known as the standard error of the mean.

If the distribution of the sampled population is not normal, we can still use this sampling distribution, provided the sample size is sufficiently large. This is possible because the **central limit theorem** states that the sampling distribution of the mean can be closely approximated by the normal distribution, regardless of the distribution of the population from which the sample is drawn, provided that the sample size is large. Although the theorem itself is an asymptotic result (being exactly true only if n goes to infinity), the approximation is usually very good for moderate sample sizes.

The definition of the sampling distribution of the mean is used to construct the statistic,

$$z = \frac{\bar{y} - \mu}{\sqrt{\dfrac{\sigma^2}{n}}},$$

which is normally distributed with mean of zero and unit variance. Probabilities associated with this distribution can be found in Appendix Table A.1 or can be obtained with computer programs.

Notice that this statistic has two parameters, as does the normal distribution. If we know σ^2, then we can use this statistic to make inferences on μ. If we do not know the population variance, then we use a statistic of the same form as the z, but with an estimate of σ^2 in its place. This distribution is known as the **t distribution.**

The estimated variance is computed by the familiar formula[1]:

$$s^2 = \frac{\Sigma(y - \bar{y})^2}{n - 1}.$$

---

[1] Formulas for more convenient computation exist but will not be presented.

For future reference it is important to note that this formula is evaluated using two distinct steps:

1. Calculating the sum of squares. The numerator of this equation, $\Sigma(y - \bar{y})^2$, is the sum of squared deviations of the observed values from the point estimate of the mean. This quantity is called the **sum of squares**[2] and is denoted by SS or S_{yy}.
2. Calculating the mean square. The **mean square** is an "average" square deviation and is calculated by dividing the sum of squares by the degrees of freedom and is denoted by MS. The **degrees of freedom** are the number of elements in the sum of squares minus the number of point estimates of parameters used in that sum. In this case there is only one such estimate, $\bar{y}$ (the estimate of μ); hence the degrees of freedom are $(n - 1)$. We will frequently use the notation MS instead of s^2.

We now substitute the mean square for σ^2 in the statistic, resulting in the expression

$$t(v) = \frac{\bar{y} - \mu}{\sqrt{\dfrac{\text{MS}}{n}}}.$$

This statistic has the Student t or simply the t distribution. This sampling distribution depends on the degrees of freedom used in computing the mean square, which is denoted by v in the equation. The necessary values for doing statistical inference can be obtained from Appendix Table A.2 and are automatically provided in most computer outputs. When the variance is computed as shown, the degrees of freedom are $(n - 1)$, but we will see that this is not applicable in all situations. Therefore, the appropriate degrees of freedom must always be specified when computing probabilities. As we will see in Section 1.4, the normal or the t distribution is also used to describe sampling distributions for the difference between two sample means.

Sampling Distribution of the Variance

Consider a sample of n independently drawn sample values from the Z (standard normal) distribution. Call these values z_i, $i = 1, 2, \ldots, n$. The sample statistic,

$$X^2 = \Sigma z_i^2,$$

is also a random variable whose distribution we call χ^2 (the Greek letter "chi").

Like the t distribution, the chi-square distribution depends on its degrees of freedom, the number of z-values in the sum of squares. Thus, the variable X^2 described earlier would have a χ^2 distribution with degrees of freedom equal to n. As in the t distribution, the degrees of freedom are denoted by the Greek letter v, and the distribution usually denoted by $\chi^2(v)$. A few important characteristics of the χ^2 distribution are as follows:

---

[2] We will use the second notation in formulas, as it conveniently describes the variable(s) involved in the computations. Thus, for example, $S_{xy} = \Sigma(x - \bar{x})(y - \bar{y})$.

1. χ^2 values can not be negative since they are sums of squares.
2. The shape of the χ^2 distribution is different for each value of v; hence, a separate table is needed for each value of v. For this reason, tables giving probabilities for the χ^2 distribution give values for only a selected set of probabilities. Appendix Table A.3 gives probabilities for the χ^2 distribution. However, tables are not often needed because probability values are available in most computer outputs.

The χ^2 distribution is used to describe the distribution of the sample variance. Let $y_1, y_2, \ldots, y_n$ be a random sample from a normally distributed population with mean μ and variance σ^2. Then the quantity

$$\frac{\Sigma(y_i - \bar{y})^2}{\sigma^2} = \frac{SS}{\sigma^2}$$

is a random variable whose distribution is described by a χ^2 distribution with $(n - 1)$ degrees of freedom. Notice that the sample variance, s^2, is the sum of squares divided by $n - 1$. Therefore, the χ^2 distribution is readily useful for describing the sampling distribution of s^2.

Sampling Distribution of the Ratio of Two Variances

A sampling distribution that occurs frequently in statistical methods is one that describes the distribution of the ratio of two estimates of σ^2. Assume two independent samples of size n_1 and n_2 from normally distributed populations with variances σ_1^2 and σ_2^2, respectively. The statistic

$$F = \frac{s_1^2/\sigma_1^2}{s_2^2/\sigma_2^2},$$

where s_1^2 and s_2^2 represent the usual variance estimates, is a random variable having the F distribution. The F distribution has two parameters, v_1 and v_2, called degrees of freedom, and is denoted by $F(v_1, v_2)$. If the variances are estimated in the usual manner, the degrees of freedom are $(n_1 - 1)$ and $(n_2 - 1)$, respectively, but this is not always the case. Also, if both populations have equal variance, that is, $\sigma_1^2 = \sigma_2^2$, the F statistic is simply the ratio s_1^2/s_2^2. A few important characteristics of the F distribution are as follows:

1. The F distribution is defined only for nonnegative values.
2. The F distribution is not symmetric.
3. A different table is needed for each combination of degrees of freedom. Fortunately, for most practical problems only a relatively few probability values are needed.
4. The choice of which variance estimate to place in the numerator is somewhat arbitrary; hence, the table of probabilities of the F distribution always gives the right tail value, that is, it assumes that the larger variance estimate is in the numerator.

Appendix Table A.4 gives probability values of the F distribution for selected degrees of freedom combinations for right tail areas.

Relationships among the Distributions

All of the sampling distributions presented in this section start with normally distributed random variables; hence, they are naturally related. The following relationships are not difficult to verify and have implications for many of the methods presented later in this book:

(1) $\qquad$ $t(\infty) = z$

(2) $\qquad$ $z^2 = \chi^2(1)$

(3) $\qquad$ $F(1, v_2) = t^2(v_2)$

(4) $\qquad$ $F(v_1, \infty) = \chi^2(v_1)/v_1.$

1.3 Inferences on a Single Population Mean

If we take a random sample of size n from a population described by a normal distribution with mean μ and standard deviation σ, then we can use the resulting data to make inferences on the unknown population mean μ in two ways. The first method uses the standard approach using the sampling distribution of an estimate of μ; the second uses the concept of a linear model. As we shall see, both give exactly the same result.

Inferences Using the Sampling Distribution of the Mean

The best single-valued or point estimate of the population mean is the sample mean. Denote the sample observations by y_i, $i = 1, 2, \ldots, n$, where n is the sample size. Then the sample mean is defined

$$\bar{y} = \frac{\Sigma y_i}{n}.$$

For the purpose of making inferences on the population mean, the sample mean is the maximum likelihood estimator. We have already noted that the sampling distribution of $\bar{y}$ has mean μ and standard deviation $\sigma/\sqrt{n}$.

We use the sampling distribution of the sample mean to make inferences on the unknown value μ. Usually, inferences on the mean take two forms. One form consists of establishing the reliability of our estimation procedure by constructing a confidence interval. The other is to test hypotheses on the unknown mean, μ.

The $(1 - \alpha)$ **confidence interval** on the unknown value μ is the interval contained by the endpoints defined by the formula

$$\bar{y} \pm z_{\alpha/2}\sqrt{\frac{\sigma^2}{n}},$$

where $z_{\alpha/2}$ is the $\alpha/2$ percentage point of the standard normal distribution. This interval includes the true value of the population mean with reliability $(1 - \alpha)$. In other words, we say that we are $(1 - \alpha)$ confident that the true value of the population mean is inside the computed interval. The level of confidence is often expressed as a percentage. That is, we often say that we are $(1 - \alpha) \times 100\%$ confident that the true value of the population mean is inside the computed interval.

Usually the population variance used in the preceding inference procedures is not known, so we estimate it and use the t distribution. The endpoints of the $(1 - \alpha)$ confidence interval using the estimated variance are computed by

$$\bar{y} \pm t_{\alpha/2}(n - 1)\sqrt{\frac{\text{MS}}{n}},$$

where $t_{\alpha/2}(n - 1)$ is the $\alpha/2$ percentage point of the t distribution with $(n - 1)$ degrees of freedom. The interpretation is, of course, the same as if the variance were known.

A **hypothesis test** can be conducted to determine if a hypothesized value of the unknown mean is reasonable given the particular set of data obtained from the sample. A statistical hypothesis test on the mean takes the following form. We test the null hypothesis

$$H_0: \mu = \mu_0$$

against the alternate hypothesis,

$$H_1: \mu \neq \mu_0,$$

where μ_0 is a specified value.[3] To test this hypothesis we use the sampling distribution of the mean to obtain the probability of getting a sample mean as far (or farther) away from the null hypothesis value as the one obtained in the sample. If the probability is smaller than some specified value, called the significance level, the evidence against the null hypothesis is deemed sufficiently strong to reject it. If the probability is larger than the significance level, there is said to be insufficient evidence to reject. A significance level of 0.05 is frequently used, but other levels may be used.

The hypothesis test is performed by computing the test statistic

$$z = \frac{\bar{y} - \mu_0}{\sqrt{\dfrac{\sigma^2}{n}}}.$$

If the null hypothesis is true, this test statistic has the standard normal distribution and can be used to find the probability of obtaining this sample mean or one farther away from the null hypothesis value. This probability is called the p-value. If the p-value is less than the significance level, the null hypothesis is rejected.

Again, if the variance is not known, we use the t distribution with the test statistic

$$t = \frac{\bar{y} - \mu_0}{\sqrt{\dfrac{\text{MS}}{n}}}$$

and compare it with values from the t distribution with $n - 1$ degrees of freedom.

---

[3] One-sided alternatives, such as $H_0: \mu > \mu_0$, are possible but will not be explicitly covered here.

Inferences Using the Linear Model

In order to explain the behavior of a random variable, y, we can construct a *model* in the form of an algebraic equation that involves the *parameter(s)* of the distribution of that variable (in this case, μ and σ). If the model is a statistical model, it also contains a component that represents the variation of an individual observation on y from the parameter(s). We will use a **linear model** where the model is a linear or additive function of the parameters.

For inferences on the mean from a single population, we use the linear model:

$$y_i = \mu + \epsilon_i,$$

where

> y_i is the ith observed value[4] of the *response* or *dependent* variable in the sample, $i = 1, 2, \ldots, n$
> μ is the population mean of the response variable
> $\epsilon_i, i = 1, 2, \ldots, n$, are a set of n independent and normally distributed random variables with mean zero and standard deviation σ

This model effectively describes the n observed values of a random sample from a normally distributed population having a mean of μ and standard deviation of σ.

The portion, μ, of the right-hand side of the model equation is the **deterministic** portion of the model. That is, if there were no variation ($\sigma = 0$), all observed values would be μ, and any one observation would exactly describe or determine the value of μ. Because the mean of ϵ is zero, it is readily seen that the mean or expected value of y is the deterministic portion of the model.

The ϵ_i make up the stochastic or **random** component of the model. It can be seen that these are deviations from the mean and can be expressed as $(y_i - \mu)$. That is, they describe the variability of the individual values of the population about the mean. It can be said that this term, often referred to as the "error" term, describes how well the deterministic portion of the model describes the population. Small values of ϵ_i imply that the mean provides a lot of information about the population mean, whereas larger values imply that, because there is considerable variation among the values, the mean provides less information about the population mean. The population parameter, σ^2, is the variance of the ϵ and is a measure of the magnitude of the dispersion of the error terms. A small variance implies that most of the error terms are near zero and the mean is "close" to the observed value y_i, and is therefore a measure of the "fit" of the model. A small variance implies a "good fit."

Notice that with this formulation of the model, the variance σ^2 can be expressed as

$$\sigma^2 = \Sigma \frac{\epsilon^2}{n}.$$

Using this model, we can perform statistical inferences using sample observations. The first task of the statistical analysis is to find a single point estimate

---

[4] The subscript, i, is usually omitted unless it is necessary for clarification.

of the parameter μ, which is the deterministic portion of the model. The idea is to find an estimate of μ, call it $\hat{\mu}$, that causes the model to best "fit" the observed data. A very convenient, and indeed the most popular, criterion for goodness of fit is the magnitude of the sum of squared differences, called **deviations**, between the observed values and the estimated mean. Consequently, the estimate that best fits the data is found by using the principle of **least squares**, which results in the estimate for which the sum of squared deviations is minimized.

Define

$$\hat{\epsilon}_i = y_i - \hat{\mu}$$

as the ith deviation (often called the ith residual). That is, the deviation is the difference between the ith observed sample value and the estimated mean. The least squares criterion requires that the value of $\hat{\mu}$ minimize the sum of squared deviations, that is, it minimizes

$$SS = \Sigma\hat{\epsilon}^2 = \Sigma(y_i - \hat{\mu})^2.$$

The estimate is obtained by using calculus to minimize SS. (See Appendix C for a discussion of this procedure.) Notice that this actually minimizes the variance of the error terms. This procedure results in the following equation:

$$\Sigma y - n\hat{\mu} = 0.$$

The solution to this equation, obviously, is

$$\hat{\mu} = \frac{\Sigma y}{n} = \bar{y},$$

which is the same estimate we obtained using the sampling distribution approach.

If we substitute this estimate for $\hat{\mu}$ in the formula for SS above, we obtain the minimum SS:

$$SS = \Sigma(y - \bar{y})^2,$$

which is the numerator portion of s^2, the estimate of the variance we used in the previous section. Now the variance is simply the sum of squares divided by degrees of freedom; hence the sample mean, $\bar{y}$, is that estimate of the mean that minimizes the variation about the model. In other words, the least squares estimate provides the estimated model that best fits the sample data.

Hypothesis Testing

As before, we test the null hypothesis,

$$H_0: \mu = \mu_0,$$

against the alternate hypothesis,

$$H_1: \mu \neq \mu_0.$$

We have noted that the variance is an indicator of the effectiveness of the model in describing the population. It then follows that if we have a choice among models, we can use the relative magnitudes of variances of the different models as a criterion for choice.

The hypothesis test statements above actually define two competitive models as follows:

1. The null hypothesis specifies a model where the mean is μ_0, that is,

$$y_i = \mu_0 + \epsilon_i.$$

   This model is referred to as the **restricted** model, since the mean is restricted to the value specified by the null hypothesis.
2. The alternate hypothesis specifies a model where the mean may take any value. This is referred to as the **unrestricted** model, which allows any value of the unknown parameter μ.

The sum of squares using the restricted model,

$$SSE_{restricted} = \Sigma(y - \mu_0)^2,$$

is called the restricted error sum of squares, since it is the sum of squares of the random error when the mean is restricted by the null hypothesis. This sum of squares has n degrees of freedom because it is computed from deviations from a quantity (μ_0) that is not computed from the data.[5]

The sum of squares for the unrestricted model is

$$SSE_{unrestricted} = \Sigma(y - \bar{y})^2$$

and represents the variability of observations from the best-fitting estimate of the model parameter. It is called the error sum of squares for the unrestricted model. As we have seen, it has $(n - 1)$ degrees of freedom and is the numerator of the formula for the estimated variance. Since the parameter is estimated by least squares, we know that this sum of squares is as small as it can get. This result ensures that

$$SSE_{restricted} \geq SSE_{unrestricted}.$$

The magnitude of this difference is used as the basis of the hypothesis test.

It would now appear logical to base the hypothesis test on a comparison between these two sums of squares. However, it turns out that a test based on a *partitioning* of sums of squares works better. An exercise in algebra provides the following relationship:

$$\Sigma(y - \mu_0)^2 = \Sigma(y - \bar{y})^2 + n(\bar{y} - \mu_0)^2.$$

This formula shows that $\Sigma(y - \mu_0)^2 = SSE_{restricted}$, the restricted error sum of squares, can be partitioned into two parts:

1. $\Sigma(y - \bar{y})^2$, the unrestricted model error sum of squares ($SSE_{unrestricted}$), which has $(n - 1)$ degrees of freedom.
2. $n(\bar{y} - \mu_0)^2$, which is the increase in error sum of squares due to the restriction imposed by the null hypothesis. In other words, it is the increase in the error sum of squares due to imposing the restriction that the null hypothesis

[5] The corresponding mean square is rarely calculated.

is true, and is denoted by $SS_{hypothesis}$. This sum of squares has one degree of freedom because it shows the decrease in the error sum of squares when going from a model with no parameters estimated from the data (the restricted model) to a model with one parameter, μ, estimated from the data. Equivalently, it is the sum of squares due to estimating one parameter.

Thus, the relationship can be written

$$SSE_{restricted} = SSE_{unrestricted} + SS_{hypothesis},$$

that is, the restricted sum of squares is partitioned into two parts. This *partitioning of sums of squares* is the key element in performing hypothesis tests using linear models.

Just as the preceding expression shows a partitioning of sums of squares, there is an equivalent partitioning of the degrees of freedom:

$$df_{restricted} = df_{unrestricted} + df_{hypothesis},$$

that is,

$$n = (n - 1) + 1.$$

Furthermore, we can now compute mean squares:

$$MS_{hypothesis} = SS_{hypothesis}/1, \text{ and}$$
$$MSE_{unrestricted} = SSE_{unrestricted}/(n - 1).$$

It stands to reason that as $SS_{hypothesis}$ increases relative to the other sums of squares, the hypothesis is more likely to be rejected. However, in order to use these quantities for formal inferences, we need to know what they represent in terms of the model parameters.

Remember that the mean of the sampling distribution of a sample statistic, called its expected value, tells us what the statistic estimates. It is, in fact possible to derive formulas for the means of the sampling distributions of mean squares. These are called **expected mean squares** and are denoted by E(MS). The expected mean squares of $MS_{hypothesis}$ and $MSE_{unrestricted}$ are as follows:

$$E(MS_{hypothesis}) = \sigma^2 + n(\mu - \mu_0)^2,$$

and

$$E(MSE_{unrestricted}) = \sigma^2.$$

Recall that in the discussion of sampling distributions that the F distribution describes the distribution of the ratio of two independent estimates of the same variance. If the null hypothesis is true, that is, $\mu = \mu_0$, or, equivalently $(\mu - \mu_0) = 0$, the expected mean squares[6] show that both mean squares are estimates of σ^2. Therefore, if the null hypothesis is true, then

$$F = \frac{MS_{hypothesis}}{MSE_{unrestricted}}$$

will follow the F distribution with 1 and $(n - 1)$ degrees of freedom.

---

[6] The fact that these estimates are independent is not proved here.

However, if the null hypothesis is not true, then $(\mu - \mu_0)^2$ will be a positive quantity[7]; hence, the numerator of the sample F statistic will tend to become larger. This means that calculated values of this ratio falling in the right-hand tail of the F distribution will favor rejection of the null hypothesis. The test of the hypothesis that $\mu = \mu_0$ is thus performed by calculating these mean squares and rejecting that hypothesis if the calculated value of the ratio exceeds the $(1 - \alpha)$ right-tail value of the F distribution with 1 and $(n - 1)$ degrees of freedom.

It is not difficult to show that the square root of the ratio used for the F distribution is indeed identical to the t statistic used for hypothesis testing using the sampling distribution of the mean. In fact, we have already noted that the square root of any percentage point of the F distribution with one degree of freedom in the numerator is the same as the percentage point of the t distribution with the same denominator degrees of freedom, remembering only that both the positive and negative tails of the t distribution go to the right tail of the F distribution. Therefore, the t and F tests will give identical results.

If both tests give the same answer, then why use the F test? Actually, for one-parameter models, t tests are preferred, and also have the advantage that they are easily converted to confidence intervals and may be used for one-sided alternative hypotheses. The purpose of presenting this method is to illustrate the principle for a situation where the derivations of the formulas for the linear model approach are easily understood.

EXAMPLE 1.1   Consider the following set of 10 observations shown in Table 1.1. The response variable is y. We will use the variable DEV later in the example.

Inferences using the sampling distribution. The quantities needed can easily be calculated:

$$\bar{y} = 11.55$$
$$s^2 = 27.485/9 = 3.0539,$$

TABLE 1.1

DATA FOR EXAMPLE 1.1

OBS	y	DEV
1	13.9	3.9
2	10.8	0.8
3	13.9	3.9
4	9.3	−0.7
5	11.7	1.7
6	9.1	−0.9
7	12.0	2.0
8	10.4	0.4
9	13.3	3.3
10	11.1	1.1

---

[7] Note that this occurs without regard to the sign of $(\mu - \mu_0)$. Hence, this method is not directly useful for one-sided alternative hypotheses.

and the estimated standard error of the mean,

$$\sqrt{s^2/n},$$

is 0.55262.

The 0.95 confidence interval can now be calculated:

$$11.55 \pm (2.262)(0.55262),$$

where 2.262 is the two-sided 0.05 tail value of the t distribution with nine degrees of freedom from Table A.2. The resulting interval contains the values from 10.30 to 12.80. That is, based on this sample, we are 95% confident that the interval (10.30 to 12.80) contains the true value of the mean.

Assume we want to test the null hypothesis

$$H_0: \mu = 10$$

with the alternative hypothesis

$$H_1: \mu \neq 10.$$

The test statistic is

$$t = \frac{11.55 - 10}{0.55262} = 2.805.$$

The 0.05 two-sided tail value for the t distribution is 2.262; hence, we reject the null hypothesis at the 0.05 significance level. A computer program will provide a p-value of 0.0206.

Inferences using the linear model. For the linear model test of $H_0: \mu = 10$, we need the following quantities:

1. The restricted model sum of squares: $\Sigma(y - 10)^2$. The individual values of these differences are the variable DEV in Table 1.1, and the sum of squares of this variable is 51.51.
2. The unrestricted model error sum of squares, 27.485, was obtained as an intermediate step in computing the estimated standard error of the mean for the t test earlier. The corresponding mean square is 3.054 with nine degrees of freedom.
3. The difference $51.51 - 27.485 = 24.025$, which can also be calculated directly as $10(\bar{y} - 10)^2$, is the $SS_{\text{hypothesis}}$. Then the F ratio is $24.025/3.054 = 7.867$. From Appendix Table A.4, the 0.05 tail value of the F distribution with (1,9) degrees of freedom is 5.12; hence, the null hypothesis should be rejected. Note that the square root of 7.867 is 2.805, the quantity obtained from the t test, and the square root of 5.12 is 2.262, the value from the t distribution needed to reject for that test. ◆

Although a confidence interval can be constructed using the linear model approach, the procedure is quite cumbersome. Recall that a confidence interval and the rejection region for a hypothesis test are related. That is, if the hypothesized value of the mean, μ_0, is not in the $1 - \alpha$ confidence interval, then we will reject the null hypothesis with a level of significance α. We could use this concept to go from the hypothesis test just given to a confidence interval on the mean, and that interval would be identical to that given using the sampling distribution of the mean.

1.4 Inferences on Two Means Using Independent Samples

Assume we have two populations with means μ_1 and μ_2 and variances σ_1^2 and σ_2^2, respectively, and with distributions that are approximately normal. Independent random samples of n_1 and n_2, respectively, are drawn from the two populations. We are interested in inferences on the means, specifically on the difference between the two means, that is $(\mu_1 - \mu_2) = \delta$, say. Note that although we have two means, the focus of inference is really on the single parameter, δ. As in the previous section, we first present inferences using the sampling distribution of the means and then using a linear model and the partitioning of sums of squares.

Assume that we have random samples taken independently from two populations. The measurement of interest is y_{ij}, where $i = 1, 2$ and $j = 1, \ldots, n_i$. The sample means are $\bar{y}_1$ and $\bar{y}_2$.

Inferences Using the Sampling Distribution

Since the point estimates of μ_1 and μ_2 are $\bar{y}_1$ and $\bar{y}_2$, the point estimate of δ is $(\bar{y}_1 - \bar{y}_2)$. A generalization of the sampling distribution of the mean shows that the sampling distribution of $(\bar{y}_1 - \bar{y}_2)$ tends to be normally distributed with a mean of $(\mu_1 - \mu_2)$ and variance of $(\sigma_1^2/n_1 + \sigma_2^2/n_2)$.

The statistic

$$z = \frac{(\bar{y}_1 - \bar{y}_2) - \delta}{\sqrt{\dfrac{\sigma_1^2}{n_1} + \dfrac{\sigma_2^2}{n_2}}}$$

has a standard normal distribution. If the variances are known, this statistic is used for confidence intervals and hypothesis tests on the difference between the two unknown population means.

The $(1 - \alpha)$ confidence interval for δ is the interval between endpoints defined as:

$$(\bar{y}_1 - \bar{y}_2) \pm z_{\alpha/2}\sqrt{\frac{\sigma_1^2}{n_1} + \frac{\sigma_2^2}{n_2}},$$

which states that we are $(1 - \alpha)$ confident that the true mean difference is within the interval defined by these endpoints.

For hypothesis testing, the null hypothesis is

$$H_0: (\mu_1 - \mu_2) = \delta_0.$$

In most applications, δ_0 is zero for testing the hypothesis that $\mu_1 = \mu_2$. The alternative hypothesis is

$$H_1: (\mu_1 - \mu_2) \neq \delta_0.$$

The test is performed by computing the test statistic

$$z = \frac{(\bar{y}_1 - \bar{y}_2) - \delta_0}{\sqrt{\dfrac{\sigma_1^2}{n_1} + \dfrac{\sigma_2^2}{n_2}}}$$

and comparing the resulting value with the appropriate percentage point of the standard normal distribution.

As for the one-population case, this statistic is not overly useful, since it requires the values of the two usually unknown population variances. Simply substituting estimated variances is not useful, since the resulting statistic does not have the t distribution because the denominator contains independent estimates of two variances. One way to adjust the test statistic so that it does have the t distribution is to assume that the two population variances are equal and find a mean square that serves as an estimate of that common variance. That mean square, called the pooled variance, is computed as follows:

$$s_p^2 = \frac{\Sigma_1(y - \bar{y}_1)^2 + \Sigma_2(y - \bar{y}_2)^2}{(n_1 - 1) + (n_2 - 1)},$$

where Σ_1 and Σ_2 represent the summation over samples 1 and 2. Using the convention of denoting sums of squares by SS, the pooled variance can be written

$$s_p^2 = \frac{SS_1 + SS_2}{n_1 + n_2 - 2},$$

where SS_1 and SS_2 are the sum of squares calculated separately for each sample. This formula explicitly shows that the estimate of the variance is of the form

$$\frac{\text{Sum of squares}}{\text{Degrees of freedom}},$$

and the degrees of freedom are $(n_1 + n_2 - 2)$ because two estimated parameters, $\bar{y}_1$ and $\bar{y}_2$, are used in computing the sum of squares.[8]

Substituting the pooled variance for both population variances in the test statistic provides

$$t(n_1 + n_2 - 2) = \frac{(\bar{y}_1 - \bar{y}_2) - \delta}{\sqrt{s_p^2\left(\dfrac{1}{n_1} + \dfrac{1}{n_2}\right)}},$$

which is called the "pooled t" statistic.

---

[8] Many references show the numerator as $(n_1 - 1)s_1^2 + (n_2 - 1)s_2^2$. However, this expression does not convey the fact that this is indeed a sum of squares. It is also more difficult to compute.

The $(1 - \alpha)$ confidence interval for δ is the interval between the endpoints defined by

$$(\bar{y}_1 - \bar{y}_2) \pm t_{\alpha/2}(n_1 + n_2 - 2)\sqrt{s_p^2\left(\frac{1}{n_1} + \frac{1}{n_2}\right)},$$

where $t_{\alpha/2}(n_1 + n_2 - 2)$ is the notation for the $\alpha/2$ percentage point of the t distribution with $(n_1 + n_2 - 2)$ degrees of freedom.

For hypothesis testing, the null hypothesis is

$$H_0\colon (\mu_1 - \mu_2) = \delta_0,$$

against the alternative hypothesis

$$H_1\colon (\mu_1 - \mu_2) \neq \delta_0.$$

As noted, usually $\delta_0 = 0$ for testing the null hypothesis, $H_0\colon \mu_1 = \mu_2$.

To perform the test, compute the test statistic

$$t(n_1 + n_2 - 1) = \frac{(\bar{y}_1 - \bar{y}_2) - \delta_0}{\sqrt{s_p^2\left(\frac{1}{n_1} + \frac{1}{n_2}\right)}}$$

and reject the null hypothesis if the computed statistic falls in the rejection region defined by the appropriate significance level for the t distribution with $(n_1 + n_2 - 2)$ degrees of freedom.

If the variances cannot be assumed equal for the two populations, approximate methods must be used. A discussion of this problem can be found in several texts, including Afifi and Azen (1979).

Inference for Two-Population Means Using the Linear Model

For inferences on the means from two populations, we use the linear model

$$y_{ij} = \mu_i + \epsilon_{ij},$$

where

y_{ij} represents the jth observed value from population i, $i = 1, 2,$ and $j = 1, 2, \ldots, n_i$

μ_i represents the mean of population i

ϵ_{ij} represents a normally distributed random variable with mean zero and variance σ^2

This model describes n_1 sample observations from population 1 with mean μ_1 and variance σ^2 and n_2 observations from population 2 with mean μ_2 and variance σ^2. Note that this model specifies that the variance is the same for both populations. As in the methods using sampling distributions, violations of this assumption are treated by special methods.

The null hypotheses to be tested is

$$H_0: \mu_1 = \mu_2,$$

against the alternative[9] hypothesis

$$H_1: \mu_1 \neq \mu_2.$$

Using the least squares procedures involves finding the values of $\hat{\mu}_1$ and $\hat{\mu}_2$ that minimize

$$\Sigma_{\text{all}} (y_{ij} - \hat{\mu}_i)^2$$

where Σ_{all} denotes the summation over all sample observations. This procedure yields the following equations (called the normal equations):

$$\Sigma_j y_{ij} - n\hat{\mu}_i = 0 \text{ , for } i = 1, 2.$$

The solutions to these equations are $\hat{\mu}_1 = \bar{y}_1$ and $\hat{\mu}_2 = \bar{y}_2$.

The unrestricted model error variance is computed from the sum of squared deviations from the respective sample means:

$$\text{SSE}_{\text{unrestricted}} = \Sigma_1 (y - \bar{y}_1)^2 + \Sigma_2 (y - \bar{y}_2)^2 = \text{SS}_1 + \text{SS}_2.$$

As already noted, the computation of this sum of squares requires the use of two estimated parameters, $\bar{y}_1$ and $\bar{y}_2$; hence, the degrees of freedom for this sum of squares are $(n_1 + n_2 - 2)$. The resulting mean square is indeed the pooled variance used for the pooled t statistic.

The null hypothesis is $\mu_1 = \mu_2 = \mu$, say. The restricted model, then, is

$$y_{ij} = \mu + \epsilon_{ij}.$$

The least squares estimate of μ is the overall mean of the total sample,

$$\bar{y} = \Sigma_{\text{all}} \frac{y_{ij}}{n_1 + n_2}.$$

The restricted model error sum of squares is the sum of squared deviations from this estimate, that is,

$$\text{SSE}_{\text{restricted}} = \Sigma_{\text{all}} (y - \bar{y})^2.$$

Since only one parameter estimate is used to compute this sum of squares, it has $(n_1 + n_2 - 1)$ degrees of freedom.

As before, the test of the hypothesis is based on the difference between the restricted model and unrestricted model error sums of squares. The partitioning of sums of squares is

$$\text{SSE}_{\text{restricted}} = \text{SS}_{\text{hypothesis}} + \text{SSE}_{\text{unrestricted}}.$$

An exercise in algebra provides the formula

$$\text{SS}_{\text{hypothesis}} = n_1 (\bar{y}_1 - \bar{y})^2 + n_2 (\bar{y}_2 - \bar{y})^2.$$

---

[9] The linear model approach is not normally used for the more general null hypothesis $(\mu_1 - \mu_2) = \delta$ for nonzero δ nor, as previously noted, for one-sided alternative hypotheses.

The degrees of freedom for the hypothesis sum of squares is the difference between the restricted and unrestricted model degrees of freedom. That difference is one because the unrestricted model has two parameters and the restricted model has only one, and the basis for the hypothesis test is to determine if the model with two parameters fits significantly better than the model with only one.

It is again useful to examine the expected mean squares to determine an appropriate test statistic:

$$E(MS_{hypothesis}) = \sigma^2 + \frac{n_1 n_2}{n_1 + n_2}(\mu_1 - \mu_2)^2$$

$$E(MSE_{unrestricted}) = \sigma^2.$$

The ratio of the resulting mean squares,

$$F = \frac{\left(\dfrac{SS_{hypothesis}}{1}\right)}{\left(\dfrac{SS_{unrestricted}}{n_1 + n_2 - 2}\right)} = \frac{MS_{hypothesis}}{MSE_{unrestricted}},$$

has the following properties:

1. If the null hypothesis is true, it is the ratio of two mean squares estimating the same variance and therefore has the F distribution with $(1, n_1 + n_2 - 2)$ degrees of freedom.
2. If the null hypothesis is not true, $(\mu_1 - \mu_2) \neq 0$, which means that $(\mu_1 - \mu_2)^2 > 0$. In this case, the numerator of the F statistic will tend to become large, again indicating rejection for large values of this statistic.

Another exercise in algebra gives the relationship

$$SS_{hypothesis} = n_1(\bar{y}_1 - \bar{y})^2 + n_2(\bar{y}_2 - \bar{y})^2$$

$$= (\bar{y}_1 - \bar{y}_2)^2 \frac{n_1 n_2}{n_1 + n_2}.$$

This shows that the F statistic can be expressed as

$$F = \frac{(\bar{y}_1 - \bar{y}_2)^2 \dfrac{n_1 n_2}{n_1 + n_2}}{MSE_{unrestricted}}$$

$$= \frac{(\bar{y}_1 - \bar{y}_2)^2}{MSE_{unrestricted}\left(\dfrac{1}{n_1} + \dfrac{1}{n_2}\right)}.$$

As we have seen, $MSE_{unrestricted}$ is the pooled variance; hence, the F statistic is the square of the t statistic. In other words, the pooled t test and the linear model F test are equivalent.

EXAMPLE 1.2   As before, we use some artificially generated data, consisting of 10 sample observations from population 1 and 15 from population 2. The data are shown in Table 1.2.

Inferences using the sampling distribution. For the pooled t test we compute the following quantities:

$$\bar{y}_1 = 25.1700, \bar{y}_2 = 29.3933,$$
$$SS_1 = 133.2010, SS_2 = 177.9093; \text{ hence,}$$
$$s_p^2 = (133.2010 + 177.9093)/23 = 311.1103/23 = 13.5265.$$

We want to test the hypothesis

$$H_0: \mu_1 = \mu_2$$

against the hypothesis

$$H_1: \mu_1 \neq \mu_2.$$

The pooled t statistic, then, is

$$t = \frac{25.1700 - 29.3933}{\sqrt{13.5265\left(\dfrac{1}{10} + \dfrac{1}{15}\right)}}$$
$$= 2.8128.$$

The 0.05 two-sided tail value for the t distribution with $(n_1 + n_2 - 2) = 23$ degrees of freedom is 2.069, and the null hypothesis is rejected at the 0.05 level. A computer program gives the p-value as 0.0099.

Inferences using linear models. For the linear models partitioning of sums of squares, we need the following quantities:

$$SSE_{\text{unrestricted}} = SS_1 + SS_2 = 311.1103$$
$$SSE_{\text{restricted}} = \Sigma_{\text{all}}(y - 27.704)^2 = 418.1296,$$

where 27.704 is the mean of all observations. The difference is 107.0193, which can also be calculated directly using the means:

$$SS_{\text{hypothesis}} = 10(25.1700 - 27.704)^2 + 15(29.3933 - 27.704)^2 = 107.0193.$$

The F statistic, then, is

$$F = \frac{107.0193}{\left(\dfrac{311.1103}{23}\right)} = 7.9118,$$

TABLE 1.2

DATA FOR EXAMPLE 1.2

Population 1:	25.0	17.9	21.4	26.6	29.1	27.5	30.6	25.1	21.8	26.7
Population 2:	31.5	27.3	26.9	31.2	27.8	24.1	33.5	29.6	28.3	29.3
	34.4	27.3	31.5	35.3	22.9					

which is larger than 4.28, the 0.05 upper-tail value of the F distribution with (1, 23) degrees of freedom; hence, the null hypothesis is rejected. A computer program gives the p-value of 0.0099, which is, of course, the same as for the t test. We can also see that the square of both the computed and table value for the t test is the same as the F-values for the partitioning of sums of squares test.

As in the one-sample case, the t test is more appropriately used for this application, not only because a confidence interval is more easily computed, but also because the t test allows for a hypothesis test other than that of $\mu_1 = \mu_2$. ◆

1.5 Inferences on Several Means

The extrapolation from two populations to more than two populations might, at first, seem straightforward. However, recall that in comparing two population means, we used the simple difference between them as a comparison between the two. If the difference was zero, the two means were the same. Unfortunately, we cannot use this procedure to compare more than two means. Therefore, there is no simple method of using sampling distributions to do inferences on more than two population means. Instead, the procedure is to use the linear model approach. This approach has wide applicability in comparing means from more than two populations in many different configurations.

The linear model for the analysis of any number of means is simply a generalization of the model we have used for two means. Assuming data from independent samples of n_i from each of t populations, the model is

$$y_{ij} = \mu_i + \epsilon_{ij}, \, i = 1, 2, \ldots, t, j = 1, 2, \ldots, n_i,$$

where y_{ij} is the jth sample observation from population i, μ_i is the mean of the ith population, and ϵ_{ij} is a random variable with mean zero and variance σ^2. This model is one of many that are referred to as an **analysis of variance** or **ANOVA model.** This form of the ANOVA model is called the "cell means model." As we shall see later, the model often is written in another form. Note that, as before, the linear model automatically assumes that all populations have the same variance. Inferences are to be made about the μ_i, usually in the form of a hypothesis test.

$$H_0: \mu_i = \mu_j, \text{ for all } i \neq j$$
$$H_1: \mu_i \neq \mu_j, \text{ for one or more pairs.}$$

The least square estimates for the unknown parameters μ_i are those values that minimize

$$\Sigma_{\text{all}} \, (y_{ij} - \hat{\mu}_i)^2, \, i = 1, \ldots, t.$$

The values that fit this criterion are the solutions to the t normal equations:

$$\Sigma_{\text{all}} \, y_{ij} = n_i \hat{\mu}_i, \, i = 1, \ldots, t.$$

The solutions to these equations are $\hat{\mu}_i = \overline{y}_i$, for $i = 1, \ldots, t$.

Then the unrestricted model error sum of squares is

$$\text{SSE}_{\text{unrestricted}} = \Sigma_1(y - \bar{y}_1)^2 + \Sigma_2(y - \bar{y}_2)^2 + \ldots + \Sigma_t(y - \bar{y}_t)^2,$$

which, because t sample means are used for computation, has $(N - t)$ degrees of freedom, where N is the total number of observations, $N = \Sigma n_i$.

The restricted model is

$$y_{ij} = \mu + \epsilon_{ij},$$

and the estimate of μ is the grand or overall mean of all observations:

$$\bar{y} = \Sigma_{\text{all}} y_{ij} / N.$$

Hence, the restricted error sum of squares is

$$\text{SSE}_{\text{restricted}} = \Sigma_{\text{all}} (y_{ij} - \bar{y})^2,$$

which has $(N - 1)$ degrees of freedom because only one parameter estimate, $\bar{y}$, is used.

The partitioning of sums of squares results in

$$\text{SSE}_{\text{restricted}} = \text{SS}_{\text{hypothesis}} + \text{SSE}_{\text{unrestricted}}.$$

This means that computing any two (usually $\text{SSE}_{\text{restricted}}$ and $\text{SS}_{\text{hypothesis}}$) allows the third to be computed by subtraction.[10]

The basis for the test is the difference

$$\text{SS}_{\text{hypothesis}} = \text{SSE}_{\text{restricted}} - \text{SSE}_{\text{unrestricted}},$$

which, using some algebra, can be computed directly by

$$\text{SSE}_{\text{hypothesis}} = \Sigma n_i (\bar{y}_i - \bar{y})^2$$

and has

$$(N - 1) - (N - t) = (t - 1)$$

degrees of freedom. This is because the unrestricted model estimates t parameters while the restricted model only has one.

As before, the expected mean squares provide information on the use of these mean squares. In order to make the formulas easier to understand, we will now assume that the samples from the populations are equal, that is, all $n_i = n$, say.[11] Then,

$$E(\text{MS}_{\text{hypothesis}}) = \sigma^2 + \frac{n}{t - 1}\Sigma(\mu_i - \mu)^2$$

$$E(\text{MS}_{\text{unrestricted}}) = \sigma^2.$$

Now, if the null hypothesis of equal population means is true, then $\Sigma(\mu_i - \mu)^2 = 0$ and both mean squares are estimates of σ^2. If the null hypothesis is not true, the

---

[10] Shortcut computational formulas are available, but are not of interest here.

[11] If the sample sizes are not all equal, the expression for $E(\text{MS}_{\text{hypothesis}})$ is more complicated in that it contains a weighted function of the $(\mu_i - \mu)^2$, with the weights being rather messy functions of the sample sizes, but the basic results are the same.

expected mean square for the hypothesis and consequently the F statistic will tend to become larger. Hence, the ratio of these mean squares provides the appropriate test statistic.

We now compute the mean squares:

$$MS_{hypothesis} = SS_{hypothesis} / (t - 1),$$
$$MSE_{unrestricted} = SSE_{unrestricted} / (N - t),$$

and the test statistic is

$$F = MS_{hypothesis} / MSE_{unrestricted},$$

which is to be compared to the F distribution with $[(t - 1),(N - t)]$ degrees of freedom.

EXAMPLE 1.3   The data for this example consist of weights of samples of six tubers of four varieties of potatoes grown under specific laboratory conditions. The data and some summary statistics are given in Table 1.3

The computations:

$$\bar{y} = 0.2896, \text{ then } \Sigma_{all} (y_{ij} - \bar{y})^2 = 0.5033,$$

which is $SSE_{restricted}$ with 23 degrees of freedom.

$$SSE_{unrestricted} = 0.0405 + 0.0319 + 0.0134 + 0.2335 = 0.3193,$$
$$\text{with 20 degrees of freedom}$$
$$SS_{hypothesis} = 0.5033 - 0.3193 = 0.1840,$$

or

$$SS_{hypothesis} = 6(0.1533 - 0.2896)^2 + \ldots + 6(0.2833 - 0.2896)^2,$$
$$\text{with three degrees of freedom.}$$

TABLE 1.3

DATA FOR EXAMPLE 1.3

		Variety		
	BUR	KEN	NOR	RLS
	0.19	0.35	0.27	0.08
	0.00	0.36	0.33	0.29
	0.17	0.33	0.35	0.70
	0.10	0.55	0.27	0.25
	0.21	0.38	0.40	0.19
	0.25	0.38	0.36	0.19
Mean	0.1533	0.3197	0.3300	0.2833
SS	0.0405	0.0319	0.0134	0.2335

The F statistic is

$$F = \frac{\left(\dfrac{0.1840}{3}\right)}{\left(\dfrac{0.3193}{20}\right)} = \frac{0.0613}{0.01597} = 3.84.$$

The 0.05 upper-tail percentage point of the F distribution with (3, 20) degrees of freedom is 3.10; hence, the hypothesis of equal mean weights of the four varieties may be rejected at the 0.05 level. Of course, this does not specify anything more about these means; this may be done with multiple comparison methods, which are another matter and are not presented here.

Table 1.4 gives the output of a computer program (PROC ANOVA of the SAS System) for the analysis of variance of this data set. Note that the nomenclature of the various statistics is somewhat different from what we have presented, but is probably closer to what has been presented in prerequisite courses. The equivalences are as follows:

What we have called $SSE_{restricted}$ is denoted "Corrected total." This is the sum of squares "corrected" for the mean, and since a model containing only a single mean is usually considered as having no model, this is the total variation if there is no model.

What we have called $SSE_{unrestricted}$ is simply called "Error," since this is the error sum of squares for the model specified for the analysis.

What we have called $SS_{hypothesis}$ is called "Model," since this is the decrease in the error sum of squares for fitting the model.

TABLE 1.4

COMPUTER OUTPUT FOR ANOVA

```
                      Analysis of Variance Procedure

Dependent Variable: WEIGHT
                               Sum of           Mean
Source                DF       Squares          Square      F Value   PR > F

Model                  3     0.18394583      0.06131528       3.84    0.0254
Error                 20     0.31935000      0.01596750
Corrected Total       23     0.50329583


                 Level of       -----------WEIGHT-----------
                 VAR       N       Mean                SD

                 BUR       6     0.15333333          0.09003703
                 KEN       6     0.39166667          0.07985403
                 NOR       6     0.33000000          0.05176872
                 RLS       6     0.28333333          0.21611725
```

The nomenclature used in the computer output is quite natural and easy to understand. However, it is not adequate for all inferences in more complicated models that we will encounter later.

As seen in the output, the computer program presents the sums of squares, mean squares, and F statistics and gives the p-value of 0.0254, which is, of course, less than 0.05, leading to the same conclusion reached earlier.

Below these statistics are the four variety means and the standard deviations of the observations for each variety. ◆

Another version of this model that reflects the partitioning of the sums of squares is obtained by a redefinition of the parameters, usually referred to as a **reparameterization** of the model. The reparameterization in this model consists of redefining each population mean as being conceptually composed of two parts: an overall or common mean plus a component due to the individual population. In the common application of a designed experiment with treatments randomly applied to experimental units, we are interested in the *effects* of the individual treatments. We can rewrite the model to represent this interest. The model is written

$$y_{ij} = \mu + \alpha_i + \epsilon_{ij}, \ i = 1, 2, \ldots, t, j = 1, 2, \ldots, n_i,$$

where

n_i is the number of observations in each sample or treatment group
t is the number of such populations, often referred to as levels of experimental factors or treatments
μ is the overall mean
α_i are the specific factor levels or treatment effects

In other words, this model has simply defined

$$\mu_i = \mu + \alpha_i.$$

The interpretation of the random error is as before. For more effective use of this model we add the restriction

$$\Sigma\alpha_i = 0,$$

which means that the "average" population effect is zero.[12]

The model written in this form is often called the "single-factor ANOVA model," or the "one-way classification ANOVA model."

For the reparameterized model, the equivalent hypotheses are

$$H_0: \alpha_i = 0, \text{ for all } i$$
$$H_1: \alpha_i \neq 0 \text{ for one or more } i.$$

In other words, the hypothesis of equal means translates to one of no factor effects.

---

[12] The restriction $\Sigma\alpha_i = 0$ is not absolutely necessary. See Chapter 10.

1.6 Summary

In this chapter we have briefly and without much detail reviewed the familiar one-sample and pooled t statistics and the analysis of variance procedures for inferences on means from one, two, or more populations. The important message of this chapter is that each of these methods is simply an application of the linear model and that inferences are made by comparing an unrestricted and restricted model. Although this principle may appear cumbersome for these applications, it will become more useful and, in fact, imperative to use in the more complicated models to be used later. This fact is amply illustrated by most books, which first introduce linear models for use in regression analysis where inferences cannot be made without using this approach.

EXAMPLE 1.4   Freund and Wilson (1993, p. 465) report data from an experiment done to compare the yield of three varieties of wheat tested over five subdivisions of a field. The experiment was performed in the randomized complete block design (RCBD), since the variation in subdivisions of the field was not of interest to the experimenters, but needed to be removed from the analysis of the results. The results are given in Table 1.5.

Since this experiment actually has two factors, the variety and the subdivision, we will use the "two-factor ANOVA model" with one factor considered as a block. The general model for the RCBD (with t treatments and b blocks) is written

$$y_{ij} = \mu + \alpha_i + \beta_j + \epsilon_{ij} , i = 1, 2, \ldots, t, j = 1, 2, \ldots, b,$$

where:

y_{ij} = the response from the ith treatment and the jth block
μ = the overall mean
α_i = the effect of the ith treatment
β_j = the effect of the jth block
ϵ_{ij} = the random error term

TABLE 1.5

WHEAT YIELDS

		Subdivisions				
		1	2	3	4	5
	A	31.0	39.5	30.5	35.5	37.0
Variety	B	28.0	34.0	24.5	31.5	31.5
	C	25.5	31.0	25.0	33.0	29.5

We are interested in testing the hypothesis

$$H_0: \alpha_i = 0, \text{ for all } i,$$
$$H_1: \alpha_i \neq 0, \text{ for one or more } i.$$

This means that the restricted model can be written

$$y_{ij} = \mu + \beta_j + \epsilon_{ij}.$$

Notice that this is simply the one-way ANOVA model considered previously. We can find the needed sums of squares by using PROC GLM in SAS. Table 1.6 gives the necessary sums of squares.

The appropriate sums of squares for testing the hypothesis then become

$$SSE_{hypothesis} = 112.833 - 14.400 = 98.433,$$

with $10 - 8 - 2$ degrees of freedom. The F test then becomes

$$F = \frac{98.433/2}{1.800} = 27.34.$$

This test statistic has a p-value of 0.0003. We therefore reject the null hypothesis and conclude that there is a difference in varieties.

Of course, this analysis would probably have been done using a two-way ANOVA table. Table 1.7 shows such an analysis done on PROC ANOVA in SAS.

TABLE 1.6

ANALYSIS OF EXAMPLE 1.4

ANOVA for unrestricted model:

Dependent Variable: YIELD

Source	DF	Sum of Squares	Mean Square	F Value	Pr > F
Model	6	247.333333	41.222222	22.90	0.0001
Error	8	14.400000	1.800000		
Corrected Total	14	261.733333			

ANOVA for restricted model:

Dependent Variable: YIELD

Source	DF	Sum of Squares	Mean Square	F Value	Pr > F
Model	4	148.900000	37.225000	3.30	0.0572
Error	10	112.833333	11.283333		
Corrected Total	14	261.733333			

TABLE 1.7

ANALYSIS OF VARIANCE FOR EXAMPLE 1.4

Analysis of Variance Procedure

Dependent Variable: YIELD

Source	DF	Sum of Squares	Mean Square	F Value	Pr > F
Model	6	247.333333	41.222222	22.90	0.0001
Error	8	14.400000	1.800000		
Corrected Total	14	261.733333			

Analysis of Variance Procedure

Dependent Variable: YIELD

Source	DF	ANOVA SS	Mean Square	F Value	Pr > F
BLOCK	4	148.900000	37.225000	20.68	0.0003
VARIETY	2	98.433333	49.216667	27.34	0.0003

Notice that the sums of squares for VARIETY, the F value, and the $Pr > F$ all agree with the previous analysis. ◆

CHAPTER EXERCISES

In addition to the exercises presented in this chapter, we suggest a review of exercises from prerequisite courses, redoing some of them using the linear model approach.

1. From extensive research it is known that the population of a particular freshwater species of fish has a mean length of $\mu = 171$ mm. The lengths are known to have a normal distribution. A sample of 100 fish suspected to come from this species is taken from a local lake. This sample yielded a mean length of $\bar{y} = 167$ mm with a sample standard deviation of 44 mm. Use the linear models approach and test the hypothesis that the mean length of the population of fish from the local lake is the same as that of the suspected species. Use a level of significance of 0.05.

2. M. Fogiel (*The Statistics Problem Solver*, 1978) describes an experiment in which a reading test is given to an elementary school class that consists of 12 Anglo-American children and 10 Mexican-American children. The results of the test are given in Table 1.8.

TABLE 1.8

DATA FOR EXERCISE 2

Group	Mean	Standard Deviation
Mexican-American	70	10
Anglo-American	74	8

(a) Write out an appropriate linear model to explain the data. List the assumptions made on the model. Estimate the components of the model.
(b) Using the linear models approach, test for differences between the two groups. Use a level of significance of 0.05.

3. Table 1.9 gives the results of a study of the effect of diet on the weights of laboratory rats. The data are weights in ounces of rats taken before the diet and again after the diet.

(a) Define an appropriate linear model to explain the data. Estimate the components of the model from the data.
(b) Using the linear models approach and $\alpha = 0.01$, test whether the diet changed the weight of the laboratory rats.

4. The shelf life of packaged fresh meat in a supermarket cooler is considered to be about 20 days. To determine if the meat in a local market meets this standard, a sample of 10 packages of meat were selected and tested. The data are as follows:

$$8, 24, 24, 6, 15, 38, 63, 59, 34, 39$$

(a) Define an appropriate linear model to describe this data. What assumptions would be made on this model? Estimate the components of the model.
(b) Using the linear models approach, test the hypothesis that the supermarket is in compliance. Use a level of significance of 0.05.

5. Wright and Wilson (1979) reported on a study designed to compare soil-mapping points on the basis of several properties. The study used eight contiguous sites near Albudeite in the province of Murcia, Spain. One of the properties

TABLE 1.9

DATA FOR EXERCISE 3

Rat	1	2	3	4	5	6	7	8	9	10
Before	14	27	19	17	19	12	15	15	21	19
After	16	18	17	16	16	11	15	12	21	18

of interest was clay content. Data from five randomly selected locations within each of the mapping points are given in Table 1.10.

   (a) Define an appropriate linear model for the data. What assumptions are made on this model? Estimate the components of the model.

   (b) Completely analyze the data. Assume the points are going from east to west in order of numbering. Completely explain the results.

6. A large bank has three branch offices located in a small Midwestern town. The bank has a liberal sick leave policy, and officers of the bank are concerned that employees might be taking advantage of this policy. To determine if there is a problem, employees were sampled randomly from each bank and the number of sick leave days in 1990 was recorded. The data are given in Table 1.11.

   (a) Use the linear models approach to test for differences between branch offices. Use a level of significance of 0.05.

7. Three different laundry detergents are being tested for their ability to get clothes white. An experiment was conducted by choosing three brands of washing machines and testing each detergent in each machine. The measure used was a whiteness scale, with high values indicating more "whiteness." The results are given in Table 1.12.

   (a) Define an appropriate model for this experiment. Consider the difference between washing machines a nuisance variation and not of interest to the experimenters.

TABLE 1.10

DATA FOR EXERCISE 5

Site	Clay Content				
1	30.3	27.6	40.9	32.2	33.7
2	35.9	32.8	36.5	37.7	34.3
3	34.0	36.6	40.0	30.1	38.6
4	48.3	49.6	40.4	43.0	49.0
5	44.3	45.1	44.4	44.7	52.1
6	37.0	31.3	34.1	29.7	39.1
7	38.3	35.4	42.6	38.3	45.4
8	40.1	38.6	38.1	39.8	46.0

TABLE 1.11

DATA FOR EXERCISE 6

Branch 1	Branch 2	Branch 3
15	11	18
20	15	19
19	11	23
14		

TABLE 1.12

DATA FOR EXERCISE 7

		Machine	
Solution	1	2	3
1	13	22	18
2	26	24	17
3	4	5	1

(b) Find $SSE_{unrestricted}$ and $SSE_{restricted}$.

(c) Test the hypothesis that there is no difference between detergents.

8. An experiment was conducted using two factors, A with two levels and B with two levels, in a factorial arrangement. That is, each combination of both factors received the same number of experimental units. The data from this experiment are given in Table 1.13.

The ANOVA model for this 2×2 factorial design is

$$y_{ijk} = \mu + \alpha_i + \beta_j + (\alpha\beta)_{ij} + \epsilon_{ijk}, \ i = 1, 2, j = 1, 2, \text{ and } k = 1, 2, 3.$$

y_{ijk} = the kth response from the ith level of factor A and the jth level of factor B

μ = the overall mean

α_i = the effect of factor A

β_j = the effect of factor B

$(\alpha\beta)_{ij}$ = the interaction effect between A and B

ϵ_{ijk} = the random error term

(a) The first step in the analysis is to test for interaction. Define the restricted model for testing $H_0: (\alpha\beta)_{ij} = 0$. Test the hypothesis using the linear models method.

TABLE 1.13

DATA FROM A FACTORIAL EXPERIMENT

		Factor A	
		1	2
		5.3	8.8
	1	3.6	8.9
Factor B		2.5	6.8
		4.8	3.6
	2	3.9	4.1
		3.4	3.8

(b) The next step is to test for main effects.
  (i) Define the restricted model for testing H_0: $\beta_j = 0$. Test the hypothesis using the linear models method.
  (ii) Define the restricted model for testing H_0: $\alpha_i = 0$. Test the hypothesis using the linear models method.

(c) Do a conventional ANOVA for the data and compare your results in parts (a) through (c).

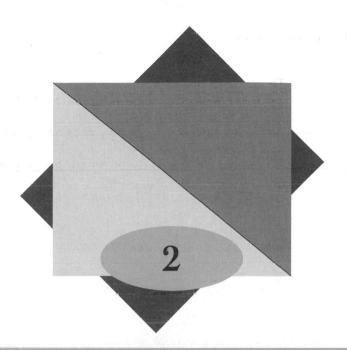

2

Simple Linear Regression

2.1 Introduction

In Chapter 1 we introduced the linear model as an alternative to standard methodology to make inferences on one or more population means where the measurement of interest (the response) was quantitative. The populations were simply identified by arbitrary labels. A schematic picture of such a model for three populations is shown in Figure 2.1, where the three populations of the variable y have means of 4, 6, and 8, respectively (as indicated on the horizontal axis), and the populations are all normally distributed.

A regression model is an application of the linear model where populations of the response variable are identified with numeric values of one or more quantitative variables that are called factor or **independent** variables. A regression model is shown in Figure 2.2. This model specifies that the mean of the independent variable (y, identified on the horizontal axis) is related to the independent variable (x, on the vertical axis) by the equation $y = 8 - 0.5x$, described by the diagonal line. Thus, for example, when $x = 0$, the mean of y is 8; when $x = 4$, the mean of y is 6; and when $x = 8$, then the mean of y is 4. This is the deterministic portion of the model.

The stochastic or statistical portion of the model specifies that the individual observations of the populations are distributed normally about these means.

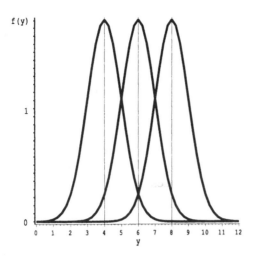

FIGURE 2.1 Schematic of Analysis of Variance

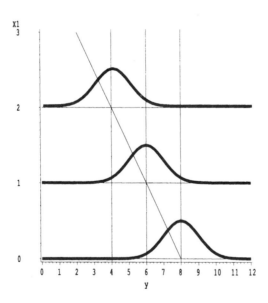

FIGURE 2.2 Schematic of Regression Model

Note that although in Figure 2.2 the distribution is shown only for populations defined by three values of x, the regression model states that for any value of x, whether or not observed in the data, there exists a population of the dependent variable that has a mean:

$$E(y) = 8 - 0.5x.$$

The purpose of a statistical analysis of a regression model is not primarily to make inferences on differences among the means of these populations, but rather to make inferences about the relationship of the mean of the response variable to the independent variables. These inferences are made through the parameters of the model, in this case the intercept of 8 and slope of -0.5. The resulting relationship can then be used to predict or explain the behavior of the response variable.

Some examples of analyses using regression models include the following:

- Estimating weight gain by the addition to children's diet of different amounts of various dietary supplements
- Predicting scholastic success (grade point ratio) based on students' scores on an aptitude or entrance test
- Estimating amounts of sales associated with levels of expenditures for various types of advertising
- Predicting fuel consumption for home heating based on daily temperatures and other weather factors
- Estimating changes in interest rates associated with the amount of deficit spending

2.2 The Linear Regression Model

The simplest regression model is the **simple linear regression model**, which is written

$$y = \beta_0 + \beta_1 x + \epsilon.$$

This model is similar to those discussed in Chapter 1 in that it consists of a deterministic part and a random part. The deterministic portion of the model,

$$\beta_0 + \beta_1 x,$$

specifies that for any value of the independent variable, x,[1] the population mean of the **dependent** or **response** variable, y, is described by the straight-line function ($\beta_0 + \beta_1 x$). Following the usual notation for the general expression for a straight line, the parameter β_0, the **intercept**, is the value of the mean of the dependent variable when x is zero, and the parameter β_1, the **slope**, is the change in the mean of the dependent variable associated with a unit change in x. These parameters are often referred to as the **regression coefficients**. Note that the intercept may not have a practical interpretation in cases where x cannot take a zero value.

As in the previously discussed linear models, the random part of the model explains the variability of the responses about the mean. We again assume that the terms (known as the error terms) have a mean zero and a constant variance, σ^2. In order to do statistical inferences we also make the assumption that the errors have a normal distribution as seen in Figure 2.2.

---

[1] In many presentations of this model the subscript "i" is associated with x and y, indicating that the model applies to the ith sample observation. For simplicity, we will not specify this subscript unless it is needed for clarity.

The fact that the regression line represents a set of means is often over-looked, a fact that often clouds the interpretation of the results of a regression analysis. This fact is demonstrated by providing a formal notation for a two-stage definition of the regression model. First, we define a linear model:

$$y = \mu + \epsilon,$$

where the standard assumptions are made on ϵ. This model states that the observed value, y, comes from a population with mean μ and variance σ^2.

For the regression model, we now specify that the mean is related to the independent variable x by the model equation

$$\mu = \mu_{y|x} = \beta_0 + \beta_1 x,$$

which shows that the mean of the dependent variable is linearly related to values of the independent variable. The notation $\mu_{y|x}$ indicates that the mean of the variable y depends on a given value of x.

A regression analysis is a set of procedures, based on a sample of n ordered pairs, (x_i, y_i), $i = 1, 2, \ldots, n$, for estimating and making inferences on the parameters, β_0 and β_1. These estimates can then be used to estimate mean values of the dependent variable for specified values of x.

EXAMPLE 2.1   One task assigned to foresters is to estimate the potential lumber harvest of a forest. This is typically done by selecting a sample of trees, making some non-destructive measures of these trees, and then using a prediction formula to estimate lumber yield. The prediction formula is obtained from a study using a sample of trees for which actual lumber yields were obtained by harvesting. Table 2.1 shows data for a such a sample of 20 trees, giving the values of the three variables representing the nondestructive measures:

HT, the height, in feet
DBH, the diameter of the trunk at breast height (about 4 feet), in inches
D16, the diameter of the trunk at 16 feet of height, in inches

and the measure obtained by harvesting the trees:

VOL, the volume of lumber (a measure of the yield), in cubic feet

Because DBH is the most easily measured nondestructive variable, it is logical to first see how well this measure can be used to estimate lumber yield. That is, we propose a regression model that uses DBH to estimate the mean lumber yield. The scatterplot shown in Figure 2.3 indicates that the two variables are indeed related and that it may be possible that a simple linear regression model can be used to estimate VOL using DBH. The deterministic portion of the model is

$$\mu_{\text{VOL|DBH}} = \beta_0 + \beta_1 \text{DBH},$$

where $\mu_{\text{VOL|DBH}}$ is the mean of a population of trees for a specified value of DBH; β_0 is the mean volume of the population of trees having zero DBH (in

TABLE 2.1

DATA FOR ESTIMATING TREE VOLUMES

OBS	DBH	D16	HT	VOL
1	10.20	9.3	89.00	25.93
2	13.72	12.1	90.07	45.87
3	15.43	13.3	95.08	56.20
4	14.37	13.4	98.03	58.60
5	15.00	14.2	99.00	63.36
6	15.02	12.8	91.05	46.35
7	15.12	14.0	105.60	68.99
8	15.24	13.5	100.80	62.91
9	15.24	14.0	94.00	58.13
10	15.28	13.8	93.09	59.79
11	13.78	13.6	89.00	56.20
12	15.67	14.0	102.00	66.16
13	15.67	13.7	99.00	62.18
14	15.98	13.9	89.02	57.01
15	16.50	14.9	95.09	65.62
16	16.87	14.9	95.02	65.03
17	17.26	14.3	91.02	66.74
18	17.28	14.3	98.06	73.38
19	17.87	16.9	96.01	82.87
20	19.13	17.3	101.00	95.71

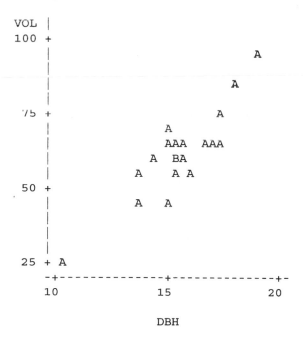

```
A=1, B=2, etc.  Plot of VOL*DBH.


VOL |
100 +
    |
    |                                      A
    |
    |                                 A
    |
 75 +                            A
    |                     A
    |                     AAA   AAA
    |                  A  BA
    |                A    A A
 50 +
    |                A    A
    |
    |
    |
 25 + A
    -+-------------+-------------+-
    10            15            20

                 DBH
```

FIGURE 2.3 Scatterplot of Volume and DBH

this example this parameter has no practical meaning); and β_1 is the increase in the mean height of trees as DBH increases by 1 inch. The complete regression model, including the error, is

$$\text{VOL} = \beta_0 + \beta_1 \text{DBH} + \epsilon.$$

First we will use the data to estimate the parameters, β_0 and β_1, the regression coefficients that describe the model, and then we will employ statistical inference methods to ascertain the significance and precision of these estimates as well as the precision of estimated values of VOL obtained by this model.   ◆

2.3   Inferences on the Parameters β_0 and β_1

We have defined the simple linear regression model

$$y = \beta_0 + \beta_1 x + \epsilon,$$

where y is the dependent variable, β_0 the intercept, β_1 the slope, x the independent variable, and ϵ the random error term. A sample of size n is taken that consists of measurements on the ordered pairs (x,y). The data from this sample is used to construct estimates of the coefficients, which are used in the following equation for estimating the mean of y:

$$\hat{\mu}_{y|x} = \hat{\beta}_0 + \hat{\beta}_1 x.$$

This is the equation of a line that is the locus of all values of $\hat{\mu}_{y|x}$, the estimate of the mean of the dependent variable, y, for any specified value of x, the independent variable. We now illustrate the method of estimating these parameters from the sample data.

Estimating the Parameters β_0 and β_1

In Section 1.3 we introduced the principle of least squares to provide an estimate for the mean. We use the same principle to estimate the coefficients in a regression equation. That is, we find those values of $\hat{\beta}_0$ and $\hat{\beta}_1$ that minimize the sum of squared deviations:

$$\text{SS} = \Sigma (y - \hat{\mu}_{y|x})^2 = \Sigma (y - \hat{\beta}_0 - \hat{\beta}_1 x)^2.$$

The values of the coefficients that minimize the sum of squared deviations for any particular set of sample data are given by the solutions of the following equations, which are called the normal equations[2]:

$$\hat{\beta}_0 n + \hat{\beta}_1 \Sigma x = \Sigma y$$

$$\hat{\beta}_0 \Sigma x + \hat{\beta}_1 \Sigma x^2 = \Sigma xy.$$

The solution for two linear equations in two unknowns is readily obtained and

---

[2] As in Chapter 1, these are obtained through an exercise in calculus; see Appendix C.

provides the estimators of these parameters as follows:

$$\hat{\beta}_1 = \frac{\sum xy - \dfrac{(\sum x)(\sum y)}{n}}{\sum x^2 - (\sum x)^2 \div n}$$

$$\hat{\beta}_0 = \bar{y} - \hat{\beta}_1 \bar{x}.$$

The estimator of β_1 can also be written

$$\hat{\beta}_1 = \frac{\sum(x - \bar{x})(y - \bar{y})}{\sum(x - \bar{x})^2}.$$

This last formula more clearly shows the structure of the estimate: It is the sum of cross products of the deviations of observed values from the means of x and y divided by the sum of squared deviations of the x values. Commonly, we call $\sum(x - \bar{x})^2$ and $\sum(x - \bar{x})(y - \bar{y})$ the corrected, or means centered, sums of squares and cross products. Since these quantities occur frequently, we will use the following notation and computational formulas:

$$S_{xx} = \sum(x - \bar{x})^2 = \sum x^2 - (\sum x)^2/n,$$

which is the corrected sum of squares for the independent variable x, and

$$S_{xy} = \sum(x - \bar{x})(y - \bar{y}) = \sum xy - \sum x \sum y/n,$$

which is the corrected sum of products of x and y. Later, we will need

$$S_{yy} = \sum(y - \bar{y})^2 = \sum y^2 - (\sum y)^2/n,$$

the corrected sum of squares of the dependent variable, y. Using this notation, we can write

$$\hat{\beta}_1 = S_{xy}/S_{xx}.$$

EXAMPLE 2.1   **Estimating the Parameters**   We illustrate the computations for Example 2.1 using the computational formulas. The preliminary computations are:

$$n = 20.$$

$$\sum x = 310.63, \text{ and } \bar{x} = 15.532.$$

$$\sum x^2 = 4889.0619.$$

We then compute:

$$S_{xx} = 4889.0619 - (310.63)^2/20 = 64.5121.$$

$$\sum y = 1237.03, \text{ and } \bar{y} = 61.852.$$

$$\sum y^2 = 80256.52, \text{ and we compute}$$

$$S_{yy} = 80256.52 - (1237.03)^2/20 = 3744.36 \text{ (which we will need later)}.$$

$$\sum xy = 19659.10, \text{ and we compute}$$

$$S_{xy} = 19659.10 - (310.63)(1237.03)/20 = 446.17.$$

The estimates of the parameters are

$$\hat{\beta}_1 = S_{xy}/S_{xx} = 446.17/64.5121 = 6.9161, \text{ and}$$

$$\hat{\beta}_0 = \bar{y} - \hat{\beta}_1\bar{x} = 61.852 - (6.9161)(15.532) = -45.566,$$

which provides the estimating equation:

$$\hat{\mu}_{VOL|DBH} = -45.566 + 6.9161(DBH).$$

The interpretation of $\hat{\beta}_1$ is that the mean volume of trees increases by 6.91 cubic feet for each 1-inch increase in DBH. The estimate of $\hat{\beta}_0$ implies that the mean volume of trees having zero DBH is -45.66. This is obviously an impossible value and reinforces the fact that for practical purposes this parameter cannot be literally interpreted in cases where a zero value of the independent variable cannot occur or is beyond the range of available data. A plot of the data points and estimated line is shown in Figure 2.4 and shows how the regression line fits the data. ◆

Inferences on β_1 Using the Sampling Distribution

Although the regression model has two parameters, the primary focus of inference is on β_1, the slope of the regression line. This is because if $\beta_1 = 0$, there is

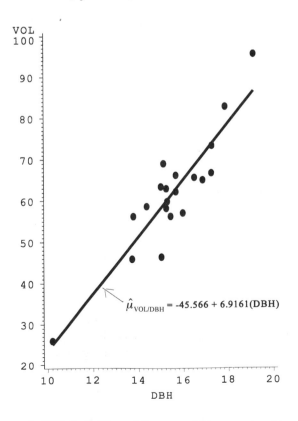

FIGURE 2.4 Plot of Data and Regression Line

no regression and the model is simply that of a single population (Section 1.3). Inferences on β_0 will be presented later.

An adaptation of the central limit theorem states that the sampling distribution of the estimated parameter $\hat{\beta}_1$ is approximately normal with mean β_1 and variance σ^2/S_{xx}, where σ^2 is the variance of the random component of the model. Consequently, the standard error of the estimated parameter is $\sigma/\sqrt{S_{xx}}$.

The standard error is a measure of the precision of the estimated parameter. We can more easily see how this is affected by the data by noting that $S_{xx} = (n-1)s_x^2$, where s_x^2 is the estimated variance computed from the values of the independent variable. Using this relationship, we can see the following:

1. The precision decreases as the standard deviation of the random error, σ, increases.
2. Holding constant s_x^2, the precision increases with larger sample size.
3. Holding constant the sample size, the precision increases with a higher degree of dispersion of the observed values of the independent variable (as s_x^2 gets larger).

The first two characteristics are the same that we observed for the sampling distribution of the mean. The third embodies a new concept that states that a regression relationship is more precisely estimated when values of the independent variable are observed over a wide range. This does make intuitive sense and will become increasingly clear (see especially Section 4.1).

We can now state that

$$z = \frac{\hat{\beta}_1 - \beta_1}{\left(\dfrac{\sigma}{\sqrt{S_{xx}}}\right)}$$

has the standard normal distribution. If the variance is known, this statistic can be used for hypothesis tests and confidence intervals.

Because the variance is typically not known, we must first obtain an estimate of that variance and use that estimate in the statistic. We have seen that estimates of variance are mean squares defined as

$$\text{Mean square} = \frac{\text{Sum of squared deviations from the estimated mean}}{\text{Degrees of freedom}}$$

When we are using a regression model, the deviations, often called *residuals*, are measured from the values of $\hat{\mu}_{y|x}$ obtained for each observed value of x. The degrees of freedom are defined as the number of elements in the sum of squares minus the number of parameters in the model used to estimate the means. For the simple linear regression model there are n terms in the sum of squares, and the $\hat{\mu}_{y|x}$ are calculated with a model having two estimated parameters, $\hat{\beta}_0$ and $\hat{\beta}_1$; hence, the degrees of freedom are $(n-2)$. The resulting mean square is the estimated variance and is denoted by $s_{y|x}^2$, indicating that it is the variance of the dependent variable, y, after fitting a regression model involving the independent variable, x. Thus,

$$s_{y|x}^2 = \text{MSE} = \frac{\text{SSE}}{n-2} = \frac{\Sigma(y - \hat{\mu}_{y|x})^2}{n-2}.$$

A shortcut formula that does not require the calculation of individual values of $\hat{\mu}_{y|x}$ is developed later in this section.

The statistic now becomes

$$t = \frac{\hat{\beta}_1 - \beta_1}{\sqrt{\dfrac{s^2_{y|x}}{S_{xx}}}},$$

which has the t distribution with $(n - 2)$ degrees of freedom. The denominator in this formula is the estimate of the standard error of the parameter estimate. For testing the hypotheses

$$H_0: \beta_1 = \beta_1^* \text{ against}$$

$$H_1: \beta_1 \neq \beta_1^*,$$

where β_1^* is any desired null hypothesis value, compute the statistic

$$t = \frac{\hat{\beta}_1 - \beta_1^*}{\sqrt{\dfrac{s^2_{y|x}}{S_{xx}}}}$$

and reject H_0 if the p-value for that statistic is less than the desired significance level. The most common hypothesis is that $\beta_1^* = 0$.

The $(1 - \alpha)$ confidence interval is calculated by

$$\hat{\beta}_1 \pm t_{\alpha \div 2}(n - 2)\sqrt{\frac{s^2_{y|x}}{S_{xx}}},$$

where $t_{\alpha/2}(n - 2)$ denotes the $(\alpha/2)$ percentage point of the t distribution with $(n - 2)$ degrees of freedom.

EXAMPLE 2.1
CONTINUED

Inferences on β_1 Using the Sampling Distribution   The first step is to compute the estimated variance. The necessary information is provided in Table 2.2.

The last column contains the residuals (deviations from the estimated means), which are squared and summed:

$$\text{SSE} = 0.9518^2 + (-3.4529)^2 + \ldots + 8.9708^2 = 658.570.$$

Dividing by the degrees of freedom:

$$s^2_{y|x} = 658.570/18 = 36.5871.$$

The estimated standard error of $\hat{\beta}_1$ is

$$\text{Standard error } (\hat{\beta}_1) = \sqrt{\frac{s^2_{y|x}}{S_{xx}}} = \sqrt{\frac{36.587}{64.512}} = 0.7531.$$

A common application is to test the hypothesis of no regression, that is,

$$H_0: \beta_1 = 0 \text{ and}$$

$$H_1: \beta_1 \neq 0,$$

TABLE 2.2

DATA FOR CALCULATING THE VARIANCE

Obs	Dependent Variable VOL (cub. ft) y	Predicted Value (cub. ft) $\hat{\mu}_{y|x}$	Residual (cub. ft) $(y - \hat{\mu}_{y|x})$
1	25.9300	24.9782	0.9518
2	45.8700	49.3229	−3.4529
3	56.2000	49.7379	6.4621
4	58.6000	53.8184	4.7816
5	63.3600	58.1756	5.1844
6	46.3500	58.3139	−11.9639
7	68.9900	59.0055	9.9845
8	62.9100	59.8355	3.0745
9	58.1300	59.8355	−1.7055
10	59.7900	60.1121	−0.3221
11	56.2000	61.1495	−4.9495
12	66.1600	62.8094	3.3506
13	62.1800	62.8094	−0.6294
14	57.0100	64.9534	−7.9434
15	65.6200	68.5498	−2.9298
16	65.0300	71.1087	−6.0787
17	66.7400	73.8060	−7.0660
18	73.3800	73.9443	−0.5643
19	82.8700	78.0249	4.8451
20	95.7000	86.7392	8.9708

for which the test statistic becomes

$$t = \frac{\hat{\beta}_1}{\text{standard error}} = \frac{6.9161}{0.7531} = 9.184.$$

The rejection criterion for a two-tailed t test with $\alpha = 0.01$ is 2.5758. The value of 9.184 exceeds this value, so the hypothesis is rejected. (The actual p-value obtained from a computer program is 0.0001.)

For the 0.95 confidence interval, we find $t_{0.025}(18) = 2.101$, and the interval is

$$6.916 \pm 2.101(0.7531), \text{ or}$$

$$6.916 \pm 1.582,$$

resulting in the interval from 5.334 to 8.498. In other words, we are 0.95 (or 95%) confident that the population mean increase in volume is between 5.334 and 8.498 cubic feet per 1-inch increase in DBH. ◆

Inferences on β_1 Using the Linear Model

The unrestricted model for simple linear regression is

$$y = \beta_0 + \beta_1 x + \epsilon.$$

The least squares estimates of the parameters are obtained as before and are used to compute the conditional means, $\hat{\mu}_{y|x}$. These are then used to compute the unrestricted model error sum of squares:

$$\text{SSE}_{\text{unrestricted}} = \Sigma(y - \hat{\mu}_{y|x})^2.$$

This is indeed the estimate we obtained earlier, and the degrees of freedom are $(n - 2)$.

The null hypothesis is

$$H_0: \beta_1 = 0;$$

hence, the restricted model is

$$y = \beta_0 + \epsilon,$$

which is equivalent to the model for a single population:

$$y = \mu + \epsilon.$$

From Section 1.2, we know that the point estimate of the parameter μ is $\bar{y}$. The restricted model error sum of squares is now the error sum of squares for that model, that is,

$$\text{SSE}_{\text{restricted}} = \Sigma(y - \bar{y})^2,$$

which has $(n - 1)$ degrees of freedom.

The hypothesis test is based on the difference between the restricted and unrestricted model error sums of squares, that is,

$$\text{SSE}_{\text{hypothesis}} = \text{SSE}_{\text{restricted}} - \text{SSE}_{\text{unrestricted}},$$

which has $[n - 1 - (n - 2)] = 1$ (one) degree of freedom. That is, we have gone from a restricted model with one parameter, μ, to the unrestricted model with two parameters, β_0 and β_1.

Note that we again have a partitioning of sums of squares, which in this case also provides a shortcut for computing $\text{SSE}_{\text{unrestricted}}$. We already know that

$$\text{SSE}_{\text{unrestricted}} = \Sigma(y - \hat{\mu}_{y|x})^2, \text{ and}$$

$$\text{SSE}_{\text{restricted}} = \Sigma(y - \bar{y})^2.$$

Now:

$$\text{SS}_{\text{hypothesis}} = \Sigma(\bar{y} - \hat{\mu}_{y|x})^2$$
$$= \Sigma(\bar{y} - \hat{\beta}_0 - \hat{\beta}_1 x)^2.$$

Substituting the least squares estimators for $\hat{\beta}_0$ and $\hat{\beta}_1$ results in some cancellation of terms and in a simplified form:

$$\text{SS}_{\text{hypothesis}} = \hat{\beta}_1 S_{xy}.$$

This quantity can also be computed by using the equivalent formulas: $\hat{\beta}_1^2 S_{xx}$ or S_{xy}^2/S_{xx}. The most convenient procedure is to compute $\text{SS}_{\text{restricted}}$ and $\text{SS}_{\text{hypothesis}}$ and obtain $\text{SS}_{\text{unrestricted}}$ by subtraction.

As before, it is useful to examine the expected mean squares to establish the test statistic. For the regression model:

$$E(MS_{hypothesis}) = \sigma^2 + \beta_1^2 S_{xx},$$
$$E(MSE_{unrestricted}) = \sigma^2.$$

If the null hypothesis, H_0: $\beta_1 = 0$, is true, both mean squares are estimators of σ^2, and the ratio

$$F = \frac{\left(\dfrac{SS_{hypothesis}}{1}\right)}{\left(\dfrac{SSE_{unrestricted}}{n-2}\right)} = \frac{MS_{hypothesis}}{MSE_{unrestricted}},$$

is indeed distributed as F with $(1,(n-2))$ degrees of freedom. If the null hypothesis is not true, the numerator will tend to increase leading to rejection in the right tail. Of course, as before, the F test is two-tailed.

Remembering that $S_{xx} = (n-1)s_x^2$, it is of interest to note that the numerator will become larger as

β_1 becomes larger,
n becomes larger,
The dispersion of x increases, and/or
$s_{y|x}^2$ becomes smaller.

Note that these are the same conditions we noted for the t test. In fact, the two tests are identical, since $t^2(n-2) = F(1, n-2)$. For this case, the t statistic may be preferable because it can be used for both one- and two-tailed tests, as well as for tests of other hypotheses, and it can be used for a confidence interval. However, as we will see later, the t statistic is not directly applicable to more complex models.

EXAMPLE 2.1
CONTINUED

Inferences on β_1 Using the Linear Model The preliminary calculations we have already used for obtaining the estimates of the parameters provide the quantities required for this test. We have

$$SSE_{restricted} = S_{yy} = 3744.36,$$
$$SS_{hypothesis} = S_{xy}^2/S_{xx} = 3085.74,$$

then by subtraction,

$$SSE_{unrestricted} = SSE_{restricted} - SS_{hypothesis} = 3744.36 - 3085.74$$
$$= 658.62.$$

The small difference from the result obtained directly from the residuals is due to roundoff. We can now compute

$$MSE_{unrestricted} = 658.62/18 = 36.59,$$

and the F statistic:

$$F = \frac{MS_{hypothesis}}{MS_{unrestricted}} = \frac{3085.74}{36.59} = 84.333.$$

The *p*-value of less than 0.0001 can be obtained from a computer and leads to rejection of the hypothesis of no regression. The square of the *t* test obtained using the sampling distribution is 84.346; again, the slight difference is due to roundoff.   ◆

Most statistical calculations, especially those for regression analyses, are performed on computers using preprogrammed computing software packages. Virtually all such packages for regression analysis are written for a wide variety of analyses of which simple linear regression is only a special case. This means that these programs provide options and output statistics that may not be useful for this simple case.

EXAMPLE 2.1
CONTINUED   **Computer Output**   We will illustrate a typical computer output with PROC REG of the SAS System. We will perform the regression for estimating tree volumes (VOL) using the diameter at breast height (DBH). The results are shown in Table 2.3. All of the quantities we have presented are available in this output. However, the nomenclature is somewhat different from what we have used and corresponds to the more conventional usage in statistical computer packages.

There are three sections of this output. The first portion of the output refers to the partitioning of the sums of squares and the test for the effect of the model, which for the simple linear regression model is equivalent to testing H_0: $\beta_1 = 0$. The column headings are self-explanatory.

TABLE 2.3

COMPUTER OUTPUT FOR TREE-VOLUME REGRESSION

Dependent Variable: VOL

Analysis of Variance

Source	DF	Sum of Squares	Mean Square	F Value	Prob > F
Model	1	3085.78875	3085.78875	84.341	0.0001
Error	18	658.56971	36.58721		
C Total	19	3744.35846			

Root MSE	6.04874	R-square	0.8241	
Dep Mean	61.85150	Adj R-sq	0.8143	
C.V.	9.77945			

Parameter Estimates

Variable	DF	Parameter Estimate	Standard Error	T for H0: Parameter=0	Prob > |T|
INTERCEP	1	-45.566252	11.77448591	-3.870	0.0011
DBH	1	6.916122	0.75308532	9.184	0.0001

The first line, labeled "Model," lists the sums of squares, mean squares, the F-value, and the p-value associated with the F test for testing the effect of the overall regression. The sum of squares, mean square, and F statistics correspond to those we computed previously, except for roundoff error. Because these can also be interpreted as the reduction in the error sum of squares due to fitting a regression, these are commonly referred to as the **regression sum of squares** (SSR), **regression mean square** (MSR), and the **F test for regression**.

The second line, labeled "Error," refers to the unrestricted model error statistics, which are generally referred to simply as the **error** statistics, (SSE and MSE). The mean square from this line is used as the denominator for the F statistic.

The third line, labeled "C Total," corresponds to what we have called the restricted error sum of squares (the corresponding mean square is not given as it is rarely used). The nomenclature "C Total" stands for "corrected total" and is the sum of squared deviations from the mean, also called "corrected" for the mean. This quantity is often simply called the **total sum of squares** (TSS), as it is a measure of the total variability of observations from the overall mean.

The second section contains some miscellaneous descriptive statistics:

"Root MSE" is the square root of the error mean square. The value of 6.05 is the standard deviation of the residuals, which we called $s_{y|x}$.
"Dep Mean" is simply $\bar{y}$.
"C.V.," the coefficient of variation, is the standard deviation divided by $\bar{y}$, expressed as a percentage.
R-square and Adj R-sq will be discussed later.

The last portion of the output contains statistics associated with the regression coefficients. The entries under the heading "Variable" identify the parameters. INTERCEP refers to β_0, the intercept, and DBH is the mnemonic computer name for the independent variable x and identifies β_1, the estimated regression parameter for that variable. The output gives the standard error and the test for zero values of both the slope and the intercept. Note that the square of the t statistic is indeed equal to the F statistic in the top portion. ◆

2.4 Inferences on the Response Variable

In addition to the inferences about the parameters of the regression model, we are also interested in how well the model estimates the behavior of the dependent variable. In other words, we want information on the reliability of the regression estimate of the dependent variable. In this context there are two different, but related, inferences:

1. *Inferences on the mean response*. In this case, we are concerned with how well the model estimates $\mu_{y|x}$, the conditional mean of the population for any specified x value.
2. *Inferences for prediction*. In this case, we are interested in how well the model predicts the value of the response variable, y, for a single randomly chosen future observation having a specified value of the independent variable, x.

The point estimate for both of these inferences is the value of $\hat{\mu}_{y|x}$ for the specified value of x. However, because the point estimate represents two different inferences, we denote them by two different symbols. Specifically, we denote the estimated mean response by $\hat{\mu}_{y|x}$, and predicted single value by $\hat{y}_{y|x}$. And, because these estimates have different implications, each has a different variance (and standard error).

For a specified value of x, say x^*, the variance for the estimated mean is

$$\text{var}(\hat{\mu}_{y|x*}) = \sigma^2 \left[\frac{1}{n} + \frac{(x^* - \bar{x})^2}{S_{xx}} \right],$$

and the variance for a single predicted value is

$$\text{var}(\hat{y}_{y|x*}) = \sigma^2 \left[1 + \frac{1}{n} + \frac{(x^* - \bar{x})^2}{S_{xx}} \right].$$

Both of these variances vary with different values of x^* and both are at their minimum value when $x^* = \bar{x}$. In other words, the response is estimated with greatest precision when the independent variable is at its mean, with the variance of both estimates increasing as x^* deviates from $\bar{x}$. It is also seen that var$(\hat{y}_{y|x*}) >$ var$(\hat{\mu}_{y|x*})$ because a mean is estimated with greater precision than is a single value. Finally, it is of interest to note that when x^* takes the value $\bar{x}$, the estimated conditional mean is $\bar{y}$ and the variance of the estimated mean is indeed σ^2/n, the familiar variance of the mean.

Substituting the mean square error, MSE, for σ^2 provides the estimated variance. The square root is the corresponding standard error used in hypothesis testing or (more commonly) interval estimation using the appropriate value from the t distribution with $(n - 2)$ degrees of freedom.

The variance of the intercept, β_0, can be found by letting $x^* = 0$ in the variance of $\hat{\mu}_{y|x}$. Thus, the variance of $\hat{\beta}_0$ is

$$\text{var}(\hat{\beta}_0) = \sigma^2 \left[\frac{1}{n} + \frac{(\bar{x})^2}{S_{xx}} \right] = \sigma^2 \left[\frac{\sum x^2}{n S_{xx}} \right].$$

Substituting MSE for σ^2 and taking the square root provides the estimated standard error, which can be used for hypothesis tests and confidence intervals. As we have noted, in most applications β_0 represents an extrapolation and is thus not a proper candidate for inferences. However, since a computer does not know if the intercept is a useful statistic for any specific problem, most computer programs do provide that standard error as well as the test for the null hypothesis that $\beta_0 = 0$.

EXAMPLE 2.1
CONTINUED

Inferences for the Response   We illustrate the calculations for the confidence interval for the mean volume (VOL) for x = DBH = 10.20 inches in Example 2.1. Putting the value x = 10.20 in the regression equation, we get $\hat{\mu}_{y|x}$ = 24.978 cubic feet. From previous calculations we have

$$\bar{x} = \text{mean DBH} = 15.5315,$$

$$S_{xx} = 64.5121,$$

$$\text{MSE}_{\text{unrestricted}} = 36.5872.$$

Using these quantities we have

$$\text{var}(\hat{\mu}_{y|x}) = 36.5872\ [0.05 + (10.20 - 15.5315)^2/\ 64.5121]$$

$$= 36.5872\ (0.05 + 0.4406)$$

$$= 17.950.$$

The square root of 17.950 = 4.237 is the standard error of the estimated mean. The 95% confidence interval for the estimated mean, using t = 2.101 for α = 0.05 and 18 degrees of freedom, is

$$24.978 \pm (2.101)\ (4.237),$$

or from 16.077 to 33.879 cubic feet. This interval means that, using the regression model, we are 95% confident that the true mean height of the population of trees with DBH = 10.20 inches is between 16.077 and 33.879 cubic feet. The width of this interval may be taken as evidence that the estimated model may not have sufficient precision to be very useful. A plot of the actual values, the estimated regression line, and the locus of all 0.95 confidence intervals are shown in Figure 2.5. The minimum width of the intervals at the mean of the independent variable is evident.

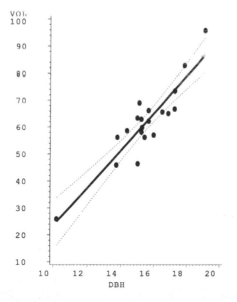

FIGURE 2.5 Plot of Confidence Intervals

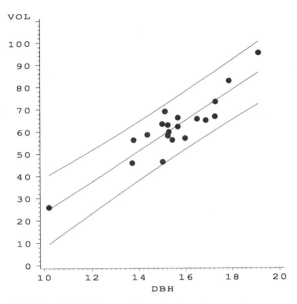

FIGURE 2.6 Plot of Prediction Intervals

The computations for the prediction interval are similar and will, of course, produce wider intervals. This is to be expected since we are predicting individual observations rather than estimating means. Figure 2.6 shows the 0.95 prediction intervals along with the original observations and the regression line. Comparison with Figure 2.4 shows that the intervals are indeed much wider, but both do have the feature of being narrowest at the mean. In any case, the width of these intervals may suggest that the model is not adequate for very reliable prediction. ◆

At this point it is important to note that both estimation and prediction is only valid within the range of the sample data. In other words, extrapolation is typically *not* valid. Extrapolation and other potential misuses of regression will be discussed in Section 2.8.

2.5 Correlation and the Coefficient of Determination

The purpose of a regression analysis is to estimate or explain a response variable (y) for a specified value of a factor variable (x). This purpose implies that the variable x is chosen or "fixed" by the experimenter (hence, the term independent or factor variable) and the primary interest of a regression analysis is to make inferences about the dependent variable using information from the independent variable. However, this is not always the case. For example, in a study of the relation between the height of a person and that person's weight, a random sample of individuals would be taken, and the height and weight of each individual recorded. The objective of the study might be to simply examine the relationship between the two variables. Examining the strength of the relationship between the two variables would be done using a **correlation model**. A correlation model consists of variables, all of which are random. Unlike a regression model, a correlation model specifies the joint distribution of both variables, not just the conditional distribution of y with a fixed x.

The correlation model most often used is the normal correlation model. This model specifies that the two variables (x, y) have what is known as the bivariate normal distribution. This distribution is defined by five parameters: the means of x and y, the variances of x and y, and the **correlation coefficient**, ρ. The correlation coefficient measures the strength of a linear (straight-line) relationship between the two variables. The correlation coefficient has the following properties:

1. Its value is between $+1$ and 1 inclusive. A positive correlation coefficient implies a direct relationship while a negative coefficient implies an inverse relationship.
2. Values of $+1$ and -1 signify an *exact* direct and inverse relationship, respectively, between the variables. That is, a plot of the values of x and y exactly describe a straight line with a positive or negative slope.
3. A correlation of zero indicates there is no linear relationship between the two variables. This condition does not necessarily imply that there is no relationship, because correlation only measures the strength of a straight line relationship.
4. The correlation coefficient is symmetric with respect to the two variables. It is thus a measure of the strength of a linear relationship between any two variables, even if one is an independent variable in a regression setting.

Because correlation and regression are related concepts, they are often confused, and it is useful to repeat the basic definitions of the two concepts:

DEFINITION

The **regression model** describes a linear relationship where an independent or factor variable is used to estimate or explain the dependent or response variable. In this analysis, one of the variables, x, is "fixed," or chosen at particular values. The other, y, is the only variable subject to a random error.

DEFINITION

The **correlation model** is used to describe the strength of a linear relationship between two variables, where both are random variables.

The parameter ρ of the normal correlation model can be estimated from a sample of n pairs of observed values of two variables x and y using the following estimator[3]:

$$\hat{\rho} = r = \frac{\Sigma(x - \bar{x})(y - \bar{y})}{\sqrt{\Sigma(x - \bar{x})^2 \Sigma(y - \bar{y})^2}} = \frac{S_{xy}}{\sqrt{S_{xx} S_{xy}}}.$$

The value r, called the *Pearson product moment correlation coefficient*, is the sample correlation between x and y and is a random variable. The sample correlation coefficient has the same four properties as the population correlation coefficient. Since we will be interested in making inferences about the population correlation coefficient, it seems logical to use this sample correlation.

The hypothesis that is usually of interest is

$$H_0: \rho = 0, \text{ vs}$$

$$H_1: \rho \neq 0.$$

The appropriate test statistic is

$$t_{(n-2)} = r \frac{\sqrt{n-2}}{\sqrt{1-r^2}},$$

where $t(n - 2)$ is the t distribution with $n - 2$ degrees of freedom.

To construct a confidence interval on ρ is not so simple. The problem is that the sampling distribution of r is very complex for nonzero values of ρ and therefore does not lend itself to standard confidence interval construction techniques. Instead, this task is performed by an approximate procedure. The Fisher z transformation states that the random variable

$$z' = \tfrac{1}{2} \log_e \left[\frac{1 + r}{1 - r} \right]$$

is an approximately normally distributed variable with mean

$$\text{Mean} = \tfrac{1}{2} \log_e \left[\frac{1 + \rho}{1 - \rho} \right], \text{ and}$$

$$\text{Variance} = \frac{1}{n - 3}.$$

The use of this transformation for hypothesis testing is quite straightforward: The computed z' statistic is compared to percentage points of the normal distribution. A confidence interval is obtained by first computing the interval for z':

$$z' \pm z_{\alpha/2} \sqrt{\frac{1}{n - 3}}.$$

Note that this formula provides a confidence interval in terms of z', which must be converted to an interval in terms of ρ. This conversion requires the solution of

[3] These estimators are obtained by using maximum likelihood methods (discussed in Appendix C).

a nonlinear equation and is therefore more efficiently performed with the aid of a table. An example of such a table can be found in Neter *et al.* (1989), Table A.7.

EXAMPLE 2.2   A study is being performed to examine the correlation between scores on a traditional aptitude test and scores on a final exam given in a statistics course. A random sample of 100 students is given the aptitude test, and upon completing the statistics course, given a final exam. The data resulted in a sample correlation coefficient value of 0.65. We first test to see if the correlation coefficient is significant. If so, we will then construct a 95% confidence interval on ρ.

The hypotheses of interest are

$$H_0: \rho = 0, \text{ vs}$$

$$H_1: \rho \neq 0.$$

The test statistic is

$$t = \frac{(0.65)\sqrt{98}}{\sqrt{1 - (0.65)^2}} = 20.04.$$

The p-value for this statistic is less than 0.0001, indicating that the correlation is significant.

For the confidence interval, substituting 0.65 for r in the formula for z' gives the value 0.775. The variance of z' is given by $1/97 = 0.0103$; the standard deviation is 0.101. Since we want a 95% confidence interval, $z_{\alpha/2} = 1.96$. Substituting into the formula for the confidence interval on z gives us 0.576 to 0.973. Using the table from Neter *et al.* (1989), we obtain the corresponding values of ρ, which are 0.52 and 0.75. Thus, we are 0.95 confident that the true correlation between the scores on the aptitude test and the final exam is between 0.52 and 0.75.   ◆

Although statistical inferences on the correlation coefficient are only strictly valid when the correlation model fits (that is, when the two variables have the bivariate normal distribution), the concept of correlation also has application in the traditional regression context. Since the correlation coefficient measures the strength of the linear relationship between the two variables, it follows that the correlation coefficient between the two variables in a regression equation should be related to the "goodness of fit" of the linear regression equation to the sample data points. In fact, this is true. The sample correlation coefficient is often used as an estimate of the "goodness of fit" of the regression model. More often, however, the square of the correlation coefficient, called the **coefficient of determination**, is used for this effort.

It is not difficult to show that

$$r^2 = SSR/TSS,$$

where SSR is the sum of squares due to regression, and TSS is the corrected total sum of squares in a simple regression analysis. The coefficient of determination, or r-square, is a descriptive measure of the relative strength of the corresponding regression. In fact, as can be seen from the foregoing relationship, r^2 is the proportional reduction of total variation associated with the regression of y on x and is therefore widely used to describe the effectiveness of a linear regression model. This is the statistic labeled R-SQUARE in the computer output (Table 2.3). It can also be shown that

$$F = \frac{MSR}{MSE} = \frac{(n-2)r^2}{(1-r^2)},$$

where F is the computed F statistic from the test for the hypothesis that $\beta_1 = 0$. This relationship shows that large values of the correlation coefficient generate large values of the F statistic, both of which imply a strong linear relationship. This relationship also shows that the test for a zero correlation is identical to the test for no regression, that is, the hypothesis test of $\beta_1 = 0$. {Remember that $[t(v)]^2 = F(1, v)$.}

We illustrate the use of r^2 using the data in Example 2.1. The correlation coefficient is computed using the quantities available from the regression analysis:

$$r = \frac{S_{xy}}{\sqrt{S_{xx}S_{xy}}}$$

$$= \frac{446.17}{\sqrt{(64.5121)(3744.36)}}$$

$$= \frac{446.17}{491.484} = 0.908.$$

Equivalently, from Table 2.3, the ratio of SSR to TSS is 0.8241, the square root is 0.908, which is the same result. Furthermore, $r^2 = 0.8241$, as indicated by R-SQUARE in Table 2.3, which means that approximately 82% of the variation in tree volumes can be attributed to the linear relationship of volume to DBH.

2.6 Regression through the Origin

In some applications it is logical to assume that the regression line goes through the origin, that is, $\hat{\mu}_{y|x} = 0$ when $x = 0$. For example, in Example 2.1, it can be argued that when DBH $= 0$, there is no tree and therefore the volume must be zero. If this is the case, the model becomes

$$y = \beta_1 x + \epsilon,$$

where y is the response variable, β_1 is the slope, and ϵ is a random variable with mean zero and variance σ^2. Unfortunately, the naive use of this model may lead to confusing inferences.

Regression through the Origin Using the Sampling Distribution

The least squares principle is used to obtain the estimator for the coefficient:

$$\hat{\beta}_1 = \frac{\sum xy}{\sum x^2}.$$

The resulting estimate $\hat{\beta}_1$ has a sampling distribution with mean β_1 and variance

$$\text{Variance } (\hat{\beta}_1) = \frac{\sigma^2}{\Sigma x^2}.$$

The error sum of squares and the corresponding mean square can be calculated directly[4]:

$$\text{MSE} = \frac{\Sigma(y - \hat{\mu}_{y|x})^2}{n - 1}.$$

Notice that the degrees of freedom are $(n - 1)$ because the model contains only one parameter to be estimated. This mean square can be used for the t test of the hypothesis H_0: $\beta_1 = 0$:

$$t = \frac{\hat{\beta}_1}{\sqrt{\dfrac{\text{MSE}}{\Sigma x^2}}},$$

which is compared to the t distribution with $(n - 1)$ degrees of freedom.

EXAMPLE 2.1
REVISITED

Regression through the Origin   We will use the data from Example 2.1 and assume that the regression line goes through the origin. The preliminary calculations have already been presented and provide

$$\hat{\beta}_1 = \frac{\Sigma xy}{\Sigma x^2} = \frac{19659.1}{4889.06} = 4.02104.$$

In other words, our estimated regression equation is

$$\text{Estimated volume} = 4.02104(\text{DBH})$$

Using this equation, we compute the individual values of $\hat{\mu}_{y|x}$ and the residuals $(y - \hat{\mu}_{yx})$ (computations not shown). These are used to compute the error sum of squares and the error mean square, which is our estimate of σ^2:

$$s_{y|x}^2 = \text{MSE} = \frac{\Sigma(y - \hat{\mu}_{x|y})^2}{n - 1} = \frac{1206.51}{19} = 63.500.$$

The variance of the sampling distribution of $\hat{\beta}_1$ is $\sigma^2/\Sigma x^2$. Using the estimated variance, we compute the estimated variance of $\hat{\beta}_1$:

$$\text{Variance } \hat{\beta}_1 = \frac{\text{MSE}}{\Sigma x^2} = \frac{63.500}{4889.06} = 0.01299.$$

Finally, the t statistic for testing the hypothesis $\beta_1 = 0$:

$$t = \frac{\hat{\beta}_1}{\sqrt{\dfrac{\text{MSE}}{\Sigma x^2}}} = \frac{4.02104}{0.11397} = 35.283.$$

The hypothesis that there is no regression is easily rejected.

[4] The computational form is presented in the next section.

Regression through the Origin Using Linear Models

The least squares estimator is the same we have just obtained. The restricted model for $H_0: \beta_1 = 0$ is

$$y = \epsilon.$$

In other words, the restricted model specifies $\mu_{y|x} = 0$; hence, the restricted, or total, sum of squares is $\sum y^2$, which has n degrees of freedom since its formula does not require any sample estimates.

For this model, the shortcut formula for the hypothesis sum of squares is

$$SS_{hypothesis} = \frac{[\sum xy]^2}{\sum x^2} = \hat{\beta}_1 \sum xy,$$

and the unrestricted model error sum of squares is obtained by subtraction from the restricted model error sum of squares, $\sum y^2$.

We now compute the required sums of squares:

$$SS_{restricted} = TSS = \sum y^2 = 80256.52$$

$$SS_{hypothesis} = \hat{\beta}_1 \sum xy = 4.02104 \cdot 19659.1 = 79050.03$$

$$SSE_{unrestricted} = SSE_{restricted} - SS_{hypothesis} = 1206.49.$$

The unrestricted error sum of squares has 19 degrees of freedom; hence, the mean square is 63.500, which is the same we obtained directly. The F ratio for the test of $\beta_1 = 0$ is

$$F = \frac{79050.03}{63.500} = 1244.89,$$

which leads to rejection. Again, note that the F value is the square of the t statistic obtained previously.

It is of interest to compare the results of the models estimated with and without the intercept as shown by abbreviated computer outputs in Table 2.4. We can immediately see that the coefficient for DBH is smaller when there is no intercept and the error mean square is considerably larger, implying that the no intercept model provides a poorer fit.

The reason for this can be seen by a plot of the actual values and both regression lines as shown in Figure 2.7. It is apparent that the no-intercept regression line (dotted line) does not fit the data as well as the with intercept line (solid line).

However, the statistics for the significance tests suggest that the no-intercept regression is "stronger," since the F (or t) statistics as well as R-square are larger in value than they are for the with-intercept regression. The reason for these apparently contradictory results is readily understood by reviewing the unrestricted and restricted models for the two tests.

In the with-intercept regression, the test compares the error sum of squares for the model

$$\hat{\mu}_{y|x} = \hat{\beta}_0 + \hat{\beta}_1 x,$$

TABLE 2.4

REGRESSION WITH AND WITHOUT INTERCEPT

Dependent Variable: VOL, Regression with intercept

Analysis of Variance

Source	DF	Sum of Squares	Mean Square	F Value	Prob>F
Model	1	3085.78875	3085.78875	84.341	0.0001
Error	18	658.56971	36.58721		
C Total	19	3744.35846			

Parameter Estimates

Variable	DF	Parameter Estimate	Standard Error	T for H0: Parameter=0	Prob > |T|
INTERCEP	1	-45.566252	11.77448591	-3.870	0.0011
DBH	1	6.916122	0.75308532	9.184	0.0001

Dependent Variable: VOL, Regression without intercept

Analysis of Variance

Source	DF	Sum of Squares	Mean Square	F Value	Prob>F
Model	1	79050.01113	79050.01113	1244.873	0.0001
Error	19	1206.50837	63.50044		
U Total	20	80256.51950			

Parameter Estimates

Variable	DF	Parameter Estimate	Standard Error	T for H0: Parameter=0	Prob > |T|
DBH	1	4.021038	0.11396608	35.283	0.0001

which is 658.6, with the error sum of squares for the line

$$\hat{\mu} = \bar{y},$$

which is 3744.4. On the other hand, the test for the no-intercept regression compares the error sum of squares for the line

$$\hat{\mu}_{y|x} = \hat{\beta}_1 x,$$

which is 1206.5, to that for the line represented by

$$\hat{\mu} = 0,$$

which is 80246.5. Now the error sum of squares for the intercept model is indeed the smaller of the two, but the restricted or total sum of squares is very much larger in the for the no intercept model; hence, the larger value of the test statistic and R-square.

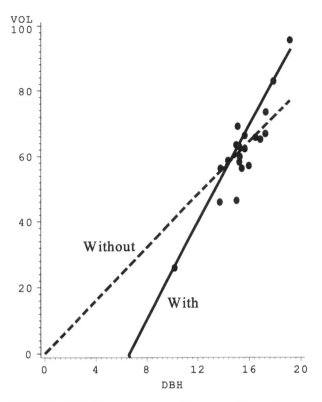

FIGURE 2.7 Regressions with and without Intercept

In fact, completely unreasonable results can be obtained if the no-intercept model is inappropriate. This is illustrated with an example.

EXAMPLE 2.3   Table 2.5 shows a small data set for which the estimated model with intercept is

$$\hat{\mu}_{yx} = 9.3167 - 0.830x,$$

with an error sum of squares of 0.606 and a t statistic for the test of no regression of -21.851.

The estimated regression line for the no-intercept regression is

$$\hat{\mu}_{y|x} = 0.6411x,$$

with an error sum of squares of 165.07 and a t statistic of 2.382 ($p = 0.0444$). Not only does the coefficient have the wrong sign, but in spite of an apparently significant regression, the error sum of squares is *larger* than the total sum of squares used for the model with intercept.

The reason for these unusual results is shown by the plot of actual data and the no-intercept model predictions shown in Figure 2.8. The positively sloping line does indeed fit better than the line

$$y = 0,$$

TABLE 2.5

INAPPROPRIATE DATA FOR
A NO-INTERCEPT
REGRESSION

Obs	x	y
1	1	8.5
2	2	7.8
3	3	6.9
4	4	5.5
5	5	5.1
6	6	4.8
7	7	3.3
8	8	2.9
9	9	1.7

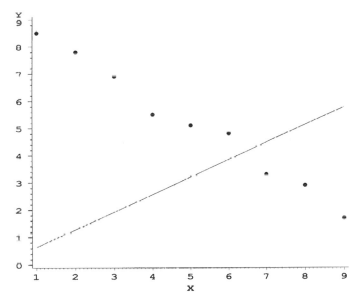

FIGURE 2.8 Plot of Inappropriate No-Intercept Model

which is the line with $\beta_1 = 0$ in the no-intercept model (which explains why the t test rejected that hypothesis). However, it does not fit as well as the line

$$y = \bar{y},$$

which is the line with $\beta_1 = 0$ in the regression model with intercept.   ◆

This is certainly an extreme example that is unlikely to be proposed by anyone familiar with the process that gave rise to the data. However, it does illustrate the problems occurring when forcing the regression through the intercept. For this reason, we strongly recommend that the results of any no-intercept regression be examined very critically, and that the diagnostic methods discussed in the next chapter be done any time a no-intercept model is applied.[5]

When using the no-intercept model, care must be used in interpreting the residuals because they do not sum to zero. Another important caution when using regression through the origin is, as Example 2.3 illustrated, the fact that the sums of squares due to error may be larger than the total sums of squares (as typically defined). This means that the coefficient of determination, RSQUARE, may turn out to negative!

Finally, it should be stated that the regression through the origin model is not often used in practical applications and only should be used when it really is appropriate. Even though there are many processes that have zero-valued response when the value of the independent variable is zero, that does not necessarily mean that the regression line that best fits the data will go through the origin. This is especially true for problems where data is not measured at $x = 0$. Therefore, it is usually recommended that the intercept term be included in the analysis, if only as a "placeholder."

2.7 Assumptions on the Simple Linear Regression Model

In Section 2.2 we briefly listed the assumptions underlying the regression model. Obviously, the validity of the results of a regression analysis requires that these assumptions be satisfied. We can summarize these assumptions as follows:

1. The model adequately describes the behavior of the data.
2. The random error, ϵ, is an independently and normally distributed random variable, with mean zero and variance σ^2.

These assumptions may look familiar,[6] as they should, since virtually all statistical methods presented in statistical methodology courses are based on linear models. Therefore, in a regression analysis it is advisable to make an effort to determine if violations of assumptions may have occurred.

In this section we discuss the following:

How these violations may occur
How existence of these violations may affect the results
Some tools for detecting violations

If the model is not correctly specified, the analysis is said to be subject to specification error. This error most often occurs when the model should contain

---

[5] Another way of evaluating a no-intercept regression is to perform the restricted regression that is illustrated in Freund and Littell, 1991, *The SAS System for Regression*, Section 2.4.5.

[6] Not discussed here is the assumption that x is fixed and measured without error. Although this is an important assumption, it is not very frequently violated to the extent as to greatly influence the results of the analysis. Also diagnostic and remedial methods for violations of this assumption are beyond the scope of this book (C.F. Seber, 1977).

additional parameters and/or additional independent variables. It can be shown that a specification error causes coefficient and variance estimates to be biased, and since the bias is a function of the unknown additional parameters, the magnitude of the bias is not known. A common example of a specification error is for the model to describe a straight line when a curved line should be used.

Assumption 2 just listed actually contains three conditions that may occur independently or in any combination. These three are (a) independence of the random errors, (b) normality of the random errors, and (c) constant variance of the random errors. The assumption of independence is often related to the specification of the model in the sense that if an important independent variable is omitted, the responses, and hence the errors, may be related to that variable. This relationship will often result in correlated error terms. An example of this is when responses are measured over time and changes in the response are a result of time. A remedial measure for this violation is to use the variable time in a model that allows for correlated errors.

The assumption of equal variances is one that is frequently violated in practice, and unequal variances often come in conjunction with nonnormality. When the error variance varies in a systematic fashion, we can use the method of weighted least squares discussed in Chapter 4 to obtain the estimators of the regression parameters. Frequently, nonnormality and unequal variances take the form of increasing skewness and increasing variability as the mean of the response variable increases (or decreases). In other words, the size of the variance and/or the amount of skewness will be related to the mean of the response variable. Fortunately, it is often the case that one transformation on the response variable will correct both violations. Transformations are discussed in detail in Chapters 4 and 8.

Outliers or unusual observations may be considered a special case of unequal variances. The existence of outliers can cause biased estimates of coefficients as well as incorrect estimates of the variance. It is, however, very important to emphasize that simply discarding observations that appear to be outliers is not good statistical practice. Since any of these violations of assumptions may cast doubt on estimates and inferences, it is important to see if such violations may have occurred (see Chapter 4).

A popular tool for detecting violation of assumptions is an analysis of the residuals. Recall that the residuals are the differences between the observed y-values and the estimated conditional means, $\hat{\mu}_{y|x}$, that is, $(y - \hat{\mu}_{y|x})$. An important part of an analysis of residuals is a residual plot, which is a scatterplot featuring the individual residual values $(y - \hat{\mu}_{y|x})$ on the vertical axis and either the predicted values $(\hat{\mu}_{y|x})$ or x values on the horizontal axis. Occasionally, residuals may also be plotted against possible candidates for additional independent variables.

A regression model that has no violations of assumptions will have a residual plot that appears as a roughly horizontal band around zero. This band will have about the same width for all values on the horizontal axis. Specification errors will show up as a distinct, recognizable pattern other than the horizontal band. Violations of the common variance assumption may show up as a fan-shaped pattern. The most frequently occurring pattern is for the point of the fan facing left, that is, the larger residuals occur with the larger values of $\hat{\mu}_{y|x}$. Outlier may show up as points that lie far beyond the scatter of the remaining residuals. The key word here is "may," because violations are not always easily detected by residual plots and may have occurred even when plots look quite well behaved.

Additional analyses of residuals consist of using descriptive methods, especially the exploratory data analysis techniques such as stem and leaf or box plots. Virtually all computer programs for regression provide for the relatively easy implementation of such analyses.

Residual Plots for Example 2.1   Figure 2.9 contains two residual plots. The left plot is the plot of the residuals from the model using the intercept, while the right one contains the residual plot using the no-intercept model. In both cases, the residuals are plotted against the independent variable (DBH).

The residuals from the model with the intercept appear to be randomly scattered in both directions, suggesting that there are no serious violations of assumptions. The residuals from the no-intercept model show a definite trend, suggesting a model deficiency and reinforcing the inadequacy of the model we have already noted.   ◆

The examination of residuals using residual plots is a very subjective method. We will see later that, in spite of the residual plot looking as if the model is perfectly adequate, simply using DBH is not good enough. That is, adding the

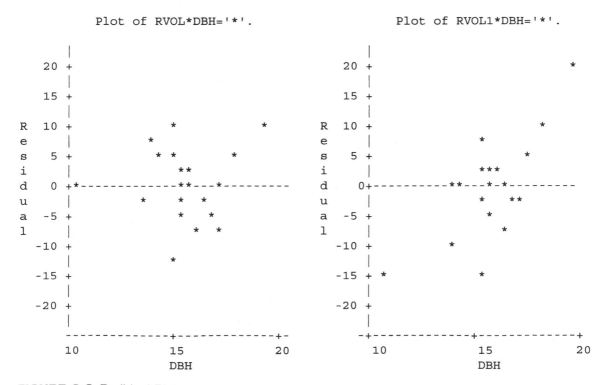

FIGURE 2.9 Redidual Plots

other variables shown in Table 2.1 does indeed provide a better model, as we will see in Chapter 3. In other words, the apparent lack of recognizable patterns in the residual plot is no guarantee that there are no violations of assumptions.

The detection of violations of assumptions can be a very technical exercise, and although it usually becomes more difficult as the complexity of the model increases, the general strategy is the same. For this reason, the complete presentation of this topic is postponed until Chapter 4, following the coverage of multiple regression.

2.8 Uses and Misuses of Regression

As we have noted, the validity of the results of the statistical analysis requires fulfillment of certain assumptions about the data. However, even if all assumptions are fulfilled, there are some limitations of regression analysis:

- The fact that a regression relationship has been found to exist does *not*, by itself, imply that *x causes y*. For example, it has been clearly demonstrated that smokers have more lung cancer (and other diseases) than do non-smokers, but this relationship does not *by itself* prove that smoking causes lung cancer. Basically, to prove cause and effect, it must also be demonstrated that no other factor could cause that result.
- It is not advisable to use an estimated regression relationship for extrapolation. That is, the estimated model should not be used to make inferences on values of the dependent variable beyond the range of observed *x*-values. Such extrapolation is dangerous, because although the model may fit the data quite well, there is no evidence that the model is appropriate outside the range of the existing data.

2.9 Inverse Predictions

At times, it is desirable to use a regression of *y* on *x* to make predictions of the value of *x* that resulted in a new observation of *y*. This procedure is known as *inverse prediction* or the *calibration problem*. This method has application when inexpensive, quick, or approximate measurements (y) are related to precise, often expensive, or time-consuming measurements (x). A sample of *n* observations can be used to determine the estimated regression equation:

$$\hat{\mu}_{y|x} = \hat{\beta}_0 + \hat{\beta}_1 x,$$

and the equation solved for *x*:

$$x = \frac{\hat{\mu}_{y|x} - \hat{\beta}_0}{\hat{\beta}_1}.$$

Suppose that we had a new observation, $\hat{y}_{y|x\ (new)}$, and we want to estimate the value of x that gave rise to this new observation, say $\hat{x}_{(new)}$. We then get an estimate for this value as:

$$\hat{x}_{(new)} = \frac{\hat{y}_{y|x(new)} - \hat{\beta}_0}{\hat{\beta}_1}.$$

A 95% confidence interval on $\hat{x}_{(new)}$ is given by:

$$\hat{x}_{(new)} \pm t_{\alpha/2}s(\hat{x}_{new}),$$

where

$$s^2(\hat{x}_{new}) = \frac{MSE}{\hat{\beta}_1^2}\left[1 + 1 \div n + \frac{(\hat{x}_{(new)} - \bar{x})^2}{\Sigma(x_i - \bar{x})^2}\right],$$

and $t_{\alpha/2}$ has $n - 2$ degrees of freedom.

EXAMPLE 2.4   Prices of homes for sale are usually based on several characteristics of the home, including the amount of living space. A certain real estate agent wanted to examine the relationship between the asking price of homes and the living space. To establish an estimate of this relationship, a sample of 15 homes listed for sale in the agent's home town was obtained. The asking price (y) was recorded in thousands of dollars and the living space (x) determined in thousands of square feet. The data are given in Table 2.6. Since the determination of the actual living space is quite tedious, the agent wanted to be able to estimate the living space based on the asking price for the next house to be listed for sale.

The least squares equation for predicting price on living space is

$$\hat{\mu}_{y|x} = -12.064 + 32.879x.$$

Using this equation, the agent wanted to estimate the living space for a house being offered for sale at $47,000 with 95% confidence. Using the data in Table 2.6, the following values were computed:

TABLE 2.6

SAMPLE OF HOUSES FOR SALE

Home	Space	Price	Home	Space	Price
1	1.326	27.90	9	1.708	41.50
2	1.391	33.50	10	1.529	35.50
3	1.000	19.00	11	2.234	65.50
4	1.542	31.00	12	1.607	48.55
5	0.735	18.90	13	1.648	41.50
6	1.444	38.30	14	1.608	46.50
7	1.796	45.00	15	1.020	18.00
8	1.770	43.50			

1. MSE = 21.27
2. $\sum(x_i - \bar{x})^2 = 1.8997$
3. $\bar{x} = 1.49$

Using $\hat{x}_{(new)} = (47 + 12.064)/32.879 = 1.796$, we can estimate the living space for the $47,000 house as about 1800 square feet. We get the following value:

$$s^2(\hat{x}_{(new)}) = \frac{21.27}{(32.879)^2}\left[1 + \frac{1}{15} + \frac{(1.796 - 1.49)^2}{1.8997}\right] = 0.022.$$

From Table A.2, we get $t_{.025}(13) = 2.1604$. Therefore, the desired 95% confidence interval will be

$$1.796 \pm 2.1604(\sqrt{0.022}) = 1.796 \pm 0.32, \text{ or}$$

$$1.476 \text{ to } 2.116.$$

This interval may be too wide to be very meaningful to the real estate agent. This is to be expected given that only 15 houses were sampled and given the large MSE from the regression analysis. ◆

2.10 Summary

This chapter provides a detailed description of the simple linear regression model,

$$y = \beta_0 + \beta_1 x + \epsilon,$$

which is used as a basis for establishing the nature of a relationship between values of an independent or factor variable x, and the values of a dependent or response variable y. The chapter begins with the presentation and interpretation of the model and continues with the estimation of the model parameters, based on a sample of n pairs of observations. We next discuss statistical inferences on these parameters and the estimated or predicted values of the response variable.

This is followed by the related concept of correlation, which establishes the strength of the relationship between two variables. The correlation model is based on the linear correlation coefficient. Next is a short introduction to the use of restricted models, specifically the presentation of what happens when the model does not include the intercept. Final sections cover possible violations of assumptions and misuses of regression methods and inverse prediction.

CHAPTER EXERCISES

1. For those who want to exercise their calculators, use the following small data set for practice.

x	1	2	3	4	5	6	7	8
y	2	5	5	8	9	7	9	10

(a) Calculate the least squares estimates of the coefficients of the regression line that predicts y from x.

(b) Test the hypothesis that $\beta_1 = 0$. Construct a 95% confidence interval on β_1.

(c) Calculate R^2. Explain it.

(d) Construct a 95% prediction interval on the value of y when $x = 5$.

(e) Calculate the residuals and plot them against x. Explain the plot.

(f) Completely explain the relation between x and y.

2. A sample of 22 bass were caught by a fisheries scientist who measured length (TL) in mm and weight (WT) in grams. The data are shown in Table 2.7 and are available as File REG02P02 on the data diskette.

(a) Use a regression model to see how well weight can be estimated by length.

(b) Use the residuals to check for violations of assumptions.

(c) Calculate a 95% confidence interval on the mean weight of a 300-mm bass.

3. Table 2.8 gives data on gas mileage (MPG) and size of car (WT in pounds) for a selection of 32 different cars. The data are also available as file REG02P03 on the data diskette.

(a) Calculate the linear regression equation that relates MPG (y) to WT (x).

(b) Use the inverse prediction method of Section 2.9 to predict the weight of a car that gets 25 mpg with 95% confidence.

4. The data in Table 2.9 show the grades for 15 students on the midterm examination and the final average in a statistics course.

(a) Do the regression analysis to predict the final average based on the midterm examination score.

(b) Estimate, using a 90% confidence interval, the value of a midterm score for a student whose final average will be 70.

(c) Fit a regression through the origin and compare it with part (a). Which model seems best?

5. The 1995 *Statistical Abstract of the United States* lists the breakdown of the Consumer Price Index by major groups for the years 1960 to 1994. The Consumer Price Index reflects the buying patterns of all urban consumers. Table 2.10 on page 70 and file REG02P05 on the data diskette list the major groups energy and transportation.

(a) Perform a correlation analysis to determine the relationship between transportation and energy. Calculate the confidence interval on the correlation coefficient. Explain the results.

(b) Perform separate regression analyses using year as the independent variable and transportation and energy, respectively, as dependent variables. Use residual plots to check assumptions. Explain the results.

TABLE 2.7

BASS DATA FOR EXERCISE 2

TL	WT
387	720
366	680
421	1060
329	480
293	330
273	270
268	220
294	380
198	108
185	89
169	68
102	28
376	764
375	864
374	718
349	648
412	1110
268	244
243	180
191	84
204	108
183	72

TABLE 2.8

MILEAGE DATA
FOR EXERCISE 3

WT	MPG
2620	21.0
2875	21.0
2320	22.8
3215	21.4
3440	18.7
3460	18.1
3570	14.3
3190	24.4
3150	22.8
3440	19.2
3440	17.8
4070	16.4
3730	17.3
3780	15.2
5250	10.4
5424	10.4
5345	14.7
2200	32.4
1615	30.4
1835	33.9
2465	21.5
3520	15.5
3435	15.2
3840	13.3
3845	19.2
1935	27.3
2140	26.0
1513	30.4
3170	15.8
2770	19.7
3570	15.0
2780	21.4

TABLE 2.9

DATA FOR EXERCISE 4

Student	Midterm	Final Average
1	82	76
2	73	83
3	95	89
4	66	76
5	84	79
6	89	73
7	51	62
8	82	89
9	75	77
10	90	85
11	60	48
12	81	69
13	34	51
14	49	25
15	87	74

6. A political scientist suspects that there is a relationship between the number of promises a political candidate makes and the number of promises that are fulfilled once the candidate is elected. Table 2.11 on page 71 lists the "track record" of 10 politicians.

(a) Calculate the correlation coefficient between promises made and kept.
(b) Test the hypothesis that $\rho = 0$.
(c) Interpret the results.

7. One way of evaluating a testing instrument is to do what is called an item analysis of the exam. One part of the item analysis is to examine individual questions relative to how they affect the score made on the test. This is done by correlating the score with "right" or "wrong" on each question. The resulting correlation coefficient is known as a point biserial correlation coefficient. It is calculated by correlating the score on the test with a variable that has a value 0 for "wrong" and 1 for "right." The higher the correlation, the more the question contributes toward the score. Use the equation given in Section 2.5 to calculate the point biserial correlation coefficient between the score and the variable x (which has 1 for "right" and 0 for "wrong") for the data on 30 students given in Table 2.12 on page 71. Test it for significance and explain the results. (This is for one question only; in evaluating a testing instrument, this would be done for all questions.)

8. In Exercise 6 we attempted to quantify the strength of the relationship between promises made and promises kept by politicians.

(a) Use the data in Table 2.11 to construct a regression equation that predicts the number of promises kept based on the number made.
(b) Do the inferences on the coefficients, including a 95% confident interval on β_1. Explain the results. Do they agree with the results of Exercise 6?

TABLE 2.10

DATA FOR EXERCISE 5

Year	Energy	Transportation
60	22.4	29.8
61	22.5	30.1
62	22.6	30.8
63	22.6	30.9
64	22.5	31.4
65	22.9	31.9
66	23.3	32.3
67	23.8	33.3
68	24.2	34.3
69	24.8	35.7
70	25.5	37.5
71	26.5	39.5
72	27.2	39.9
73	29.4	41.2
74	38.1	45.8
75	42.1	50.1
76	45.1	55.1
77	49.4	59.0
78	52.5	61.7
79	65.7	70.5
80	86.0	83.1
81	97.7	93.2
82	99.2	97.0
83	99.9	99.3
84	100.9	103.7
85	101.6	106.4
86	88.2	102.3
87	88.6	105.4
88	89.3	108.7
89	94.3	114.1
90	102.1	120.5
91	102.5	123.8
92	103.0	126.5
93	104.2	130.4
94	104.6	134.3

(c) Predict the number of promises kept for a politician making 45 promises. Use 0.95.

9. It has been argued that many cases of infant mortality are caused by teenage mothers who, for various reasons, do not receive proper prenatal care. Table 2.13 on page 72 lists data from the *Statistical Abstract of the United States* on the teenage birth rate per 1000 (TEEN) and the infant mortality rate per 1000 live births (MORT) for the 48 contiguous states.

(a) Perform a regression to estimate MORT using TEEN as the independent variable. Do the results confirm the stated hypothesis?

TABLE 2.11

DATA FOR EXERCISE 6

Politician	Promises Made	Promises Kept
1	21	7
2	40	5
3	31	6
4	62	1
5	28	5
6	50	3
7	55	2
8	43	6
9	61	3
10	30	5

TABLE 2.12

DATA FOR EXERCISE 7

Score	x	Score	x	Score	x
75	1	60	0	79	1
60	0	51	0	69	1
89	1	77	1	70	0
90	1	70	1	68	1
94	1	76	1	63	0
55	0	65	1	66	0
22	0	32	0	45	1
25	0	43	0	69	1
54	0	64	1	72	1
65	1	63	0	77	1

(b) Is the regression model significant at the 0.05 level?

(c) Construct a 95% confidence interval on β_1. Explain the result.

(d) Use residual plots to verify all assumptions on the model.

10. An engineer is interested in calibrating a pressure gauge using a set of standard pressure tanks. The gauge is used to measure pressure in five tanks with known pressure ranging from 50 psi to 250 psi. Each tank is measured three times using the gauge. The results are given in Table 2.14 on page 73.

(a) Use the inverse regression procedures of Section 2.9 to determine a calibration equation for the new gauge.

(b) Find a 95% confidence interval on the true pressure if the gauge reads 175.

TABLE 2.13

DATA FOR EXERCISE 9

STATE	TEEN	MORT
AL	17.4	13.3
AR	19.0	10.3
AZ	13.8	9.4
CA	10.9	8.9
CO	10.2	8.6
CT	8.8	9.1
DE	13.2	11.5
FL	13.8	11.0
GA	17.0	12.5
IA	9.2	8.5
ID	10.8	11.3
IL	12.5	12.1
IN	14.0	11.3
KS	11.5	8.9
KY	17.4	9.8
LA	16.8	11.9
MA	8.3	8.5
MD	11.7	11.7
ME	11.6	8.8
MI	12.3	11.4
MN	7.3	9.2
MO	13.4	10.7
MS	20.5	12.4
MT	10.1	9.6
NB	8.9	10.1
NC	15.9	11.5
ND	8.0	8.4
NH	7.7	9.1
NJ	9.4	9.8
NM	15.3	9.5
NV	11.9	9.1
NY	9.7	10.7
OH	13.3	10.6
OK	15.6	10.4
OR	10.9	9.4
PA	11.3	10.2
RI	10.3	9.4
SC	16.6	13.2
SD	9.7	13.3
TN	17.0	11.0
TX	15.2	9.5
UT	9.3	8.6
VA	12.0	11.1
VT	9.2	10.0
WA	10.4	9.8
WI	9.9	9.2
WV	17.1	10.2
WY	10.7	10.8

TABLE 2.14

CALIBRATION DATA FOR EXERCISE 10

Pressure in Tank	50	100	150	200	250
Gauge readings	48	100	154	200	247
	44	100	154	201	245
	46	106	154	205	146

Multiple Linear Regression

3.1 Introduction

Multiple linear regression is in some ways a relatively straightforward extension of simple linear regression allowing for more than one independent variable. The objective of multiple regression is the same as that of simple regression; that is, we want to use the relationship between a response (dependent) variable and factor (independent) variables to predict or explain the behavior of the response variable. This chapter will illustrate the similarities and the differences between simple and multiple linear regression, as well as develop the methodology necessary to use the multiple regression model.

Some examples of analyses using the multiple regression model include the following:

- Estimating weight gain of children using various levels of a dietary supplement, exercise, and behavior modification
- Predicting scholastic success (GPA) of college freshmen based on scores on an aptitude test, high school grades, and IQ level
- Estimating changes in sales associated with increased expenditures on advertising, increased number of sales personnel, increased number of management personnel, and various types of sales strategies

- Predicting daily fuel consumption for home heating based on daily temperature, daily humidity, daily wind velocity, and previous day's temperature
- Estimating changes in interest rates associated with the amount of deficit spending, value of the GNP, value of the CPI, and inflation rate

3.2 The Multiple Linear Regression Model

The multiple linear regression model is written as a straightforward extension of the simple linear model given in Section 2.2. The model is specified as

$$y = \beta_0 + \beta_1 x_1 + \beta_2 x_2 + \ldots + \beta_m x_m + \epsilon,$$

where

y is the dependent variable
$x_j, j = 1, 2, \ldots, m$, represent m different independent variables
β_0 is the intercept (value when all the independent variables are 0)
$\beta_j, j = 1, 2, \ldots, m$, represent the corresponding m regression coefficients
ϵ is the random error, usually assumed to be normally distributed with mean zero and variance σ^2

Although the model formulation appears to be a simple generalization of the model with one independent variable, the inclusion of several independent variables creates a new concept in the interpretation of the regression coefficients. For example, if multiple regression is to be used in estimating weight gain of children, the effect of each of the independent variables—dietary supplement, exercise, and behavior modification—depends on what is occurring with the other independent variables. In multiple regression we are interested in what happens when each variable is varied one at a time, while not changing values of any others. This is in contrast to performing several simple linear regressions, using each of these variables in turn, but where each regression ignores what may be occurring with the other variables. Therefore, in multiple regression, the coefficient attached with each independent variable should measure the average change in the response variable associated with changes in that independent variable, while all other independent variables remain fixed. This is the standard interpretation for a regression coefficient in a multiple regression model.

DEFINITION   Formally, a coefficient, β_j, in a multiple regression model is defined as a **partial regression coefficient**, whose interpretation is the change in the mean response, $\mu_{y|x}$, associated with a unit change in x_j *holding constant* all other variables.

In contrast, if m separate simple regressions are performed, the regression coefficient for the simple linear regression involving, say, x_j, is called the **total regres-**

sion coefficient and is interpreted as the change in the mean response, $\mu_{y|x}$, associated with a unit change in x_j, *ignoring* the effect of any other variables.

In other words, if we use a multiple regression model to estimate the weight gain of children, we obtain partial regression coefficients. If, on the other hand, we use three simple regressions, one for each independent variable, we will obtain total regression coefficients.

In most regression analyses the partial and total regression coefficients will have different values and, as we will see, obtaining partial coefficients is sometimes hampered by the lack of relevant data. For example, in the illustration where we want to predict scholastic success of college freshmen, most high school students who have high IQs and high scores on an aptitude test tend also to have high grades. The resulting lack of such students with low grades makes it difficult to estimate the partial coefficient due to grade.[1]

On the other hand, if we perform a regression using only students' grades, say, we imply that the other variables do not affect scholastic success. Therefore, it is not appropriate to use simple linear regressions when there exist other variables that should be included in a model. We will, in fact, see that the two types of coefficients may provide vastly different estimates and inferences.

EXAMPLE 3.1   To illustrate the differences between total and partial coefficients, we return to the forestry problem used for Example 2.1, as shown in Table 2.1. We will use three easily measured tree characteristics, DBH, D16, and HT, to estimate tree volume, VOL.

We first perform the three simple linear regressions (MODELS 1, 2, and 3) for which the estimated total regression coefficients are shown in the first portion of Table 3.1. We then perform the multiple regression (MODEL4) using all three independent variables,[2] resulting in the partial regression coefficients.

TABLE 3.1

TOTAL AND PARTIAL REGRESSION COEFFICIENTS

MODEL	INTERCEP	DBH	HT	D16
		TOTAL COEFFICIENTS		
MODEL1	−45.566	6.91612	.	.
MODEL2	−104.790	.	1.74408	.
MODEL3	−53.433	.	.	8.28789
		PARTIAL COEFFICIENTS		
MODEL4	−108.576	1.62577	0.69377	5.67140

[1] This is extensively discussed in Chapter 5.

[2] Procedures for obtaining these estimates will be presented in the next section.

It is readily seen that the total coefficients are not the same as the partial coefficients. For example, the total coefficient for DBH (MODEL1) is 6.916, whereas the partial coefficient (MODEL4) is 1.626. The difference becomes more clear if we again review the interpretation of these coefficients:

> The **total** coefficient says that the mean volume increases 6.916 units for each unit increase in DBH, ignoring the possible effect of other independent variables.
>
> The **partial** coefficient says that in a subpopulation containing trees of a specified value of HT and a specified value of D16, the volume will increase 1.626 units for each unit increase in DBH.

One feature of the independent variables in this example is that they vary together. That is, as DBH gets larger, so does D16, and to some lesser degree so does HT. In statistical terms, pairwise correlations among these variables are strong and positive. This means, for example, that the total coefficient for DBH also indirectly measures the effects of D16 and HT, whereas the partial coefficient measures only the effect of DBH for a population of trees having constant values of HT and D16. Therefore, the total coefficients are not readily useful.   ◆

From this discussion we may infer that if the independent variables do not vary together, that is, if they are uncorrelated, the partial and total coefficient will be the same. This is indeed true, but is a situation that rarely occurs in practice. However, it is true that when the relationships among independent variables are weak, then differences between partial and total regression coefficients are small and their interpretation becomes easier. The existence of correlations among independent variables is called **multicollinearity**, which is extensively discussed in Chapter 5.

We will return to the interpretation of the regression coefficients in Section 3.4 after we have discussed the estimation procedures for the multiple regression model.

3.3 Estimation of Coefficients

The multiple regression model presented in Section 3.2 has the form

$$y = \beta_0 + \beta_1 x_1 + \beta_2 x_2 + \ldots + \beta_m x_m + \epsilon,$$

where the terms in the model were defined in that section. To estimate the regression coefficients, we use a set of n observed values on the $(m + 1)$-tuple $(x_1,$

..., x_m, y) and use the least squares principle to obtain the following equation for estimating the mean of y:

$$\hat{\mu}_{y|x} = \hat{\beta}_0 + \hat{\beta}_1 x_1 + \ldots + \hat{\beta}_m x_m.$$

The least squares principle specifies that the estimates, $\hat{\beta}_i$, minimize the error sum of squares:

$$\text{SSE} = \Sigma(y - \hat{\beta}_0 - \hat{\beta}_1 x_1 - \hat{\beta}_2 x_2 - \ldots - \hat{\beta}_m x_m)^2.$$

For convenience we redefine the model:

$$y = \beta_0 x_0 + \beta_1 x_1 + \beta_2 x_2 + \ldots + \beta_m x_m + \epsilon,$$

where x_0 is a variable that has the value 1 for all observations. Obviously, the model is not changed by this definition, but the redefinition makes β_0 look like any other coefficient, which simplifies the computations in the estimation procedure.[3] The error sum of squares to be minimized is now written

$$\text{SSE} = \Sigma(y - \hat{\beta}_0 x_0 - \hat{\beta}_1 x_1 - \hat{\beta}_2 x_2 \ldots - \hat{\beta}_m x_m)^2.$$

The least squares estimates are provided by the solution to the following set of $(m + 1)$ linear equations in the $(m + 1)$ unknown parameters, $\beta_0, \beta_1, \ldots, \beta_m$ (see Appendix C for details). The solutions to these **normal equations** provide the least squares estimates of the coefficients, which we have already denoted by $\hat{\beta}_0, \hat{\beta}_1, \ldots, \hat{\beta}_m$.

$$
\begin{aligned}
\beta_0 n &+ \beta_1 \Sigma x_1 &+ \beta_2 \Sigma x_2 &+ \ldots + \beta_m \Sigma x_m &= \Sigma y \\
\beta_0 \Sigma x_1 &+ \beta_1 \Sigma x_1^2 &+ \beta_2 \Sigma x_1 x_2 &+ \ldots + \beta_m \Sigma x_1 x_m &= \Sigma x_1 y \\
\beta_0 \Sigma x_2 &+ \beta_1 \Sigma x_2 x_1 &+ \beta_2 \Sigma x_2^2 &+ \ldots + \beta_m \Sigma x_2 x_m &= \Sigma x_2 y \\
&\quad\cdot &\cdot &\quad\cdot \\
&\quad\cdot &\cdot &\quad\cdot \\
\beta_0 \Sigma x_m &+ \beta_1 \Sigma x_m x_1 &+ \beta_2 \Sigma x_m x_2 &+ \ldots + \beta_m \Sigma x_m^2 &= \Sigma x_m y
\end{aligned}
$$

Because of the large number of equations and variables, it is not possible to obtain simple formulas that directly compute the estimates of the coefficients as we did for the simple linear regression model in Chapter 2. In other words, the system of equations must be specifically solved for each application of this method. Although procedures are available for performing this task with hand-held or desk calculators, the solution is almost always obtained by computers using methods that are beyond the scope of this book. We will, however, have need to represent symbolically the solutions to the set of equations. This is done with matrices and matrix notation.[4]

---

[3] A model that does not include the x_0 variable as defined here will fit a regression through the origin. This is not commonly done with multiple regression models, because situations where there exist zero values for all independent variables rarely occur. Comments made in Section 2.6 regarding the interpretations of such models apply equally to multiple regression.

[4] Appendix B contains a brief introduction to matrix algebra. This appendix may also serve as a refresher for those who have had some exposure to matrices.

We now show the solution procedure using matrix notation for the general case and numerically for Example 3.1.

Define the matrices X, Y, E, and B as follows:[5]

$$X = \begin{bmatrix} 1 & x_{11} & x_{12} & \cdots & x_{1m} \\ 1 & x_{21} & x_{22} & \cdots & x_{2m} \\ \cdot & \cdot & \cdot & \cdots & \cdot \\ \cdot & \cdot & \cdot & \cdots & \cdot \\ 1 & x_{n1} & x_{n2} & \cdots & x_{nm} \end{bmatrix}, \quad Y = \begin{bmatrix} y_1 \\ y_2 \\ \cdot \\ \cdot \\ y_n \end{bmatrix}, \quad E = \begin{bmatrix} \epsilon_1 \\ \epsilon_2 \\ \cdot \\ \cdot \\ \epsilon_n \end{bmatrix}, \quad \text{and } B = \begin{bmatrix} \beta_0 \\ \beta_1 \\ \beta_2 \\ \cdot \\ \cdot \\ \beta_m \end{bmatrix},$$

where x_{ij} represents the ith observation of the jth independent variable, $i = 1, \ldots, n$, and $j = 1, \ldots, m$.

Using these matrices, the model equation for all observations,

$$y = \beta_0 + \beta_1 x_1 + \beta_2 x_2 + \ldots + \beta_m x_m + \epsilon,$$

can be expressed as

$$Y = XB + E.$$

For the data for Example 3.1, the matrices X and Y are

$$X = \begin{bmatrix} 1 & 10.20 & 89.00 & 9.3 \\ 1 & 19.13 & 101.00 & 17.3 \\ 1 & 15.12 & 105.60 & 14.0 \\ 1 & 17.28 & 98.06 & 14.3 \\ 1 & 15.67 & 102.00 & 14.0 \\ 1 & 17.26 & 91.02 & 14.3 \\ 1 & 15.28 & 93.09 & 13.8 \\ 1 & 14.37 & 98.03 & 13.4 \\ 1 & 15.43 & 95.08 & 13.3 \\ 1 & 15.24 & 100.80 & 13.5 \\ 1 & 17.87 & 96.01 & 16.9 \\ 1 & 16.50 & 95.09 & 14.9 \\ 1 & 15.67 & 99.00 & 13.7 \\ 1 & 15.98 & 89.02 & 13.9 \\ 1 & 15.02 & 91.05 & 12.8 \\ 1 & 16.87 & 95.02 & 14.9 \\ 1 & 13.72 & 90.07 & 12.1 \\ 1 & 13.78 & 89.00 & 13.6 \\ 1 & 15.24 & 94.00 & 14.0 \\ 1 & 15.00 & 99.00 & 14.2 \end{bmatrix}, \quad Y = \begin{bmatrix} 25.93 \\ 95.71 \\ 68.99 \\ 73.38 \\ 66.16 \\ 66.74 \\ 59.79 \\ 58.60 \\ 56.20 \\ 62.91 \\ 82.87 \\ 65.62 \\ 62.18 \\ 57.01 \\ 46.35 \\ 65.03 \\ 45.87 \\ 56.20 \\ 58.13 \\ 63.36 \end{bmatrix}$$

Note that the first column of the matrix X is a column of ones, used as the "variable" corresponding to the intercept. Using matrix notation, the normal equations can be expressed as

$$(X'X)\hat{B} = X'Y,$$

[5]We use the convention that matrices are denoted by the capital letters of the elements of the matrix. Unfortunately, the capital letters corresponding to β, ϵ, and μ are B, E, and M, respectively.

where $\hat{\beta}$ is a vector of least squares estimates of B.

The solution to the matrix equation is written

$$\hat{B} = (X'X)^{-1}X'Y.$$

Note that these expressions are valid for a multiple regression with any number of independent variables. That is, for a regression with m independent variables, the X matrix has n rows and $(m + 1)$ columns. Consequently, the matrices B and $X'Y$ are of order $\{(m + 1) \times 1\}$, and $X'X$ and $(X'X)^{-1}$ are of order $\{(m + 1) \times (m + 1)\}$.

The procedure for obtaining the estimates of the parameters of a multiple regression model is a straightforward application of matrix algebra for the solution of a set of linear equations. To apply the procedure, first compute the $X'X$ matrix:

$$X'X = \begin{bmatrix} n & \Sigma x_1 & \Sigma x_2 & \cdots & \Sigma x_m \\ \Sigma x_1 & \Sigma x_1^2 & \Sigma x_1 x_2 & \cdots & \Sigma x_1 x_m \\ \Sigma x_2 & \Sigma x_2 x_1 & \Sigma x_2^2 & \cdots & \Sigma x_2 x_m \\ \cdot & \cdot & \cdot & \cdots & \cdot \\ \Sigma x_m & \Sigma x_m x_1 & \Sigma x_m x_2 & \cdots & \Sigma x_m^2 \end{bmatrix},$$

that is, the matrix of sums of squares and cross products of all the independent variables. Next, compute the $X'Y$ matrix:

$$X'Y = \begin{bmatrix} \Sigma y \\ \Sigma x_1 y \\ \Sigma x_2 y \\ \cdot \\ \cdot \\ \cdot \\ \Sigma x_m y \end{bmatrix}.$$

For Example 3.1, $X'X$ and $X'Y$ are, using the output from PROC IML of the SAS System:[6]

(X'X)				(X'Y)
20	310.63	1910.94	278.2	1237.03
310.63	4889.0619	29743.659	4373.225	19659.105
1910.94	29743.659	183029.34	26645.225	118970.19
278.2	4373.225	26645.225	3919.28	17617.487

Note that for this example, the first element in the first row is 20, which is the sum of squares of the 20 ones corresponding to the intercept variable and is simply n, the number of observations. Some of the other quantities may be verified by the reader.

---

[6] Most computer programs have options to print many of these matrices.

The next step is to compute the inverse of $X'X$. As we have indicated earlier, we do not present here a procedure for this task. The inverse, obtained by PROC IML, is

---

$(X'X)^{-1}$

```
 20.874129  -0.180628 -0.216209   0.1897516
 -0.180628   0.109857  0.0011637 -0.117671
 -0.216209   0.0011637 0.0027755 -0.004821
  0.1897516 -0.117671 -0.004821   0.1508584
```

Again the reader may wish to manually calculate some of the elements of the multiplication of this matrix with $X'X$ to verify that the resulting elements make up the identity matrix. Finally, the matrix of estimated regression coefficients is obtained by

$$\hat{B} = (X'X)^{-1} X'Y.$$

As we will frequently refer to the $X'X$ matrix and its elements, we define $C = (X'X)^{-1}$; hence we can write

$$\hat{B} = C X'Y.$$

For Example 3.1, the result is

$\hat{B}$

```
-108.5758
   1.6257654
   0.6937702
   5.6713954
```

Again, the reader may verify some elements of this result.

These results provide the estimated regression response equation:

$$\text{VÔL} = -108.58 + 1.6258(\text{DBH}) + 0.6938(\text{HT}) + 5.6714(\text{D16}).$$

These are indeed the partial coefficients shown in Table 3.1. Thus, the coefficient for DBH, 1.6258, estimates an increase of 1.6258 in the average volume associated with a unit increase in DBH, holding constant all other variables. This means that for a subpopulation with a specified height and the diameter of the trunk at 16 feet constant, increasing DBH by 1 inch increases the average yield by 1.6258 cubic feet. Note also that the intercept is negative, an apparently impossible result. Actually, this is a meaningless number, because it estimates the volume for an impossible "tree" having zero DBH, HT, and D16. This example illustrates the problem with extrapolating beyond the range of the observed data: There are no observations of trees having even remotely near these zero values. As we shall see, this does not affect the ability of the model to adequately predict the volume of timber.

3.4 Interpreting the Partial Regression Coefficients

We have emphasized that the partial regression coefficients in a multiple regression model have a unique interpretation that differs from that of the total regression coefficients for a given set of independent variables. This can often be a problem when interpreting the results of a least squares analysis in regression, especially when it is desired to provide some real interpretations for individual coefficients. In order to explore this difference and to try to understand the interpretation of the partial regression coefficients, we will examine the following:

1. A contrived example where the reason for the different partial and total coefficients is readily apparent
2. An alternative approach to computing partial regression coefficients that, although not useful in practice, may help to explain the nature of the coefficients

EXAMPLE 3.2   In Table 3.2 are contrived data for 12 observations on two independent variables x_1, x_2, and the response variable y (ignore the last column for now) .

The estimated total regression coefficients for x_1 and x_2 as well as the partial coefficients for the model involving both variables are shown in Table 3.3, which is similar to Table 3.1. That is, the simple regression involving only y and x_1 is MODEL1, and so forth.

The differences between the partial and the total coefficients are more pronounced than they were for Example 3.1. The most interesting difference occurs for x_2, where the total coefficient is 0.7810, whereas the partial coefficient is −1.2855. In other words, the total coefficient implies a positively sloping relationship, whereas the partial coefficient implies a negatively sloping one. At first

TABLE 3.2

DATA FOR EXAMPLE OF DIFFERENT PARTIAL AND TOTAL COEFFICIENTS

Obs	x_1	x_2	y	RX2
1	0	2	2	−0.73585
2	2	6	3	0.60377
3	2	7	2	1.60377
4	2	5	7	−0.39623
5	4	9	6	0.94340
6	4	8	8	−0.05660
7	4	7	10	−1.05660
8	6	10	7	−0.71698
9	6	11	8	0.28302
10	6	9	12	−1.71698
11	8	15	11	1.62264
12	8	13	14	−0.37736

TABLE 3.3

ESTIMATES FOR TOTAL AND PARTIAL COEFFICIENTS

MODEL	INTERCEP	X1	X2
	TOTAL COEFFICIENTS		
MODEL1	1.85849	1.30189	.
MODEL2	0.86131	.	0.78102
	PARTIAL COEFFICIENTS		
MODEL3	5.37539	3.01183	-1.28549

glance, these two results seem to contradict one another. A closer look, however, explains this apparent contradiction.

The scatterplot of y against x_2 is shown in Figure 3.1. The initial impression reinforces the positive value of the total regression coefficient for x_2. However, at this point we have *ignored* the effect of x_1, which is what we do when we estimate a total regression coefficient. We can show the effect of x_1 in this plot by using the values of x_1 as the plotting symbol. That is, data points represented by the symbol "2" in Figure 3.1 indicate observations for which $x_1 = 2$, and so forth.

Remember that the partial coefficient is the relationship of y to x_2, *for constant values of* x_1. That means that data points for constant values of x_1 are the

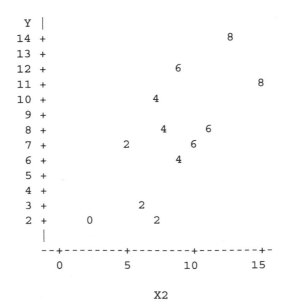

Plot of Y*X2.   Symbol is value of X1.

```
  Y |
 14 +                               8
 13 +
 12 +                    6
 11 +                              8
 10 +              4
  9 +
  8 +              4       6
  7 +         2        6
  6 +              4
  5 +
  4 +
  3 +         2
  2 +    0         2
    |
    --+---------+---------+---------+-
      0         5        10        15

                  X2
```

FIGURE 3.1 Plot Showing Total and Partial Coefficient

basis for the partial regression coefficient. The negative relationship now becomes more obvious: For example, the three data points for $x_1 = 4$ definitely show the negatively sloping relationship between y and x_2, and similar relationships hold for the other values of x_1. ◆

Estimating Partial Coefficients Using Residuals

In standard practice, the estimates of the partial regression coefficients are obtained through the solution of the normal equations as shown in Section 3.3. An alternative procedure for obtaining these estimates, which is computationally cumbersome and therefore not used in practice, may provide additional insight into the properties of the partial regression coefficients.

Assume the usual regression model:

$$y = \beta_0 + \beta_1 x_1 + \beta_2 x_2 + \ldots + \beta_m x_m + \epsilon.$$

We can estimate the partial regression coefficient for, say, x_j, using this alternative method by following these three steps:

1. Perform a least squares "regression" using the variable x_j as the dependent variable and all other x's as independent variables.
2. Compute the predicted values $\hat{\mu}_{xj}$, and the residuals $d_{xj} = (x_j - \hat{\mu}_{xj})$. These residuals measure that portion of the variability in x_j *not* explained by the variability among the other variables. Notice that we have used the letter d to denote these residuals to avoid confusion with the residuals discussed in Section 2.7
3. Perform a *simple* linear regression with the observed values of y as the dependent variable and the residuals, d_{xj}, as the independent variable. This regression measures the relationship of the response to the variation in x_j not explained by a linear regression involving the other variables. The resulting estimated coefficient is indeed the estimate of the partial regression coefficient β_j in the original model involving all m independent variables.[7]

EXAMPLE 3.2
REVISITED

To illustrate this method, we will calculate estimates of the regression coefficients for x_2 from Example 3.2. Since there are only two independent variables, "all other" x-variables consist of x_1. Then the "regression" of x_2 on x_1 provides the estimated equation:

$$\hat{x}_2 = 2.7358 + 1.3302 x_1.$$

The predicted values, the $\hat{x}_2$, from this regression are used to obtain residuals that are given under the column labeled RX2 in Table 3.2. The difference between the total and partial regressions involving x_2 can be seen in the plots of x_2 and RX2 against y as shown in Figure 3.2.

---

[7] The sum of squares for the coefficient is also the same as that for the partial coefficient. However, the error sum of squares is not useful, since it does not reflect the contribution of the other variables. See Section 3.5.

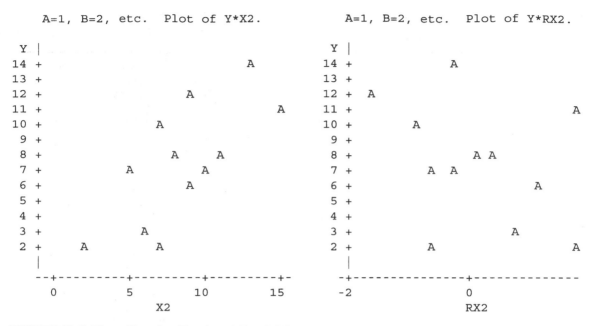

FIGURE 3.2 Plots Showing Total and Partial Regression

The plot of Y against X2 is the same as in Figure 3.1 and shows the positively sloping relationship between Y and X2 when X1 is ignored, whereas the plot of Y against RX2 indicates the negative relationship between Y and X2 when X1 is held fixed. Using the values of Y and RX2 in a simple linear regression will indeed provide the same value of -1.28549 as an estimate of β_2. ◆

EXAMPLE 3.1
REVISITED
We will again illustrate regression with residuals by obtaining the partial coefficient for DBH in Example 3.1. The data are reproduced in Table 3.4. Recall from Table 3.1 that the total coefficient was 6.9161, whereas the partial coefficient was 1.6258.

We first perform the regression using DBH as the dependent and HT and D16 as independent variables. The estimated model equation is

$$\hat{\text{DBH}} = \text{PDBH} = 1.6442 - 0.01059(\text{HT}) + 1.0711(\text{D16}).$$

This equation is used to obtain the predicted and residual values for DBH. In Table 3.4 the predicted values are given in the column labeled PDBH and the residual values under RDBH. We then estimate the simple linear regression equation using VOL and RDBH, which provides the partial regression coefficient.

The plots of the actual values (•) and the estimated regression line for the regressions of VOL on DBH (the total regression) and VOL on residual (the partial regression) are shown in Figure 3.3. The steeper slope of the total regression is evident. The reader may want to verify that performing the regression of VOL on DBH will provide the total regression coefficient, and that the regression of VOL on RDBH does provide the partial regression coefficient. ◆

TABLE 3.4

OBTAINING RESIDUAL VALUES

OBS	DBH	HT	D16	VOL	PDBH	RDBH
1	10.20	89.00	9.3	25.93	10.6629	-0.46295
2	19.13	101.00	17.3	95.71	19.1049	0.02514
3	15.12	105.60	14.0	68.99	15.5214	-0.40141
4	17.28	98.06	14.3	73.38	15.9226	1.35738
5	15.67	102.00	14.0	66.16	15.5595	0.11045
6	17.26	91.02	14.3	66.74	15.9972	1.26281
7	15.28	93.09	13.8	59.79	15.4397	-0.15970
8	14.37	98.03	13.4	58.60	14.9589	-0.58892
9	15.43	95.08	13.3	56.20	14.8831	0.54694
10	15.24	100.80	13.5	62.91	15.0367	0.20331
11	17.87	96.01	16.9	82.87	18.7293	-0.85927
12	16.50	95.09	14.9	65.62	16.5968	-0.09676
13	15.67	99.00	13.7	62.18	15.2700	0.40001
14	15.98	89.02	13.9	57.01	15.5899	0.39007
15	15.02	91.05	12.8	46.35	14.3902	0.62982
16	16.87	95.02	14.9	65.03	16.5975	0.27250
17	13.72	90.07	12.1	45.87	13.6508	0.06923
18	13.78	89.00	13.6	56.20	15.2688	-1.48880
19	15.24	94.00	14.0	58.13	15.6443	-0.40429
20	15.00	99.00	14.2	63.36	15.8056	-0.80555

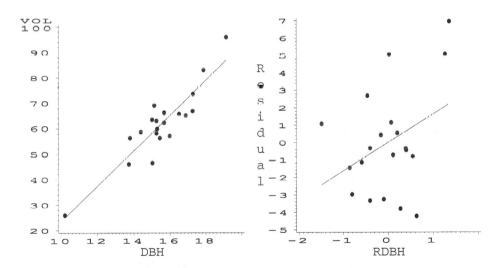

FIGURE 3.3 Total and Partial Regression Using Residuals

3.5 Inferences on the Parameters

It should now be obvious that in general we do not get the correct estimates of the partial coefficients in multiple parameter models by performing simple linear regressions using the individual independent variables. Likewise, we cannot obtain the appropriate inferences for the *partial* coefficients by direct application of the procedures for making inferences for simple linear regression coefficients.

However, inference procedures based on the comparison of unrestricted and restricted models do provide the proper framework for inferences on partial coefficients. The most common inferences consist of tests of the null hypothesis that one or more coefficients are zero. That is, the null hypothesis is that the corresponding variables are not needed in the model.

> For a test on the null hypothesis that one or more coefficients are zero, we compare the unrestricted model that contains all coefficients to the restricted model that does not contain the coefficient(s) being tested.

Note that both the unrestricted and restricted models contain all of the *other* coefficients; this corresponds to holding all other variables constant. The tests are based on the reduction in effectiveness of the restricted model as measured by the increase in the error sum of squares due to imposing the restriction.

Formally, the testing procedure is implemented as follows:

1. Divide the full set of coefficients, denoted by B, into two sets represented by submatrices B_1 and B_2 as follows:

$$B = \begin{bmatrix} B_1 \\ \cdots \\ B_2 \end{bmatrix}.$$

   Since the ordering of elements in the matrix of coefficients is arbitrary, either submatrix may contain any desired subset of the entire set of coefficients. We want to test the hypotheses

$$H_0 : B_2 = 0,$$

$$H_1 : \text{At least one element in } B_2 \text{ is not zero.}[8]$$

   Denote the number of coefficients in B_1 by q and the number of coefficients in B_2 by p. Note that $p + q = m + 1$. Since we rarely consider hypotheses concerning β_0, this coefficient is normally included in B_1.

2. Perform the regression using all coefficients, that is, using the unrestricted model $Y = XB + E_{\text{unrestricted}}$. The error sum of squares for this model is $SSE_{\text{unrestricted}}$. This sum of squares has $(n - m - 1)$ degrees of freedom.

3. Perform the regression containing only the coefficients in B_1. The model is $Y = X_1 B_1 + E_{\text{restricted}}$, which is the result of imposing the restriction $B_2 = 0$ as specified by H_0. The error sum of squares for this model is $SSE_{\text{restricted}}$ and has $(n - q)$ degrees of freedom.

4. The difference $SSE_{\text{restricted}} - SSE_{\text{unrestricted}}$ is the increase in the error sum of squares due to the deletion of B_2. As we did in Chapter 2, we will refer

---

[8] Since there is no "capital" zero, we must understand that the zero here represents a $p \times 1$ matrix of zeros.

to this as $SS_{\text{hypothesis}}$. The degrees of freedom are $(m + 1) - q = p$, which corresponds to the fact that there are p coefficients in B_2. Dividing $SS_{\text{hypothesis}}$ by its degrees of freedom provides the mean square that can be used as the numerator in the test statistic.[9]

5. The ratio of the hypothesis mean square to the error mean square of the unrestricted model is the test statistic that, under the null hypothesis, will have the F distribution with $(p, n - m - 1)$ degrees of freedom.

Obviously, to do the hypothesis test discussed earlier, we need to calculate error sums of squares for various multiple regression models. For the unrestricted model,

$$y = \beta_0 + \beta_1 x_1 + \beta_2 x_2 + \ldots + \beta_m x_m + \epsilon,$$

the error mean square is defined as

$$SSE_{\text{unrestricted}} = \Sigma(y - \hat{\mu}_{y|x})^2,$$

where the values of $\hat{\mu}_{y|x}$ are obtained by using the least squares estimates of the regression coefficients. As was the case for simple linear regression, the procedure for the partitioning of sums of squares provides for computational formulas. The partitioning of sums of squares used for this model is

$$\Sigma y^2 = \Sigma \hat{\mu}_{y|x}^2 + \Sigma(y - \hat{\mu}_{y|x})^2.$$

Note that, unlike the partitioning of sums of squares for simple linear regression, the left-hand side is the uncorrected sum of squares for the dependent variable.[10] Consequently, the term corresponding to the regression sum of squares includes the contribution of the intercept and is therefore not normally used for inferences. As in simple linear regression, the shortcut formula is the one for the sum of squares due to regression, which can be computed by any of three equivalent formulas:

$$\Sigma \hat{\mu}_{y|x}^2 = \hat{B}'X'Y = Y'X(X'X)^{-1}Y'X = \hat{B}'X'X\hat{B}.$$

The first one is the most convenient formula for manual computation and represents the following algebraic expression:

$$\Sigma \hat{\mu}_{y|x}^2 = \hat{\beta}_0 \Sigma y + \hat{\beta}_1 \Sigma x_1 y + \ldots + \hat{\beta}_m \Sigma x_m y.$$

Note that the individual terms appear to be similar to the formula for SSR for the simple linear regression model; however, these *individual* terms have *no* practical interpretation. The error sum of squares is obtained by subtracting the regression sum of squares from the total sum of squares, Σy^2.

The restricted error sum of squares is computed in the same manner by using B_1 instead of B and denoting by X_1 the columns of X corresponding to those in B_1.

---

[9] A formula for computing $SS_{\text{hypothesis}}$ directly will be presented later.

[10] This way of defining these quantities corresponds to the way most current computer programs for regression are written. References that do not rely heavily on the use of such programs may define TSS and consequently SSR in a manner analogous to that presented in Section 1.3. These different definitions cause minor modifications in computational procedures, but the ultimate results are the same.

EXAMPLE 3.1
REVISITED

We illustrate the procedure by testing the null hypothesis:

$$H_0: \beta_{\text{HT}} = 0, \text{ and } \beta_{\text{D16}} = 0.$$

Computing the Unrestricted SSE   The actual, predicted, and residual values for the unrestricted model, labeled VOL, PVOL, and RVOL, respectively, are shown in Table 3.5.

The sum of squares of the RVOL values is the $\text{SSE}_{\text{unrestricted}}$ and has a value of 153.3007. The corresponding error degrees of freedom is $(20 - 3 - 1) = 16$; hence, the error mean square is 9.581.

The quantities required to compute this quantity by the partitioning of sums of squares are available from the computations for the regression coefficients that were given in Example 3.1. For our purposes we need $X'Y$ and $\hat{B}$. Using the output from PROC IML of the SAS System:

XY	B
1237.030	-108.5758
19659.105	1.6257654
118970.190	0.6937702
17617.487	5.6713954

SSR	
80103.219	

TABLE 3.5

COMPUTING SSE DIRECTLY

OBS	VOL	PVOL	RVOL
1	25.93	22.4965	3.43351
2	95.71	90.7110	4.99902
3	68.99	68.6674	0.32261
4	73.38	68.6494	4.73056
5	66.16	67.0640	-0.90399
6	66.74	63.7328	3.00722
7	59.79	59.1142	0.67583
8	58.60	58.7934	-0.19339
9	56.20	57.9029	-1.70294
10	62.91	62.6967	0.21331
11	82.87	82.9320	-0.06204
12	65.62	68.7237	-3.10368
13	62.18	63.2813	-1.10126
14	57.01	57.9957	-0.98570
15	46.35	51.6048	-5.25479
16	65.03	69.2767	-4.24665
17	45.87	44.8414	1.02858
18	56.20	52.7037	3.49627
19	58.13	60.8148	-2.68475
20	63.36	65.0277	-1.66770

Multiplying $\hat{B}$ by $X'Y$:

$$(1237.03)(-108.5758) + \ldots + (17617.487)(5.6713954),$$

gives SSR $= 80103.219$ as shown. From the initial matrix computations we have $Y'Y = \Sigma y^2 = 80{,}256.52$, and by subtraction,

$$\text{SSE} = 80256.520 - 80103.219$$

$$= 153.301.$$

The result agrees with the directly calculated error sum of squares, as it must.   ◆

Computing the Hypothesis SSE

We illustrate the procedure by testing the following hypothesis:

$$H_0: \beta_{\text{HT}} = 0, \text{ and } \beta_{\text{D16}} = 0.$$

For this test the unrestricted model includes the variables DBH, HT, and D16, whereas the restricted model includes only DBH.

We have already calculated the unrestricted model sum of squares to be 153.30 with 16 degrees of freedom. The restricted model is the one that was used to illustrate simple linear regression in Example 2.1, where the error sum of squares was calculated to be 658.62 with 18 degrees of freedom. We now compute the hypothesis sum of squares:

$$\text{SS}_{\text{hypothesis}} = \text{SSE}_{\text{restricted}} - \text{SSE}_{\text{unrestricted}}$$

$$\text{SS}_{\text{hypothesis}} = 658.62 - 153.30 = 505.32,$$

with $18 - 16 - 2$ degrees of freedom, and the resulting mean square is 252.66.

The Hypothesis Test

Using the error mean square from the unrestricted model gives the F statistic

$$F - \frac{252.66}{9.581} - 26.371,$$

with 2 and 16 degrees of freedom, which is to be compared with the tabled value of 6.23 for rejection at the 0.01 significance level. Thus, we conclude that these two variables contribute significantly to the model over and above the effect of DBH. This test does not, however, give any indication of the relative contributions of the two individual parameters.

Commonly Used Tests

The above procedure can be used to test hypotheses for any subset of coefficients to be zero. It is, however, not generally desirable to perform all possible tests. There are two sets of tests commonly performed in the initial inference stage of a regression analysis:

1. The test that all coefficients (except β_0) are zero. The hypothesis statement is

$$H_0: \beta_1 = 0, \beta_2 = 0, \ldots, \beta_m = 0.$$

This test is referred to as the test for the model.
2. m separate tests:

$$H_{0j}: \beta_j = 0,$$

that is, a set of separate tests that each individual partial coefficient is zero.

Because these tests are performed and presented automatically in computer programs for multiple regression, we will present them in some detail.

The Test for the "Model"

This test compares the unrestricted model with the restricted model,

$$y = \beta_0 + \epsilon,$$

which is equivalent to

$$y = \mu + \epsilon,$$

that is, the model for the mean of a single population (Section 1.2). The estimate of μ is $\bar{y}$, and the error sum of squares is the familiar

$$SSE_{restricted} = (n - 1)s^2 = \Sigma(y - \bar{y})^2 = \Sigma y^2 - \frac{(\Sigma y)^2}{n},$$

which has $(n - 1)$ degrees of freedom.

The hypothesis sum of squares for this test is referred to as the model sums of squares and is $SSE_{restricted}$ minus $SSE_{unrestricted}$:

$$SS_{model} = SSE_{restricted} - SSE_{unrestricted}$$

$$= \left[\Sigma y^2 - \frac{(\Sigma y)^2}{n}\right] - [\Sigma y^2 - \hat{B}'X'Y]$$

$$= \hat{B}'X'Y - \frac{(\Sigma y)^2}{n},$$

which has m degrees of freedom. The resulting mean square provides the numerator for the F statistic, with the usual unrestricted model error mean square as the denominator.

For Example 3.1 we have already computed:

$$\hat{B}'X'Y = 80103.219$$

$$\Sigma y = 1237.03,$$

hence,

$$(\Sigma y)^2/n = 76512.161.$$

The sum of squares due to the model:

$$SS_{model} = 80103.219 - 76512.161 = 3591.058,$$

and the mean square

$$MS_{model} = 3591.058/3 = 1197.019.$$

We have previously computed $MSE_{unrestricted} - 9.581$, hence the test statistic:

$$F = 1197.019/9.581 = 124.937,$$

which obviously leads to rejection of the hypothesis that the model does not fit the data better than the mean. This result was to be expected since the test for the model with only DBH was highly significant, and it would be highly unusual for a model containing a variable whose total coefficient was significant not to also be significant.

Tests for Individual Coefficients

The restricted model for each test of a coefficient (or variable) is the error sum of squares for an $(m - 1)$ parameter model excluding that variable. This procedure requires the computation of m such regressions to get all the required tests. Fortunately, there exists a shortcut[11] formula. Remembering that we have previously defined $C = (X'X)^{-1}$, then it can be shown that the partial sum of squares for $\hat{\beta}_j$ is

$$SS_{\beta_j} = \frac{\hat{\beta}_j^2}{c_{ij}},$$

where c_{jj} is the jth diagonal element of C. These sums of squares are computed for each $\hat{\beta}_j$, and have one degree of freedom each. The unrestricted model error mean square is used for the denominator of the F statistic.

The parameter estimates and elements of C are available in the PROC IML output in Section 3.3. Using these quantities, we have

$$SS_{\beta_{DBH}} = \frac{\hat{\beta}_{DBH}^2}{c_{DBH,DBH}} = \frac{1.6258^2}{0.1099} = 24.0512, \text{ and } F = \frac{24.0512}{9.5813} = 2.510$$

$$SS_{\beta_{HT}} = \frac{\hat{\beta}_{HT}^2}{c_{HT,HT}} = \frac{0.6938^2}{0.002776} = 173.400, \text{ and } F = \frac{173.400}{9.5813} = 18.0978$$

$$SS_{\beta_{D16}} = \frac{\hat{\beta}_{D16}^2}{c_{D16,D16}} = \frac{5.6714^2}{0.1508} = 213.2943, \text{ and } F = \frac{213.2943}{9.5813} = 22.262.$$

The 0.01 right-tail value for the F distribution for $(1,16)$ degrees of freedom is 8.53; hence, we can conclude that HT contributes significantly to the regression model when DBH and D16 are fixed and that D16 contributes significantly to the regression model when DBH and HT are fixed. We cannot say that DBH is a significant contributor when HT and D16 are fixed.

[11] See "Testing a General Linear Hypothesis" in Section 3.6.

Recall from the discussion of sampling distributions that the square root of any F statistic with one degree of freedom in the numerator is equal to a t statistic with the corresponding denominator degrees of freedom. Taking the square root of the F ratio for, say, β_j, and rearranging some elements, produces

$$\sqrt{F} = t = \frac{\hat{\beta}_j}{\sqrt{c_{jj}(\text{MSE}_{\text{unrestricted}})}}.$$

Recall also that the standard formulation for a t statistic has the parameter estimate (minus the null hypothesis value) in the numerator and the estimated standard error of that estimate in the denominator. Hence, we see that the estimated standard error of $\hat{\beta}_j$ is

$$\text{Standard error } (\hat{\beta}_j) = \sqrt{c_{jj}(\text{MSE}_{\text{unrestricted}})}$$

For Example 3.1, the standard error for the DBH coefficient is

$$\text{Standard error } (\hat{\beta}_{\text{DBH}}) = \sqrt{(9.5813)(0.10986)}$$

$$= 1.0259,$$

which can be used for hypothesis tests and confidence intervals.

To test the hypothesis of a zero coefficient for DBH, calculate

$$t = \frac{1.6258}{1.0259} = 1.585,$$

which is less than 2.120, the 0.05 two-tail value for t with 16 degrees of freedom; hence, the hypothesis is not rejected, as was expected. Note that $t^2 = 1.585^2 = 2.512$, which is the F-value obtained using the partial sum of squares computed from the partitioning of sums of squares.

The preceding results are based on an application of the central limit theorem that states that the sampling distribution of an estimated partial regression coefficient, $\hat{\beta}_j$, is a random variable that is normally distributed with

$$\text{Mean } (\hat{\beta}_j) = \beta_j, \text{ and}$$

$$\text{Variance } (\hat{\beta}_j) = \sigma^2 c_{jj}.$$

Furthermore, the estimated coefficients are not independent. In fact, the **covariance** between two estimated coefficients, say, $\hat{\beta}_i$ and $\hat{\beta}_j$, is

$$\text{cov } (\hat{\beta}_i, \hat{\beta}_j) = \sigma^2 c_{ij}.$$

Although we will not directly use this definition in this chapter, we will use it in Section 9.2.

Substituting MSE for σ^2 provides the statistic

$$t = \frac{\hat{\beta}_j - \beta_j}{\sqrt{c_{jj}\text{MSE}}},$$

which has the t distribution with $(n - m - 1)$ degrees of freedom. Using this statistic we can test hypotheses that the coefficients are some specified value and construct confidence intervals.

For example, the 0.95 confidence interval for DBH uses the t value of 2.120 ($\alpha = 0.05$, df $= 16$) to obtain

$$\hat{\beta}_{DBH} \pm t_{\alpha/2}(\text{Standard error } (\hat{\beta}_{DBH})), \text{ or}$$

$$1.6258 \pm (2.120)(1.0259),$$

which provides the interval from -0.5491 to 3.8007. This interval does include zero, confirming the failure to reject the zero value with the hypothesis test.

Simultaneous Inference

In doing the analysis of a multiple regression equation, there are occasions when we require a series of estimates or tests where we are concerned about the correctness of the entire set of estimates or tests. We call this set of estimates the family of estimates. To evaluate a family of estimates we must use what is called joint or simultaneous inference. For example, a family confidence coefficient indicates the level of confidence for all the estimates in the family. If we constructed a 95% confidence interval on one parameter, using the methods previously discussed, we would expect 95% of all intervals constructed in the same manner to contain the single unknown parameter. On the other hand, if we constructed a 95% confidence interval on a collection of four unknown parameters, we would expect 95% of intervals constructed this way to contain all four of the parameters. For 5% of these intervals one or more of the four would not be in the interval. The problem of the correct confidence (or significance) levels for simultaneous inferences requires special methodology (see Neter *et al.,* 1996). One such method is to use what is known as a Bonferroni procedure. The **Bonferroni approach** gives confidence limits as follows:

$$\hat{\beta}_i \pm t_{\alpha/2r}(\text{Standard error}(\hat{\beta}_i)),$$

where r is the number of intervals to be computed. In other words, the Bonferroni approach simply adjusts the confidence coefficient for the simultaneous inferences. If in our example we wish to make simultaneous 90% intervals for the three coefficients, we would use $t_{0.0167}(16)$. Since few tables have such percentage points, the value may be obtained with a computer program or by interpolation. Interpolating gives us a value of $t_{0.0167}(16) \approx 2.377$. For the DBH coefficient, the interval is

$$1.6258 \pm (2.377)(1.0259),$$

or from -0.8128 to 4.0643. Confidence intervals for the other coefficients are calculated in a similar manner.

EXAMPLE 3.1   Computer Output   The tests for the model and individual coefficients are the ones printed by default by virtually all regression computer outputs. Table 3.6 reproduces the output for Example 3.1 as provided by PROC REG of the SAS System.

The format of this output is virtually the same as that presented in Table 2.3, because computer programs for regression treat all models alike and consider

TABLE 3.6

COMPUTER OUTPUT FOR EXAMPLE 3.1

Analysis of Variance

Source	DF	Sum of Squares	Mean Square	F Value	Prob>F
Model	3	3591.05774	1197.01925	124.933	0.0001
Error	16	153.30071	9.58129		
C Total	19	3744.35846			

Root MSE	3.09537	R-square	0.9591	
Dep Mean	61.85150	Adj R-sq	0.9514	
C.V.	5.00451			

Parameter Estimates

Variable	DF	Parameter Estimate	Standard Error	T for H0: Parameter=0	Prob > |T|
INTERCEP	1	-108.575847	14.14217723	-7.677	0.0001
DBH	1	1.625765	1.02594965	1.585	0.1326
HT	1	0.693770	0.16307253	4.254	0.0006
D16	1	5.671395	1.20225588	4.717	0.0002

simple linear regression as just a special case of the multiple regression model. The top portion of the output contains the test for the model, giving the results we have already obtained. The descriptive statistics presented in the next portion of the printout are the same ones given in Table 2.3 and described in Section 2.3. The only statistic not yet discussed is the one called R-square (and Adj R-sq). These values play an important role in evaluating the adequacy of the model and are discussed in some detail in Section 3.8.

Finally, the section of the printout under the heading Parameter Estimates gives the information about the regression coefficients. Each estimate of a regression coefficient is identified by its corresponding variable name. Notice that all the coefficient estimates show a positive relation between the independent variable and VOL, while, as previously noted, the negative value of the intercept has no meaning.

The estimates are followed by their standard errors and the results of the t test for $H_0: \beta_j = 0$. The results are identical to those obtained earlier in this section. We again see the apparent contradiction between the test for DBH in the simple linear model and the test for DBH when HT and D16 are included in the model.   ◆

The Test for a Coefficient Using Residuals

In Section 3.4 we saw that a partial regression coefficient can be computed as a total coefficient using residuals. The test for a coefficient, say β_j, can be computed by the same principles as follows:

1. Perform the regression of x_j on all other independent variables in the unrestricted model and compute the residuals, r_{xj}.
2. Perform the regression of y on all other independent variables in the unrestricted model and compute the residuals, $r_{y.xj}$.
3. Then the simple linear regression of $r_{y.xj}$ on r_{xj} will provide the partial regression coefficient *and* the test for the hypothesis that the coefficient is zero.

Remember that in Section 3.4 we showed that the partial regression coefficient can be estimated by regressing y, the observed values of the dependent variable, on the r_{xj}. However, the test for the partial coefficient also involves the use of r_{xj}, but requires the use of the residuals $r_{y.xj}$. This is because the test must reflect the partial effect *on* y after accounting for the relationship with the other independent variables.

EXAMPLE 3.1   Testing a Coefficient by Residuals   Table 3.7 shows the quantities required to obtain the various sums of squares for performing the test on the partial coefficient for DBH. The variables VOL and DBH are the originally observed data, whereas RVOL and RDBH are the residuals from the regressions using HT and D16.

The reader may verify that

$$\hat{\beta}_{DBH} = S_{RVOL,RDBH} / S_{RDBH,RDBH} = 14.7989/9.1027 = 1.6258,$$

TABLE 3.7

RESIDUALS FOR INFERENCE ON THE DBH COEFFICIENT

OBS	VOL	DBH	RVOL	RDBH
1	25.93	10.20	2.68087	-0.46295
2	95.71	19.13	5.03989	0.02514
3	68.99	15.12	-0.33000	-0.40141
4	73.38	17.28	6.93734	1.35738
5	66.16	15.67	-0.72442	0.11045
6	66.74	17.26	5.06024	1.26281
7	59.79	15.28	0.41619	-0.15970
8	58.60	14.37	-1.15084	-0.58892
9	56.20	15.43	-0.81374	0.54694
10	62.91	15.24	0.54384	0.20331
11	82.87	17.87	-1.45901	-0.85927
12	65.62	16.50	-3.26099	-0.09676
13	62.18	15.67	-0.45093	0.40001
14	57.01	15.98	-0.35154	0.39007
15	46.35	15.02	-4.23085	0.62982
16	65.03	16.87	-3.80363	0.27250
17	45.87	13.72	1.14112	0.06923
18	56.20	13.78	1.07583	-1.48880
19	58.13	15.24	-3.34203	0.40429
20	63.36	15.00	-2.97733	-0.80555

and further, the sums of squares for the test of the coefficient:

$$\text{SSR}_{\text{DBH}} = (S_{\text{RVOL,RDBH}})^2 / S_{\text{RDBH,RDBH}} = 24.0595,$$

$$\text{TSS} = S_{\text{RVOL,RVOL}} = 177.3601, \text{ and}$$

$$\text{SSE} = \text{TSS} - \text{SSR} = 153.3005,$$

which is indeed the error sum of squares for the unrestricted model. Dividing by the appropriate degrees of freedom provides the mean squares for the test. The standard error of the partial coefficient is obtained by using $S_{\text{RDBH,RDBH}}$ as S_{xx} in the formula for the standard error of the simple linear regression coefficient.[12]

Figure 3.4 shows the plot of the original variables and residuals. The plot readily shows the strength of both the total and partial relationship between VOL and DBH. It is important to note the difference in the scales of the plots: The residuals show a much smaller range of values, showing how much of the variability on both variables is accounted for by the other variables. It is apparent that the total relationship appears strong, while the partial one is not.

The plot using residuals is called the **partial residual** or **leverage** plot and will be reintroduced later as a useful tool for diagnosing problems with the data.

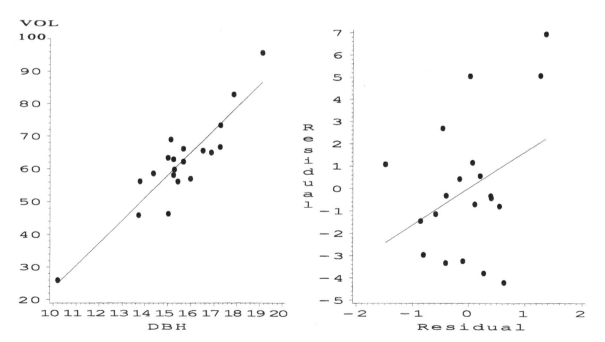

FIGURE 3.4 Original and Partial Residuals Regression

[12] Using the variables RVOL and RDBH in a computer program for regression produces the correct sums of squares. However, the program will assume that these are actual variables rather than residuals; hence, the degrees of freedom and consequently all mean squares will not be correct.

3.6 Testing a General Linear Hypothesis (Optional Topic)

In the preceding sections we have limited our discussion to hypotheses of the form H_0: $B_2 = 0$, where B_2 is some subset of the regression coefficients in B. Occasionally, however, there is a need for a more general type of hypothesis. For example, it may be desired to test hypotheses such as

$$H_0: \beta_i = \beta_j$$

or

$$H_0: \Sigma \, \beta_i = 1.$$

Such hypotheses imply a more general type of restricted model. (Recall that if any coefficients are assumed zero, the corresponding independent variable is simply removed from the model.) The implementation of tests on a more general class of restrictions is provided by the test for a general linear hypothesis.

The general linear hypothesis is stated in matrix form,

$$H_0: HB \quad K = 0,$$

where

H is a $k \times (m + 1)$ matrix of coefficients for a specified set of k restrictions. Each row of this matrix describes one set of restrictions.

B is the matrix of $(m + 1)$ coefficients of the unrestricted model.

K is a matrix of constants. In many applications, K is a null matrix (having all elements equal zero).

For example, assume a four variable model:

$$y = \beta_0 + \beta_1 x_1 + \beta_2 x_2 + \beta_3 x_3 + \beta_4 x_4 + \epsilon,$$

and we want to test

$$H_0: \beta_1 = 0.$$

Remembering that the first element of B is β_0, then for this hypothesis,

$$H = [\, 0 \ 1 \ 0 \ 0 \ 0 \,], \text{ and } K = [0].$$

The hypothesis

$$H_0: \beta_1 + \beta_2 - 2\beta_3 = 1$$

is restated as

$$H_0: \beta_1 + \beta_2 - 2\beta_3 - 1 = 0,$$

then

$$H = [\, 0 \ 1 \ 1 \ -2 \ 0 \,], \text{ and } K = [1].$$

For the composite hypothesis:

$$H_0: (\beta_1, \beta_2, \beta_3, \beta_4) = 0,$$

$$H = \begin{bmatrix} 0\ 1\ 0\ 0\ 0 \\ 0\ 0\ 1\ 0\ 0 \\ 0\ 0\ 0\ 1\ 0 \\ 0\ 0\ 0\ 0\ 1 \end{bmatrix} \text{ and } K = \begin{bmatrix} 0 \\ 0 \\ 0 \\ 0 \end{bmatrix}.$$

Using matrix notation, the hypothesis sum of squares, $SS_{\text{hypothesis}}$, is obtained by

$$SS_{\text{hypothesis}} = [H\hat{B} - K]' \, [H[X'X]^{-1} \, H']^{-1} \, [H\hat{B} - K],$$

where $\hat{B}$ is the set of estimated regression coefficients.

The fact that the entire $X'X^{-1}$ matrix is involved in this equation results from the existence of correlations among the estimated coefficients.

Although not often done in this manner, we can also obtain the estimates of the regression coefficients directly for any model subject to the restriction $HB = K$. Denoting such a set of coefficients by $\hat{B}_H$, they can be obtained by

$$\hat{B}_H = \hat{B} - [X'X]^{-1} \, H'[HX'XH']^{-1} \, [H\hat{B} - K],$$

where $\hat{B}$ is the matrix of unrestricted model parameter estimates.

All of the tests previously presented in this section are simply special cases of this general linear hypothesis testing procedure. For example, to test the hypothesis $H_0: \beta_1 = 0$ in a multiple regression model with m independent variables, $H = [0\ 1\ 0\ \ldots\ 0]$ and $K = [0]$. For this case:

$$H\hat{B} = \hat{\beta}_1 \text{ and}$$

$$H[X'X]^{-1}H' = c_{11},$$

the second[13] diagonal element of $[X'X]^{-1}$.

The inverse of the scalar, c_{11}, is its reciprocal; hence,

$$SS_{\text{hypothesis}} = \hat{\beta}_1^2 / c_{11},$$

which is the previously obtained result.

EXAMPLE 3.1 The two independent variables, DBH and D16, are rather similar measures of
REVISITED trunk diameter, yet the estimates of the coefficients of these two variables have different magnitudes: $\hat{\beta}_{\text{DBH}} = 1.626$, whereas $\hat{\beta}_{\text{D16}} = 5.671$. It may be interesting to speculate what would happen if we restricted the two coefficients to have the same value. This is obtained by implementing the restriction

$$\hat{\beta}_{\text{DBH}} - \hat{\beta}_{\text{D16}} = 0.$$

We perform the analysis by using the RESTRICT option in PROC REG of the SAS System, which performs the computations with the formulas given above. The results are shown in Table 3.8, and we can compare them with those of Table 3.6.

The analysis of variance portion now shows only two degrees of freedom for the model, since with the restriction there are effectively only two parameters.

---

[13] Remember that the first row and column correspond to the intercept.

TABLE 3.8

ESTIMATING A RESTRICTED MODEL

Note: Restrictions have been applied to parameter estimates.
Dependent Variable: VOL

Analysis of Variance

Source	DF	Sum of Squares	Mean Square	F Value	Prob > F
Model	2	3558.06335	1779.03167	162.342	0.0001
Error	17	186.29511	10.95854		
C Total	19	3744.35846			

Root MSE	3.31037	R-square	0.9502	
Dep Mean	61.85150	Adj R-sq	0.9444	
C.V.	5.35212			

Parameter Estimates

Variable	DF	Parameter Estimate	Standard Error	T for H0: Parameter=0	Prob > |T|
INTERCEP	1	-111.596504	15.02396025	-7.428	0.0001
DBH	1	3.481386	0.24541635	14.186	0.0001
HT	1	0.742575	0.17211637	4.314	0.0005
D16	1	3.481386	0.24541635	14.186	0.0001
RESTRICT	-1	-8.155564	4.70013348	-1.735	0.1008

The error mean square of 10.958 is not much larger than the unrestricted model error mean square of 9.581 found in Table 3.6, indicating that the restriction does not have much effect on the fit of the model.

The parameter estimates show the same value for the DBH and D16 coefficients (that was the restriction applied to the model). The last line is the test for the statistical significance of the restriction; the p-value of 0.1008 confirms our impression that the restriction does not seriously impair the effectiveness of the model.[14]

The results of the test may at first appear surprising. The estimated values of the coefficients of DBH and D16 in the unrestricted model were quite different, yet this difference was found not to be significant when tested. The reason for this result is that the two variables DBH and D16 are highly correlated, and therefore a linear function of the two will provide nearly equivalent prediction. In fact, the model that omits DBH entirely actually fits about as well as the model with the restriction. We will thoroughly investigate the effect of correlation among the independent variables in Chapter 5. ◆

---

[14] The RESTRICT option uses the restricted model error mean square as the denominator in the F test, resulting in a more conservative test. However, as previously noted, the restricted and the unrestricted mean squares were almost identical.

3.7 Inferences on the Response Variable in Multiple Regression

As in simple linear regression (Section 2.4), we can make two types of inferences on the response variable:

> 1. Inferences on the estimated conditional mean
> 2. Inferences on the predicted value of a single observation

For multiple regression, computing the standard errors of estimated means or predicted values is not a simple task and is normally performed with computers. The formulas for these computations are presented here in matrix format for completeness and to present the general concepts for these inferences.

We have previously denoted by $\hat{\mu}_{y|X}$ the $(n \times 1)$ matrix of the estimated mean of the response variable for each observation. The variances and covariances of these estimated means is given by the elements of the matrix:

$$\text{Variance } (\hat{\mu}_{y|X}) = \sigma^2 X [X'X]^{-1} X'.$$

This is an $n \times n$ matrix; its diagonal elements contain the variances for the estimated mean for each observation, and its off-diagonal elements the covariances of all pairs of observations. The matrix $X[X'X]^{-1}X'$ is called the *hat matrix*, which will be seen to have additional uses later. Since our primary focus is on the response for a specified set of values of the independent variables, we need the variance at that value. Denote by X_i a $1 \times (m + 1)$ matrix of values of the independent variables corresponding to that set (which may correspond to one of the observations). Then the variance of the corresponding estimated mean is

$$\text{Variance } (\hat{\mu}_{y|Xi}) = \sigma^2[X_i [X'X]^{-1} X_i'].$$

Similarly, the variance of a single predicted value at X_i is

$$\text{Variance } (\hat{y}_{y|Xi}) = \sigma^2[1 + X_i [X'X]^{-1} X_i'].$$

Substituting the unrestricted model error mean square for σ^2, we can use the t distribution with $(n - m - 1)$ degrees of freedom to calculate confidence or prediction intervals following the procedures discussed in Section 2.4. We illustrate this procedure by again referring to Example 3.1.

EXAMPLE 3.1   **Confidence and Prediction Intervals**   Because we do not normally calculate the confidence and/or prediction intervals manually, we use the data from Example 3.1 to show the confidence and prediction intervals using PROC REG of the SAS System. The results of the CLI (Confidence Limits on the Individual) and CLM (Confidence Limits on the Mean) options are presented in Table 3.9.

The first two columns in Table 3.9 list the actual and estimated values for the dependent variable, VOL. The next column contains the estimated standard error of the estimated mean (the column heading is a bit of a misnomer). The next two columns contain the output from the CLM option—the lower and upper 0.95 confidence intervals on the mean. The next two lines present the results of the CLI option—the lower and upper 0.95 prediction intervals. The last column lists the residuals for each data point, that is, $(y - \hat{\mu}_{y|x})$. Notice that both intervals are given for every observed value of the independent variables in the data set, allowing the user to pick the desired value(s).[15]

The interpretation of the confidence interval is that we are 95% confident that the true mean of the population of trees having the dimensions of the first tree (DBH = 1020, HT = 89.00, and D16 = 9.3) is between 17.84 and 27.15 cubic feet. Likewise, we are 95% confident that a single tree, picked at random from that population, will have a volume between 14.45 and 30.54 cubic feet. These intervals may be too wide to be practically useful. ◆

TABLE 3.9

CONFIDENCE AND PREDICTION INTERVALS

Obs	Dep Var VOL	Predict Value	Std Err Predict	Lower95% Mean	Upper95% Mean	Lower95% Predict	Upper95% Predict	Residual
1	25.9300	22.4965	2.197	17.8396	27.1534	14.4501	30.5429	3.4335
2	45.8700	44.8414	1.174	42.3526	47.3302	37.8234	51.8594	1.0286
3	56.2000	52.7037	1.957	48.5546	56.8529	44.9401	60.4673	3.4963
4	58.6000	58.7934	1.075	56.5142	61.0726	51.8470	65.7398	-0.1934
5	63.3600	65.0277	1.195	62.4934	67.5620	57.9935	72.0619	-1.6677
6	46.3500	51.6048	1.176	49.1117	54.0979	44.5853	58.6243	-5.2548
7	68.9900	68.6674	1.807	64.8373	72.4974	61.0696	76.2652	0.3226
8	62.9100	62.6967	1.200	60.1523	65.2411	55.6588	69.7346	0.2133
9	58.1300	60.8148	0.852	59.0086	62.6209	54.0089	67.6206	-2.6848
10	59.7900	59.1142	0.806	57.4049	60.8234	52.3334	65.8950	0.6758
11	56.2000	57.9029	0.932	55.9271	59.8787	51.0501	64.7558	-1.7029
12	66.1600	67.0640	1.248	64.4192	69.7088	59.9892	74.1388	-0.9040
13	62.1800	63.2813	1.012	61.1368	65.4257	56.3779	70.1846	-1.1013
14	57.0100	57.9957	1.328	55.1812	60.8102	50.8557	65.1357	-0.9857
15	65.6200	68.7237	0.871	66.8774	70.5699	61.9070	75.5403	-3.1037
16	65.0300	69.2767	0.913	67.3414	71.2119	62.4354	76.1179	-4.2467
17	66.7400	63.7328	1.690	60.1495	67.3161	56.2563	71.2093	3.0072
18	73.3800	68.6494	1.598	65.2612	72.0377	61.2644	76.0344	4.7306
19	82.8700	82.9320	1.815	79.0852	86.7789	75.3257	90.5384	-0.0620
20	95.7100	90.7110	1.653	87.2061	94.2159	83.2717	98.2	4.9990

Sum of Residuals	1.097789E-12
Sum of Squared Residuals	153.3007

[15] Most computer programs also have options for computing intervals for arbitrarily chosen values of the independent variables. These variables should, however, be within the range of the values in the data.

3.8 Correlation and the Coefficient of Determination

In Section 2.5 we defined the correlation coefficient as a convenient index of the strength of the linear relationship between two variables. We further showed how the square of the correlation coefficient, called the coefficient of determination, provided a useful measure of the strength of a linear regression. Equivalent statistics are available for multiple regression. There are two types of correlations that describe strengths of linear relationships among more than two variables:

> 1. **Multiple correlation**, which describes the strength of the linear relationship between one variable (usually the dependent variable) with a set of variables (usually the independent variables).
> 2. **Partial correlation**, which describes the strength of the linear relationship between two variables when all other variables are held constant or fixed. If one of these variables is the dependent variable, the partial correlation describes the strength of the linear relationship corresponding to that partial regression coefficient.

Other types of correlations not presented here are multiple-partial and part (or semipartial) correlations (Kleinbaum, Kupper, and Muller, 1988, Chapter 10). We discuss the multiple and partial correlation coefficients and the corresponding coefficient of determination in this section.

Multiple Correlation

In the multiple linear regression setting, the multiple correlation coefficient, denoted by R, is the correlation between the observed values of the response variable, Y, and the least squares estimated values, $\hat{\mu}_{y|x}$, obtained by the linear regression. Because the $\hat{\mu}_{y|x}$ are least squares estimates, it can be shown that R is also the maximum correlation obtainable between the response variable and a linear functions of the set of independent variables. In the more general setting, a multiple correlation coefficient is the maximum correlation between a linear function of a set of variables and a single variable.

Because $\hat{\mu}_{y|x}$ is the least squares estimate of the response, it follows that the multiple correlation coefficient measures the strength of the linear relationship between the response variable y and its least squares estimate, $\hat{\mu}_{y|x}$. Then, as in simple linear regression, the square of the correlation coefficient, R^2, is the ratio of the model over total sum of squares, that is, SSR/TSS.

Although the multiple correlation coefficient can be calculated by computing the $\hat{\mu}_{y|x}$ values and then calculating the correlation coefficient between $\hat{\mu}_{y|x}$ and y, it is more convenient to compute R^2 directly from the results of a regression analysis:

$$R^2 = \frac{\text{SS due to regression model}}{\text{Total SS for } y, \text{ corrected for the mean}},$$

and then R is the positive square root. Actually, R^2 is provided by all computer programs for multiple regression; in the outputs we have shown, it is denoted by "R-square." As in simple linear regression, R^2 is known as the coefficient of multiple determination and has an interpretation similar to that given r^2 in Section 2.5. That is, R^2 is the proportionate reduction in the variation of y that is directly attributable to the regression model. As in simple linear regression, the coefficient of determination must take values between and including 0 and 1, where the value 0 indicates the regression is nonexistent while the value of 1 indicates a "perfect" linear relationship. In other words, the coefficient of determination measures the proportional reduction in variability about the mean resulting from the fitting of the multiple regression model.

Also, as in simple linear regression, there is a correspondence between the coefficient of determination and the F statistic for testing for the existence of the model:

$$ F = \frac{(n - m - 1)R^2}{m(1 - R^2)} $$

The apparent simplicity of the coefficient of determination, which is often referred to as "R-square," makes it a popular and convenient descriptor of the effectiveness of a multiple regression model. This very simplicity has, however, made the coefficient of determination an often abused statistic. There is no rule or guideline as to what value of this statistic signifies a "good" regression. For some data, especially those from the social and behavioral sciences, coefficients of determination of 0.3 are often considered quite "good," whereas in fields where random fluctuations are of smaller magnitudes, as for example in engineering, coefficients of determination of less than 0.95 may imply an unsatisfactory fit. For the model analyzed in the output shown in Table 3.6, the R^2 value is given as 0.9591. Therefore, we can say that almost 96% of the variation in yield of timber (VOL) can be explained by the regression on DBH, HT, and D16. That means that only about 4% of the variation can be attributed to extraneous factors not included in the regression. However, the residual standard deviation (Root MSE in Table 3.10) of 3.10 implies that approximately 5% of observed volumes will lie more than 6.20 cubic feet (twice the standard deviation) from the estimated value, an error that may be too large for practical purposes. Furthermore, we noted that the prediction intervals were probably too wide to be useful.

An additional feature of the coefficient of determination is that when a small number of observations are used to estimate an equation, the coefficient of determination may be inflated by having a relatively large number of independent variables. In fact, if n observations are used for an $(n - 1)$ variable equation, the coefficient of determination is, by definition, unity. Therefore, the coefficient of determination can be driven to any desired value simply by adding variables! To overcome this effect, there is available an alternate statistic, called an "adjusted R-square" statistic, which indicates the proportional reduction in the mean square (rather than in the sum of squares). This quantity is also often produced automatically in computer outputs; in Table 3.6 it is denoted by "Adj R-sq." In that example, it differs little from the ordinary R-square because the number of independent variables is small compared to the sample size, which illustrates the recommendation that this statistic, although usually available in computer print-outs, is primarily useful if its value differs greatly from the ordinary R-square. The adjusted R-square statistic also has an interpretive problem due to the fact that it can assume negative values, which will occur if the F-value for the model is less than unity (and the p-value > 0.5).

Partial Correlation

The partial correlation coefficient describes the strength of a linear relationship between two variables holding constant other variables. In a sense, a partial correlation is to a simple correlation as a partial regression is to a total (or simple linear) regression.

One way to define partial correlation is to show how it can be computed from residuals. Given a set of p variables, $x_1, x_2, \ldots, x_p$, then $r_{xi,xj|\text{all others}}$, the partial correlation between x_i and x_j holding constant all other variables can be computed as follows:

1. Define $e_{i,j|\text{all others}}$ as the residuals from the regression of x_i on all other variables except x_j.
2. Define $e_{j,i|\text{all others}}$ as the residuals from the regression of x_j on all other variables except x_i.

Then the simple correlation between $e_{i,j|\text{all others}}$ and $e_{j,i|\text{all others}}$ is the partial correlation between x_i and x_j.

Because the partial correlation can be expressed as a simple correlation between two variables, it has the properties of a simple correlation: It takes a value from -1 to $+1$, with a value of 0 indicating no relationship and values of -1 and $+1$ indicating perfect linear relationship.

In the context of a regression model, the partial correlation of the dependent variable and an independent variable holding constant all other independent variables, has the following characteristics:

- There is an exact relationship between the test for the null hypothesis of a zero partial correlation and the test of the null hypothesis of a zero value for the corresponding partial regression coefficient. In this case the equivalence to the t statistic for testing whether a regression coefficient is zero is

$$|t| = \frac{|\hat{\beta}_j|}{\sqrt{c_{ij}\text{MSE}}} = \sqrt{\frac{(n - m - 1)r^2}{(1 - r^2)}}$$

where $r = r_{y,xj|\text{all other } x\text{'s}}$.
- The square of the partial correlation between y and a particular independent variable, x_j, is known as the coefficient of partial determination. The coefficient of partial determination measures the marginal contribution of adding x_j to a model containing all other independent variables. Its square

indicates the portion of the variability explained by that variable after all the other variables have been included in the model.

For example, suppose that X_1 is the age of a child, X_2 is the number of hours spent watching television, and Y is the child's score on an achievement test. The simple correlation between Y and X_2 would include the indirect effect of age on the test score and could easily cause that correlation to be positive. However, the partial correlation between Y and X_2, holding constant X_1, is the "age-adjusted" correlation between the number of hours spent watching TV and the achievement test score.

As noted, the test for the null hypothesis of no partial correlation is the same as that for the corresponding partial regression coefficient. Other inferences are made by an adaptation of the Fisher z transformations (Section 2.5), where the variance of z is $[1/(n - q - 3)]$, where q is the number of variables being held constant (usually $(m - 2)$).

Although partial correlations can be computed by residuals, there are a number of more efficient procedures for calculating these quantities. These will not be presented here (see, e.g, Kleinbaum, Kupper, and Muller, 1988, Section 10-5). The partial correlation coefficient is not widely used but has application in special situations, such as path analysis (Li, 1975). Finally, a partial correlation indicates the strength of a linear relationship between any two variables, holding constant a number of other variables, without any variables being specified as independent or dependent.

3.9 Summary and a Look Ahead
Uses and Misuses of Regression Analysis

This chapter has revolved around the use of a set of independent or factor variables to predict and/or explain the behavior of a dependent response variable. The appropriate inferences about parameters were presented under the normality assumption of Section 3.2. As in simple linear regression, there are some limitations on the uses of the results of a regression analysis. Two limitations were mentioned in Section 2.8:

1. Extrapolation is not recommended
2. The existence of a regression relationship does not, by itself, imply cause and effect

In multiple regression, these limitations have some additional features.

In avoiding extrapolation in the multiple regression setting there is the additional problem of **hidden extrapolation**. This condition occurs when values of individual independent variables are within the range of the observed values, but a **combination** of two or more variables does not occur. For example, in the data set given in Table 2.1, the range of the values of DBH is seen to be from 10 to 19 while the range of D16 is from 9 to 17. However, as shown by the data points plotted by (.) in Figure 3.5, a hypothetical tree with DBH = 14 and D16 = 18 (shown by the plotting symbol "X") is within the range of these two variables *singly*, but is definitely beyond the range of observed *combinations* of these two variables. Estimating the volume for a tree with these dimensions is an example of hidden extrapolation.

As in simple linear regression, the existence of a partial regression coefficient does not, by itself, imply that the corresponding variable causes a change in the response. As noted earlier, the reason is that there may be other variables that are causing the change. Now in multiple regression, the partial regression coefficient does indeed account for possible effects of other variables in the model, but this is no guarantee that one or more variables not in the model may still be the true cause. Of course, the more variables in the model, the greater the possibility of establishing a cause–effect relationship. However, as we will see, too many variables in a model may cause other problems.

In addition to these two obvious problems, there are many more considerations involved in the use of multiple regression. In fact, regression analyses are often afflicted by "Murphy's Law": "If anything can go wrong, it will." For this reason, the majority of the remainder of this text addresses possible problems

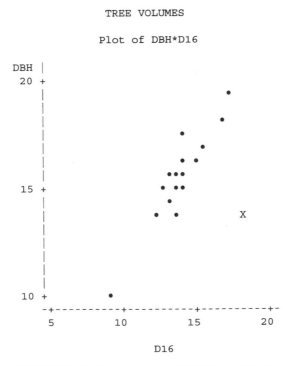

FIGURE 3.5 Relationship of DBH and D16

and the strategies used to solve them. For convenience, we have divided these potential problems into two groups—data problems and model problems.

Data Problems

Regression is mostly used with secondary data or data that are collected in a relatively random or haphazard manner, in contrast to the analysis of variance, which is primarily used for data resulting from planned experiments. This means that the results of such an analysis may be adversely affected by problems with the data. For example, outliers or unusual observations can badly bias regression model estimates. The "random" error may not be strictly random, independent, or normally distributed with variance σ^2, again causing biased estimates, incorrect standard errors, or incorrect significance levels. It is therefore advisable to make an effort to check for data problems that may cause unreliable results, and if they are found to exist, implement remedial actions. Methods to assist in these efforts are covered starting in Chapter 4.

Model Problems

Regression is a statistical method frequently used in exploratory analyses, that is, in situations where the true model is not known. Therefore, there is often a serious question as to whether the model specified is appropriate or not.

One problem is overspecification of the model—that is, including too many independent variables in the model. This does not cause biased estimates, but instead is a major cause of strong correlations among the independent variables. This phenomenon is called **multicollinearity** and causes estimated regression coefficients to have inflated standard errors. A common result when multicollinearity exists is for a regression model to be highly significant ($p = $ value < 0.05) while none of the regression coefficients approach significance (e.g., p-value > 0.05). In other words, the regression does a good job of estimating the behavior of the response, but it is impossible to tell which of the independent variables are responsible. Methods for diagnosing multicollinearity and some remedial methods are presented in Chapter 5.

On the other hand, specifying an inadequate model is known as **specification error** and may include the omission of important independent variables and/or specifying straight-line relationships when curvilinear relationships are appropriate. Making a specification error is known to cause biased estimates of model parameters. Diagnostics for specification error are presented in the text starting in Chapter 6.

Sometimes when a model appears to be overspecified, a commonly applied but not necessarily appropriate remedial method is that of discarding variables. This method, called variable selection, is presented in Chapter 6.

CHAPTER EXERCISES

The computational procedures for estimating coefficients and doing inferences on a multiple regression obviously do not lend themselves to hand calculations. Thus, it is assumed that all the exercises presented here will be analyzed using computers. Therefore, the focus of the exercises is not on the mechanics for obtaining the correct numerical results. This task is best reinforced by redoing the examples in the text. Instead the emphasis in exercises should be on the appropriate approach to the problem and the explanation of the results. For this reason we do not, in general, pose specific tasks for exercises. In general we expect the reader to do the following:

(a) Estimate all relevant parameters and evaluate the fit and the appropriateness of the model.
(b) Interpret the coefficients, and their statistical and practical significance.
(c) Examine residuals for possible violations of assumptions.
(d) Examine the predicted values and their standard errors.
(e) Summarize the results, including possible recommendations for additional analyses.

Also, for one or two selected exercises, the reader should do the following:

(f) Compute one or two coefficients and their test statistics by using residuals.

1. In Exercise 3 of Chapter 2, we related gasoline mileage to weight of some cars. The full data set included data on other variables, as follows:

WT: The weight in pounds
ESIZE: Engine power rating in cubic inches
HP: Engine horsepower
BARR: Number of barrels in the carburetor

Use a linear regression model to estimate MPG. The data are shown in Table 3.10 and are available as file REG03P01.

2. One of many organized sporting events is a "putting tour," a set of tournaments featuring professional miniature golf putters. As with all such events, there are extensive records of performances of the individuals who participate. We have such records for 32 individuals in a particular tour. The data are not reproduced here but are available on the data diskette as File REG03P02. The statistics recorded for this exercise are:

TNMT: The number of tournaments participated in by the player
WINS: The number of tournaments won
AVGMON: Average money won per tournament
ASA: The player's "adjusted" point average; the adjustment reflects the difficulties of the courses played

TABLE 3.10

MILEAGE DATA

WT	ESIZE	HP	BARR	MPG
2620	160.0	110	4	21.0
2875	160.0	110	4	21.0
2320	108.0	93	1	22.8
3215	258.0	110	1	21.4
3440	360.0	175	2	18.7
3460	225.0	105	1	18.1
3570	360.0	245	4	14.3
3190	146.7	62	2	24.4
3150	140.8	95	2	22.8
3440	167.6	123	4	19.2
3440	167.6	123	4	17.8
4070	275.8	180	3	16.4
3730	275.8	180	3	17.3
3780	275.8	180	3	15.2
5250	472.0	205	4	10.4
5424	460.0	215	4	10.4
5345	440.0	230	4	14.7
2200	78.7	66	1	32.4
1615	75.7	52	2	30.4
1835	71.1	65	1	33.9
2465	120.1	97	1	21.5
3520	318.0	150	2	15.5
3435	304.0	150	2	15.2
3840	350.0	245	4	13.3
3845	400.0	175	2	19.2
1935	79.0	66	1	27.3
2140	120.3	91	2	26.0
1513	95.1	113	2	30.4
3170	351.0	264	4	15.8
2770	145.0	175	6	19.7
3570	301.0	335	8	15.0
2780	121.0	109	2	21.4

At the end of the season each player is given a "POINT" score by the professional organization, which is supposed to indicate how good the player is considered to be.

Perform a regression to investigate how the points are related to the individual player's statistics.

3. The purpose of this study is to see how individuals' attitudes are influenced by a specific message as well as their previous knowledge and attitudes. In this study the subject is the preservation of the environment. Respondents were initially tested on their knowledge of the subject (FACT) and were given a test on their attitude (PRE): A high score indicates a pro preservation attitude. The sex (SEX) of respondents was also recorded; "1" is female and "2" is male.

The respondents were then exposed to an *anti*preservation message. After viewing the message, they recorded the number of *positive* (NUMPOS) and *negative* (NUMNEG) reactions they had to the message. It was believed that a number of positive reactions implied an antipreservation attitude and vice versa. The response variable is POST, a test on preservation attitude, where a high score indicates a propreservation attitude. The data are not shown and are available in File REG03P03.

Perform a regression to see how the POST score is related to the other variables.

4. It is generally perceived that military retirees like to retire in areas that are close to military bases and also have a pleasant climate. State data were collected to attempt to determine factors that affect the choice of states for military retirees. The variables are:

RTD: Total military retirement pay, a proxy for number of retirees (the response variable)
ACT: Total active military pay, a proxy for total military population
DOD: Total defense spending in state
POP: Total population of state
CLI: Cost of living index
LAT: Latitude (north in degrees)
PCP: Days of precipitation.

The data are in File REG03P04. Find a model that explains the decision process for the location of military retirees.

5. The data in Table 3.11 (also in file REG03P05) are consumer price index values for various commodity groups from the 1995 *Statistical Abstract of the United States.* The two groups, energy (ENERGY) and transportation (TRANS), were used in Exercise 5 of Chapter 2. Added to these two groups for this exercise are the index for medical care (MED) and the average for all items (ALL).

(a) Do the multiple regression using ALL as the dependent variable and the three others as independent variables. The interpretation of the coefficients is quite important here. What do the residuals imply here?

(b) Do separate simple regressions using ALL as the dependent variable and the other three separately as independent variables. Compare with (a). Explain.

(c) Find a simultaneous confidence region on all three coefficients in part (a) using the Bonferroni method.

(d) Find the individual confidence intervals in part (b) and compare with the results of (c).

6. A marketing research team has conducted a controlled price experiment for the purpose of studying demand relationships for three types of oranges:

1. Florida Indian River oranges, a premium Florida orange
2. Florida Interior oranges, a standard Florida orange
3. California oranges, considered superior to all Florida oranges

TABLE 3.11

DATA FOR EXERCISE 5

Year	ENERGY	TRANS	MED	ALL
60	22.4	29.8	22.3	29.6
61	22.5	30.1	22.9	29.9
62	22.6	30.8	23.5	30.2
63	22.6	30.9	24.1	30.6
64	22.5	31.4	24.6	31.0
65	22.9	31.9	25.2	31.5
66	23.3	32.3	26.3	32.4
67	23.8	33.3	28.2	33.4
68	24.2	34.3	29.9	34.8
69	24.8	35.7	31.9	36.7
70	25.5	37.5	34.0	38.8
71	26.5	39.5	36.1	40.5
72	27.2	39.9	37.3	41.8
73	29.4	41.2	38.8	44.4
74	38.1	45.8	42.4	49.3
75	42.1	50.1	47.5	53.8
76	45.1	55.1	52.0	56.9
77	49.4	59.0	57.0	60.6
78	52.5	61.7	61.8	65.2
79	65.7	70.5	67.5	72.6
80	86.0	83.1	74.9	82.4
81	97.7	93.2	82.9	90.9
82	99.2	97.0	92.5	96.5
83	99.9	99.3	100.6	99.6
84	100.9	103.7	106.8	103.9
85	101.6	106.4	113.5	107.6
86	88.2	102.3	122.0	109.6
87	88.6	105.4	130.1	113.6
88	89.3	108.7	138.6	118.3
89	94.3	114.1	149.3	124.0
90	102.1	120.5	162.8	130.7
91	102.5	123.8	177.0	136.2
92	103.0	126.5	190.1	140.3
93	104.2	130.4	201.4	144.5
94	104.6	134.3	211.0	148.2

A total of 31 price combinations (P1, P2, and P3, respectively) were selected according to an experimental design and randomly assigned to 31 consecutive days (excluding Sundays) in several supermarkets. Daily sales of the three types of oranges were recorded and are labeled as variables Q1, Q2, and Q3, respectively. For this exercise we have selected data for only one store. The data are shown in Table 3.12 and are available as File REG03P06.

Perform the three separate regressions relating the sales quantities to the three prices. Ignore the "Day" variable in the regression model, but consider it as a factor in examining the residuals. In addition to the usual interpretations, a comparison of the coefficient for the three orange types is of interest.

TABLE 3.12

DATA ON ORANGE SALES

Day	P1	P2	P3	Q1	Q2	Q3
1	37	61	47	11.32	0.00	25.47
2	37	37	71	12.92	0.00	11.07
3	45	53	63	18.89	7.54	39.06
4	41	41	51	14.67	7.07	50.54
5	57	41	51	8.65	21.21	47.27
6	49	33	59	9.52	16.67	32.64
1	37	61	71	16.03	0.00	0.00
2	45	45	63	1.37	5.15	4.12
3	41	57	51	22.52	6.04	68.49
4	45	53	55	19.76	8.71	30.35
5	65	49	59	12.55	13.08	37.03
6	61	61	71	10.40	8.88	22.50
1	41	57	67	13.57	0.00	0.00
2	49	49	59	34.88	6.98	10.70
3	49	49	43	15.63	10.45	77.24
4	61	37	71	13.91	15.65	11.01
5	57	41	67	0.71	18.58	25.09
6	41	41	67	15.60	12.57	29.53
1	57	57	67	5.88	2.26	20.36
2	45	45	55	6.65	6.01	30.38
3	53	45	55	4.72	22.05	60.63
4	57	57	51	6.14	7.62	26.78
5	49	49	75	15.95	14.36	14.36
6	53	53	63	8.07	7.02	17.19
1	53	45	63	1.45	10.87	3.26
2	53	53	55	6.50	0.00	19.49
3	61	37	47	7.06	30.88	48.58
4	49	65	59	10.29	1.20	19.86
5	37	37	47	16.34	22.99	49.24
6	33	49	59	27.05	7.79	32.79
1	61	61	47	11.43	4.29	18.57

7. Data on factors that may affect water consumption for the 48 contiguous states are available in File REG03P07. The water consumption data are from van der Leeden *et al.* (1960), and data on factors are obtained from the *Statistical Abstract of the United States* (1988) and an atlas. The variables are:

LAT: Approximate latitude of the center of the state, in degrees
INCOME: Per capita income, $1000
GAL: Per capita consumption of water, in gallons per day
RAIN: Average annual rainfall, in inches
COST: The average cost per 1000 gallons, in dollars.

Perform the regression to estimate water consumption as a function of these variables.

8. The data for this exercise concern factors considered to be influential in determining the cost of providing air service. It is desired to develop a model for estimating the cost per passenger mile so that the major factors in determining that cost can be isolated.

The data source is from a CAB report "Aircraft Operation Costs and Performance Report," August, 1972 (prior to deregulation), and is quoted in Freund and Littel (1991). The variables are:

CPM: Cost per passenger mile (cents)
UTL: Average hours per day use of aircraft
ASL: Average length of nonstop legs of flights (1000 miles)
SPA: Average number of seats per aircraft (100 seats)
ALF: Average load factor (% of seats occupied by passengers)

Data have been collected for 33 U.S. airlines with average nonstop lengths of flights greater than 800 miles. The data are available in File REG03P08. Perform the regression that estimates the cost per passenger mile.

9. The data for this exercise may be useful to ascertain factors related to gasoline consumption. The data have been extracted from Drysdale and Calef (1977). Variables are related to gasoline consumption for the 48 contiguous states with District of Columbia data incorporated into Maryland.

The variables are:

STATE: Two-character abbreviation
GAS: Total gasoline and auto diesel consumption in 10^{12} BTU
AREA: Of state in 1000 miles
POP: Population, 1970, in millions
MV: Estimated number of registered vehicles in millions
INC: Personal income in billions of dollars
VAL: Value added by manufactures in billions of dollars
Region: Codes EAST and WEST of the Mississippi river.

The data are available as File REG03P09.

Find a model that explains the factors affecting gasoline consumption. Ignore the region variable.

10. This is another contrived example of completely different total and partial regression coefficients similar to Example 3.2. Usage data in Table 3.13 perform the simple linear and the two variable multiple regressions and construct the necessary plots and/or calculations to show that these results are indeed correct.

TABLE 3.13

DATA FOR EXERCISE 3.10

OBS	x_1	x_2	y
1	3.8	3.8	3.1
2	7.9	7.4	6.9
3	4.1	3.9	3.8
4	7.2	5.2	7.5
5	4.6	3.9	3.9
6	8.8	7.8	5.4
7	1.1	2.9	0.0
8	8.4	8.5	3.4
9	8.0	7.5	3.8
10	3.4	2.5	5.4
11	3.6	4.1	3.0
12	10.0	9.0	5.1
13	5.6	6.3	1.4
14	6.4	7.2	0.5
15	1.5	1.2	4.0
16	5.9	6.2	2.0
17	1.2	3.7	–2.0
18	0.6	1.0	2.4
19	9.2	9.4	2.4
20	7.3	5.2	6.5

PART II

PROBLEMS AND REMEDIES

Part I of this book is concerned with the use of linear models to analyze the behavior of a response variable with the primary emphasis on the use of the linear regression model. Statistical analyses based on such models can provide useful results if the correct model has been chosen and other assumptions underlying the model are satisfied. In many applications these assumptions do hold, at least to a sufficient degree to ensure that the results can be used with confidence.

However, if violations of these assumptions are more substantial, the results may not reflect the true population relationships, therefore providing incorrect conclusions that may lead to recommendations or actions that do more harm than good. Unfortunately, such results are rather likely to occur in regression analyses because these are frequently used on data that are *not* the result of carefully designed experiments. For this reason it is important to scrutinize the results of regression analyses to determine if problems exist that may compromise the validity of the results, and if such problems are deemed to exist, to search for remedial measures or modify the inferences.

Part II of this book, consisting of Chapters 4, 5, and 6, addresses methods for detecting possible problems with either data or model specification and suggests possible remedies. For the purpose of this book we have divided these problems into three categories that are the basis for these three chapters.

Chapter 4 deals with how individual observations affect the results of an analysis. Because individual observations are the rows of the X and Y matrices, diagnostic tools for detecting possible problems with observations are called **row diagnostics**. We present three aspects of row diagnostics:

Outliers and influential observations are observations that in some manner appear not to "belong" and consequently may affect estimates of model parameters.

Unequal variances among observations of the response variable violate the equal variance assumption and may result in biased estimates of parameters, but more importantly, cause incorrect standard errors of estimated responses.

Correlated errors among observations of the response violate the assumption of independent errors and may cause biases in estimated standard errors of parameter as well as response estimates.

Chapter 5 deals with correlations among the independent variables. This condition, called *multicollinearity,* is not a violation of assumptions and therefore does not invalidate parameter and response estimates. However, multicollinearity does create difficulties in estimating and interpreting the partial regression coefficients. Multicollinearity is not difficult to detect; however, effective and useful remedial methods are not easy to implement and, in fact, are sometimes impossible to find.

Chapter 6 deals with variable selection. Inferences on linear models assume a correctly specified model in the sense that the model used for analysis contains the "correct" selection of independent variables. However, a regression analysis is often used to *determine* the "correct" set of independent variables by starting with an initial model that contains a large number of independent variables and then using statistical methods to select the "correct" set. By definition, this procedure is a violation of assumptions and would usually be avoided. (It has been argued that this methodology should not be presented in a textbook.) However, because these methods are intuitively attractive and readily available in all statistical software packages, they are indeed widely used (and often misused) and are, therefore, presented in this chapter.

These chapters may leave the reader somewhat frustrated, because they provide no universally "correct" procedures with "clean" answers. However, we must remember that we are here dealing largely with data and model problems, and unfortunately there is no way to make good data from bad, nor is it possible to produce a "correct" model from thin air. These are, after all, exploratory rather than confirmatory analysis procedures whose results are largely useful to guide further studies.

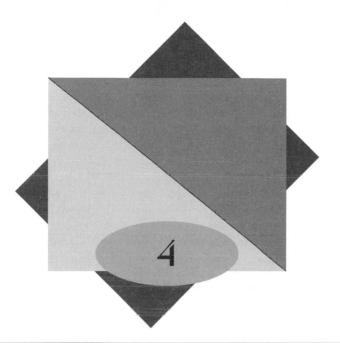

Problems with Observations

4.1 Introduction

We start the discussion of possible problems encountered with regression analyses with a look at the effect a few "extreme" observations have on the results of a statistical analysis, present some statistics that may be useful in identifying such observations, and provide some guidelines for possible remedies for dealing with such observations. We then explore the diagnostic tools necessary to detect violations of assumptions on homoscedasticity and independence of the error term. At the end of this discussion we again look at some suggested remedial action. As will be pointed out, one method of correcting for violations of assumptions is to make some form of transformation on the data. Because this topic is also important for a number of other situations, we treat transformations in a separate section in Chapter 8.

Because problems with observations involve both the independent and response variables and can occur in so many ways, this chapter is not only long but may be overwhelming. However, much of the data that we subject to statistical analysis does present us with these problems, so it is important that we cover as many of them in as comprehensive a manner as possible. To make this chapter manageable, we have divided it into two parts. The exercises at the end of the chapter are also divided in a similar manner.

Part One consists of Sections 4.1 and 4.2 and deals with the possible existence of unusual or extreme observations, often referred to as outliers. This part is quite long because, as we will see, outliers can occur in many different disguises, especially since they may exist among the independent variables as well as the response variable. For this reason there exist a number of different statistics for detection of such observations, none of which claim to be the best.

Part Two examines situations where violations of the assumptions about the variance and independence of the random errors may exist. Sections 4.3 and 4.4 address the detection of unequal variances and some suggested remedies. Section 4.5 presents a situation that often results in correlated error terms, a response variable measured over time. This type of a regression model is known as a time series and allows us to model the responses when we have time-dependent errors.

With the possible exception of time series data, we usually have no way of knowing if any one data set exhibits one or possibly more of the problems discussed in this chapter. The diagnostics presented in this chapter need not all be performed, nor do they need to be performed in the order presented. In fact, a thorough knowledge of the nature and source of the data under study may indicate that none of these techniques are needed.

▼
___

PART ONE: OUTLIERS

4.2 Outliers and Influential Observations

An observation that is markedly different from, or atypical of, the rest of the observations of a data set is known as an outlier. An observation may be an outlier with respect to the response variable and/or the independent variables. Specifically, an extreme observation in the response variable is called an **outlier**, while extreme value(s) in the x's (independent variables) are said to have high **leverage** and are often called **leverage points**.

An observation that causes the regression estimates to be substantially different from what they would be if the observation were removed from the data set is called an **influential observation**. Observations that are outliers or have high leverage are not necessarily influential, while influential observations usually are outliers and have high leverage.

In this section we start the presentation with a very simple example of a single outlier that will also show why outliers may sometimes be difficult to detect. We continue with the presentation of statistics designed to detect outliers, leverage points and influential observations. These are illustrated with an example. Unfortunately, what to do if "bad" observations are found is not a simple matter.

EXAMPLE 4.1 This artificial example consists of 10 observations generated from a simple linear regression model. The values of the observations of the independent variable, x, have the values from 1 to 10. The model is

$$y = 3 + 1.5x + \epsilon,$$

where ϵ is a normally distributed random variable with mean zero and standard deviation of three and with no outliers. The data are shown in Table 4.1.

The results of the regression analysis performed by PROC REG of the SAS System are shown in Table 4.2. The results are reasonable. The regression is certainly significant, the parameter estimates are within one standard error of the true values, and although the error mean square appears to be low, the 0.9 confidence interval does include the true value of 9.

For purposes of illustration, we have constructed two different outlier scenarios:

Scenario 1: When $x = 5$, y has been increased by 10 units
Scenario 2: When $x = 10$, y has been increased by 10 units

Note that in each scenario the observed value of the dependent variable has been increased by over three standard deviations, certainly a result that we would not expect to occur that thus qualifies the value as an outlier. The difference between the scenarios is in the location of the outlier: The first occurs in the "middle" of the range of x-values, whereas the second occurs at the (high) end. Table 4.3 gives a brief summary of results for the regression results for the original data and two outlier scenarios.

The results show a definite biasing of estimates due to the outliers, but the biases are quite different for the two scenarios. For scenario 1, $\hat{\beta}_1$ has changed little, $\hat{\beta}_0$ has increased somewhat, and the error mean square and consequently the standard errors of both coefficients are much larger. For scenario 2, $\hat{\beta}_1$ has changed rather dramatically, while the changes in $\hat{\beta}_0$ and MSE are not so marked. These results are seen in Figure 4.1, which shows the observed values and fitted

TABLE 4.1

EXAMPLE TO ILLUSTRATE OUTLIER

Obs	x	y
1	1	6.2814
2	2	5.1908
3	3	8.6543
4	4	14.3411
5	5	13.8374
6	6	11.1229
7	7	16.5987
8	8	19.1997
9	9	20.0782
10	10	19.7193

TABLE 4.2

REGRESSION RESULTS

Analysis of Variance

Source	DF	Sum of Squares	Mean Square	F Value	Prob>F
Model	1	240.87628	240.87628	56.233	0.0001
Error	8	34.26827	4.28353		
C Total	9	275.14455			

Root MSE	2.06967	R-square	0.8755
Dep Mean	13.50238	Adj R-sq	0.8599
C.V.	15.32819		

Parameter Estimates

Variable	DF	Parameter Estimate	Standard Error	T for H0: Parameter=0	Prob > |T|
INTERCEP	1	4.104439	1.41385379	2.903	0.0198
X	1	1.708717	0.22786317	7.499	0.0001

lines for the three cases. These plots show that the most visible effect on the estimated response line is due to the outlier at the upper end of the *x*-values.

The nature of these results are a direct consequence of the least squares estimating procedure. Remember, the regression line is that line for which the sum of squared distances to the data points are minimized. A physical analogue, which makes this process more understandable, is afforded by a law of physics called Hooke's law, which states: "The energy in a coil spring is proportional to the

TABLE 4.3

SUMMARY OF REGRESSIONS

Scenario	$\hat{\beta}_0$ std error	$\hat{\beta}_1$ std error	MSE
Original data	4.104	1.709	4.284
	1.414	0.228	
Scenario 1	5.438	1.648	18.469
	2.936	0.473	
Scenario 2	2.104	2.254	8.784
	2.024	0.326	

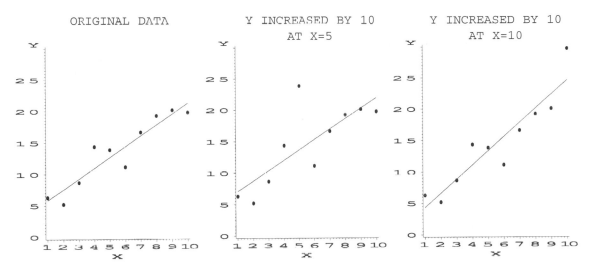

FIGURE 4.1 Effects of Outlier on Regression

square of the distance the spring is stretched." This means that if a coil spring[1] is stretched from each point to a rigid rod representing the regression line, the least squares line will be achieved by the equilibrium position of the rod.

This physical analogue allows us to visualize how the outlier in the middle exerts a balanced pull on the rod, lifting it slightly (the increase in $\hat{\beta}_0$) but leaving the slope ($\hat{\beta}_1$) essentially unchanged. On the other hand, the outlier at the upper end tends to pull the rod only on one end, which provides more leverage than the outlier in the middle; hence, the estimate of the slope ($\hat{\beta}_1$) is much more affected by that outlier. This is also the reason why the error mean square was not increased as much in scenario 2. In other words, the location of the outlier has affected the nature of the bias of the parameter estimates. This combination of outlier and leverage defines an **influential** observation.

Now that we have seen the damage that can be caused by an outlier, we need to find methods for detecting outliers. We will see that even for this very simple example it is not as easy as it may seem.

The standard tool for detecting outliers in the response variable is the **residual plot**. This is a scatterplot of the residuals ($y - \hat{\mu}_{y|x}$) on the vertical axis and the estimated values $\hat{\mu}_{y|x}$ on the horizontal axis.[2] If all is well, such a plot should reveal a random scattering of points around the zero value on the vertical axis. In such a plot outliers should be identified as points definitely not adhering to this pattern. Figure 4.2 shows the residual plots for the original data and the two outlier scenarios. In later discussions we will use various modifications of this standard residual plot to investigate other potential problems.

---

[1] Of course we assume a "perfect" spring and no gravity.

[2] For one-variable regressions it may be more useful to use the x-values on the horizontal axis.

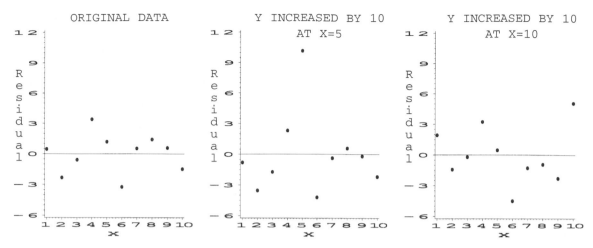

FIGURE 4.2 Effects of Outliers on Residuals

The residuals for the original data appear to satisfy the random patterns requirement. For scenario 1, the outlier definitely stands out. However, for scenario 2, the residual, although having the largest magnitude, is not much larger than at least one of the others, and therefore does not provide strong evidence of being an outlier. The reason for this result is that the high degree of leverage for this observation tends to pull the line toward itself; hence, the residual is not very large. ◆

It is apparent that additional information is needed to help detect such an outlier. Such information, in the form of a number of statistics for detecting outliers, is indeed available and is described in the rest of this section. As we will see, several different statistics will be needed because there may be several different ramifications of having an outlier present in the data. Because of the number of different statistics used in detection of outliers and measuring the effect, we present the diagnostic tools in four contexts:

1. Statistics based on *residuals*
2. Statistics designed to detect *leverage*
3. Statistics that measure the *influence* of an observation on the **estimated response**, which is a combination of residuals and leverage
4. Statistics that measure the *influence* on the **precision** of parameter estimates.

Statistics Based on Residuals

We have already noted that the actual residuals from the fitted model,

$$r_i = (y_i - \hat{\mu}_{y|x}),$$

when plotted against the predicted value, can be used in detecting outliers, especially when the observation does not have high leverage. Obviously, this is a very subjective method of identifying outliers. We can construct a slightly more rigorous method by first standardizing the residuals by dividing by their standard errors. The standard error of the residual, r_i, is

$$\sqrt{\sigma^2(1 - h_i)},$$

where h_i is the ith diagonal element of the matrix $[X(X'X)^{-1}X']$, the so-called **hat matrix**.[3] Standardized residuals have two useful properties:

1. They have zero mean and unit standard deviation. Therefore, we can determine how far an observation is from the mean in terms of standard deviation units, and, using the empirical rule, determine approximately how likely it is to obtain a suspected outlier strictly by chance.
2. By standardizing the residuals, we compensate for the leverage; therefore, they should reveal outliers regardless of the leverage of the observations.

Because we rarely know the value of σ^2, we estimate it with MSE, the mean square error of Section 3.3. The studentized residuals[4] are computed as follows:

$$\text{Stud. Res} = \frac{(y_i - \hat{\mu}_{y|x})}{\sqrt{\text{MSE}(1 - h_i)}}.$$

The residuals (Residual) and studentized residuals (Stud.Res) values for the original and two outlier scenarios in Example 4.1, as produced with PROC REG of the SAS System are shown in Table 4.4.

TABLE 4.4

RESIDUALS AND STUDENTIZED RESIDUALS

	Original Data		Scenario 1		Scenario 2	
X	Residual	Stud.Res	Residual	Stud.Res	Residual	Stud.Res
1	0.47	0.28	-0.80	-0.23	1.92	0.80
2	-2.33	-1.30	-3.54	-0.95	-1.42	-0.55
3	-0.58	-0.31	-1.73	-0.44	-0.21	-0.08
4	3.40	1.76	2.31	0.58	3.22	1.16
5	1.19	0.61	10.16	2.50	0.46	0.16
6	-3.23	-1.65	-4.20	-1.03	-4.51	-1.61
7	0.53	0.28	-0.38	-0.09	-1.28	-0.46
8	1.43	0.76	0.58	0.15	-0.94	-0.35
9	0.60	0.33	-0.19	-0.05	-2.31	-0.90
10	-1.47	-0.88	-2.20	-0.63	5.07	2.12

[3] As noted in Chapter 3, these elements were used to compute the standard error of the conditional mean. Actually,

$$\text{vâr}(\hat{\mu}_{y|x}) = \text{MSE}(h_i), \text{ and}$$
$$\text{vâr}(y - \mu_{y|x}) = \text{MSE}(1 - h_i).$$

[4] The standard error uses the estimated variance (MSE), so the appropriate distribution of that statistic is the Student t distribution; hence, the nomenclature "studentized" residuals.

Under the assumption of no outliers, the studentized residuals should be distributed as Students' t with the degrees of freedom of MSE, usually $(n - m - 1)$. Unless the degrees of freedom are extremely small, values exceeding 2.5 should rarely occur,[5] and we see that the studentized residuals do indeed point to the outliers we created. However, the outlier for scenario 2 is still not as clearly defined as that for scenario 1. Hence, the studentized residuals have helped, but are still not as sensitive as we may like them to be, so we will have to look further.

Statistics Measuring Leverage

In Example 4.1 we saw that the degree of *leverage* is an important element in studying the effect of extreme values. In order to assess the effect of leverage, we need a way to measure this effect. The most common measure of leverage is the diagonal element of the hat matrix, h_i, described earlier, which is a standardized measure of the distance between the x-values for the ith observation and the means of the x-values for all observations. An observation with high leverage measured in this way may be considered an outlier in the independent variable set. Obviously, an observation may exhibit high leverage without exerting a large influence on the model, which will generally occur when the high-leverage observation fits the model. In fact, one interpretation of an observation with high leverage is that it has the *potential* of causing a problem. In any case, it is useful to identify observations with high leverage, and then determine the influence these observations have on the model.

It now remains to specify what values of h_i imply a high degree of leverage. It can be shown that

$$\Sigma h_i = (m + 1);$$

hence the average value of the h_i,

$$\bar{h} = \frac{m + 1}{n}.$$

As a rule of thumb a value exceeding twice the average, that is, $h_i = 2\bar{h}$, is considered to indicate a high degree of leverage, but this rule is somewhat arbitrary.

In Example 4.1, as in any one-variable regression,

$$h_i = \frac{(x_i - \bar{x})^2}{\Sigma(x_i - \bar{x})^2},$$

which is indeed a measure of the relative magnitude of the squared distance of x from $\bar{x}$ for the ith observation. In Example 4.1, the two scenarios had their outlier at $x = 5$ and $x = 10$, respectively. The leverage for $x = 5$ is 0.003; for $x = 10$ it is 0.245. Although the leverage for $x = 10$ (the second scenario) is much higher than that for $x = 5$, it does not exceed twice the value $2/10 = 0.2$, and so

---

[5] The degrees of freedom in this example are 8, and the 0.05 two-tailed t-value is 2.306.

it is not considered a large leverage. We will illustrate leverage for more interesting examples later.

Statistics Measuring Influence on the Estimated Response

We saw in Example 4.1 that studentized residuals can be effective in identifying outliers regardless of their leverage, and appeared to work better for scenario 1. However, it can be argued that the outlier in scenario 2 created more bias in the parameter estimates. What has happened is that the effect of the outlier for scenario 2 is due to a combination of the magnitude of the outlier and the leverage of the independent variable. This combined effect is called the **influence** and is a measure of the effect the outlier has on the parameter estimates and hence on the estimated response, $\hat{\mu}_{y|x}$.

Recall that we previously defined an observation that causes the regression estimates to be substantially different from what they would be if the observation were removed from the data set as an influential observation. This allows us to evaluate the influence of an observation using the "leave one out" or "deleted residual" principle. This principle compares various model estimates obtained by leaving out each observation, individually, with the model obtained by using all observations. An observation for which the magnitudes of these differences are judged large is said to have a high degree of influence on the estimated model.

The most popular influence statistic is denoted by **DFFITS**, which is a mnemonic for the *DiF*ference in *FIT*, Standardized. Define $\hat{\mu}_{y|x,-i}$ as the estimated mean of the response variable using the model estimated from all *other* observations, that is, with observation i left out, and MSE$_{-i}$ as the error mean square obtained by this regression. Then

$$\text{DFFITS} = \frac{\hat{\mu}_{y|x} - \hat{\mu}_{y|x,-i}}{\text{Standard error}},$$

where "standard error" is the standard error of the numerator. Normally, a DFFITS value is calculated for each observation used in a regression. The DFFITS statistics are never calculated manually; however, the following formulas do give insight into what they measure.

$$\text{DFFITS}_i = \left(\frac{y_i - \hat{\mu}_{y|x}}{\sqrt{\text{MSE}_{-i}(1 - h_i)}}\right)\sqrt{\frac{h_i}{1 - h_i}},$$

and further,

$$\text{MSE}_{-i} = \frac{\text{SSE} - \dfrac{(y_i - \hat{\mu}_{y|x})^2}{1 - h_i}}{n - m - 1}.$$

These formulas show that the statistics are computed with quantities that are already available from the regression analysis, and are a combination of leverage and studentized residuals. We can also see that the DFFITS statistics will increase in magnitude with increases in both residuals ($y_i - \hat{\mu}_{y|x}$) and leverage (h_i).

Although this statistic is advertised as "standardized," the standard error of its distribution is not unity and is approximated by

$$\text{Standard error of DFFITS} \approx \sqrt{\frac{m + 1}{n}},$$

where m is the number of independent variables in the model.[6] It is suggested that observations with the absolute value of DFFITS statistics exceeding twice that standard error may be considered influential, but as we will see, this criterion is arbitrary.

When the DFFITS statistic has identified an influential observation, it is of interest to know which coefficients (or independent variables) are the cause of the influence. This information can be obtained by two means:

The **DFBETAS** statistics, which measure the relative differences between the least squares and deleted residual estimates of the coefficients
The use of **leverage plots**, also called partial residual plots

Using the DFBETAS Statistics

Belsley *et al.* (1980) suggest that absolute values of DFBETA exceeding $2/\sqrt{n}$ may be considered "large," and all we need to do is find those values that are large. However, since a value for DFBETA is calculated for each observation and for each independent variable, there are $n(m + 1)$ such statistics, and it would be an insurmountable task to look for "large" values among this vast array of numbers. Fortunately, the magnitude of this task can be greatly reduced by using the following strategy:

1. Examine the DFBETAS only for observations already identified as having large DFFITS.
2. Then find the coefficients corresponding to relatively large DFBETAS.

In Table 4.5 are the DFFITS and DFBETAS for the original data and two outlier scenarios of Example 4.1. Each set of three columns contain DFFITS and two DFBETAS (for β_0 and β_1, labeled DFB0 and DFB1) for the original and two outlier scenarios, with the statistics for the outliers underlined. Using the suggested guidelines, we would look for an absolute value of DFFITS larger than 0.89, then examine the corresponding DFBETAS to see if any absolute value exceeds 0.63. The DFFITS values for the outliers of both scenarios clearly exceed the value suggested by the guidelines. Further, the DFBETAS clearly show that β_0 is the coefficient affected by the outlier in scenario 1, whereas both coefficients are almost equally affected by the outlier in scenario 2.

---

[6] The original proponents of these statistics (Belsley *et al.*, 1980) consider the intercept as simply another regression coefficient. In their notation, the total number of parameters (including β_0) is p, and various formulas in their and some other books will use p rather than $(m + 1)$. Although that consideration appears to simplify some notation, it can produce misleading results (see Section 2.6 [regression through the origin].)

TABLE 4.5

DFFITS AND DFBETAS FOR EXAMPLE 4.1

| | ORIGINAL DATA | | | SCENARIO 1 | | | SCENARIO 2 | | |
OBS	DFFITS	DFB0	DFB1	DFFITS	DFB0	DFB1	DFFITS	DFB0	DFB1
1	0.19	0.19	-0.16	-0.16	-0.16	0.13	0.57	0.57	-0.48
2	-0.79	-0.77	0.61	-0.54	-0.53	0.42	-0.30	-0.30	0.23
3	-0.13	-0.12	0.09	-0.19	-0.18	0.13	-0.03	-0.03	0.02
4	0.80	0.66	-0.37	0.21	0.17	-0.10	0.46	0.37	-0.21
5	0.20	0.12	-0.03	1.68	1.02	-0.29	0.05	0.03	-0.01
6	-0.64	-0.20	-0.11	-0.35	-0.11	-0.06	-0.62	-0.19	-0.11
7	0.10	0.00	0.05	-0.03	0.00	-0.02	-0.17	0.00	-0.08
8	0.34	-0.08	0.22	0.06	-0.01	0.04	-0.15	0.04	-0.10
9	0.18	-0.07	0.14	-0.03	0.01	-0.02	-0.51	0.20	-0.40
10	-0.63	0.31	-0.53	-0.44	0.22	-0.37	2.17	-1.08	1.83

Leverage Plots

In Section 3.4 we showed that a partial regression coefficient can be computed as a simple linear regression coefficient using the residuals resulting from the two regressions of y and that independent variable on all other independent variables. In other words, these residuals are the "data" for computing that partial regression coefficient. Now, as we have seen, finding outliers and/or influential points for a simple linear regression is usually just a matter of plotting the residuals against the estimated value of y. We can do the same thing for a specified independent variable in a multiple regression by plotting the residuals as defined earlier. Such plots are called partial residual or **leverage plots**. The leverage plot gives a two-dimensional look at the hypothesis test H_0: $\beta_i = 0$. Therefore, this plot does the following:

It directly illustrates the partial correlation of x_i and y, thus indicating the effect of removing x_i from the model.

It indicates the effect of the individual observations on the estimate of that parameter. Thus, the leverage plot allows data points with large DFBETAS values to be readily spotted.

One problem with leverage plots is that in most leverage plots, individual observations are usually not easily identified. Hence, the plots may be useful for showing that there are influential observations, but they may not be readily identified. We will see later that these plots have other uses and will illustrate leverage plots at that time.[7]

Two other related statistics that are effective in identifying influential observations are Cook's D and the PRESS statistic.

---

[7] Leverage plots are, of course, not useful for one-variable regressions.

Cook's D, short for Cook's distance, is an overall measure of the impact of the ith observation on the set of estimated regression coefficients, and is thus comparable to the DFFITS statistic. Since most of these analyses are done using computer programs, we don't need the formula for calculating Cook's D (it is essentially $(DFFITS)^2/(m + 1)$). We do note that it is not quite as sensitive as DFFITS; however, since the values are squared, the potential influential observations do tend to stand out more clearly.

The **PRESS** statistic is a measure of the influence of a single observation on the residuals. First residuals are calculated from the model by the leave-one-out principle, then PRESS, a mnemonic for Prediction Error Sum of Squares:

$$\text{PRESS} = \Sigma(y_i - \hat{\mu}_{ylx,-i}).$$

The individual residuals are obviously related to the DFFITS statistic, but, because they are not standardized, they are not as effective for detecting influential observations and are therefore not often calculated.

The PRESS statistic has been found to be very useful for indicating if influential observations are a major factor in a regression analysis. Specifically, when PRESS is considerably larger than the ordinary SSE, there is reason to suspect the existence of influential observations. As with most of these statistics, "considerably larger" is an arbitrary criterion; over twice as large may be used as a start. The comparisons of SSE and PRESS for the original data and the two scenarios in Example 4.1 are below:

SCENARIO	SSE	PRESS
ORIGINAL	34.26	49.62
SCENARIO 1	147.80	192.42
SCENARIO 2	70.27	124.41

In this example, the PRESS statistics reveal a somewhat greater indicator of the influential observations[8] in scenario 2.

Statistics Measuring Influence on the Precision of Estimated Coefficients

One measure of the precision of a statistic is provided by the estimated variance of that statistic, with a large variance implying an imprecise estimate. In Section 3.5 we noted that the estimated variance of an estimated regression coefficient was

$$\hat{\text{var}}(\hat{\beta}_i) = c_{ii}\,\text{MSE},$$

where c_{ii} is the ith diagonal element of $(X'X)^{-1}$. This means that the precision of an estimated regression coefficient improves with smaller c_{ii} and/or smaller MSE and gets worse with a larger value of either of these terms. We can sum-

---

[8] The PRESS sum of squares is also sometimes used to see what effect influential observations have on a variable selection process (see Chapter 6).

marize the total precision of the set of coefficient estimates with the **generalized variance**, which is given by

$$\text{Generalized variance}(\hat{B}) = \text{MSE} \, | (X'X)^{-1} |,$$

where $| (X'X)^{-1} |$ is the determinant of the inverse of the $X'X$ matrix. The form of this generalized variance is similar to that for the variance of an individual coefficient in that a reduction in MSE and/or in the determinant of $(X'X)^{-1}$ will result in an increase in the precision. Although the determinant of a matrix is a complicated function, two characteristics of this determinant are of particular interest:

1. As the elements of $X'X$ become larger, the determinant of the inverse will tend to decrease. In other words, the generalized variance of the estimated coefficients will decrease with larger sample sizes and wider dispersions of the independent variables.
2. As correlations among the independent variables increase, the determinant of the inverse will tend to increase. Thus, the generalized variance of the estimated coefficients will tend to increase with the degree of correlation among the independent variables.[9]

A statistic that is an overall measure of how the ith observation affects the precision of the regression coefficient estimates is known as the **COVRATIO**. This statistic is the ratio of the generalized variance leaving out each observation to the generalized variance using all data. In other words, the COVRATIO statistic indicates how the generalized variance is affected by leaving out an observation. A COVRATIO near unity indicates that the observation has little effect on the precision of the parameter estimates; values greater than unity indicate that the observation increases precision of the estimates and vice versa. That is,

> Large COVRATIO: Observation increases precision
> Small COVRATIO: Observation decreases precision of the
> estimated coefficients

This statistic is defined as follows:

$$\text{COVRATIO} = \frac{|(X'X)^{-1}_{-i}|\text{MSE}_{-i}}{|(X'X)|\text{MSE}} = \frac{\text{MSE}^{m+1}_{-i}}{\text{MSE}^{m+1}}\left(\frac{1}{1-h_i}\right),$$

where the first equation is the definitional formula while the second is the computational form. In this case, the computational form is quite revealing: The magnitude of COVRATIO increases with leverage (h_i) and the relative magnitude of the deleted residual mean square.

In other words, if an observation has high leverage and leaving it out increases the error mean square, its presence has increased the precision of the

---

[9] This condition is called multicollinearity and is extensively discussed in Chapter 5.

parameter estimates (and vice versa). Of course, these two factors may tend to cancel each other to produce an "average" value of that statistic!

It has been suggested that values outside the interval

$$1 \pm \left(\frac{3(m + 1)}{n} \right)$$

suggest that the observation may be considered to affect precision. For example, a value of COVRATIO below the lower limit indicates that leaving that observation out increases precision, whereas a value greater than the upper limit indicates decreased precision by deleting the observation.

For Example 4.1, the COVRATIO values are 0.0713 and 0.3872 for the outliers in scenarios 1 and 2, respectively. These statistic confirm that both outliers caused the standard errors of the coefficients to increase (and the precision decrease), but the increase was more marked for scenario 1.

Before continuing, it is useful to discuss the implications of the arbitrary values we have used to describe "large." These criteria assume that when there are no outliers or influential observations, the various statistics are random variables having a somewhat normal distribution, and thus values more than two standard errors from the center are indications of "large." However, with so many statistics, there may very well be some "large" values even when there are no outliers or influential observations. Actually, a more realistic approach, especially for moderate-sized data sets, is to visually peruse the plots and look for "obviously" large values, using the suggested limits as rough guidelines.

The following brief summary of the more frequently used statistics may provide a useful reference:

> *Studentized residuals* are the actual residuals divided by their standard errors. Values exceeding 2.5 in magnitude may be used to indicate outliers.
>
> *Diagonals of the hat matrix, h_i,* are measures of leverage in the space of the independent variables. Values exceeding $2(m + 1)/n$ may be used to identify observations with high leverage.
>
> *DFFITS* are standardized differences between a predicted value estimated with and without the observation in question. Values exceeding $2\sqrt{(m + 1)/n}$ in magnitude may be considered "large."
>
> *DFBETAS* are used to indicate which of the independent variables contribute to large DFFITS. Therefore, these statistics are primarily useful for observations with large DFFITS values, where DFBETAS exceeding $2/\sqrt{n}$ in magnitude may be considered "large."
>
> *COVRATIO* statistics indicate how leaving out an observation affects the precision of the estimates of the regression coefficients. Values outside the bounds computed as $1 \pm 3(m + 1)/n$ may be considered "large," with values above the limit indicating less precision when leaving the observation out and vice versa for values less than the lower limit.

EXAMPLE 4.2   We again resort to some artificially generated data where we construct various scenarios and see how the various statistics identify the situations. This example has two independent variables, x_1 and x_2. We specify the model with $\beta_0 = 0$, $\beta_1 = 1$, and $\beta_2 = 1$; hence, the model for the response variable y is

$$y = x_1 + x_2 + \epsilon,$$

where ϵ is normally distributed with a mean of 0 and standard deviation of 4. Using a set of arbitrarily chosen but correlated values of x_1 and x_2, we generate a sample of 20 observations. These are shown in Table 4.6.[10] The relationship between the two independent variables creates the condition known as multi-collinearity, which we present in detail in Chapter 5, but already know to produce relatively unstable (large standard errors) estimates of the regression coefficients.

The results of the regression, produced by PROC REG of the SAS System, are shown in Table 4.7. The results are indeed consistent with the model; however, because of the multicollinearity the two regression coefficients have p-values that are much larger than that for the entire regression (see Chapter 5).

The various outlier and influence statistics are shown in Table 4.8. As expected, no values stand out; however, a few may be deemed "large" according to the suggested limits. This reinforces the argument that the suggested limits may be too sensitive.

TABLE 4.6

DATA FOR EXAMPLE 4.2

OBS	X1	X2	Y
1	0.6	−0.6	−4.4
2	2.0	−1.7	4.1
3	3.4	2.8	4.0
4	4.0	7.0	17.8
5	5.1	4.0	10.0
6	6.4	7.2	16.2
7	7.3	5.1	12.7
8	7.9	7.3	16.8
9	9.3	8.3	16.4
10	10.2	9.9	18.5
11	10.9	6.4	18.5
12	12.3	14.5	24.2
13	12.7	14.7	21.9
14	13.7	12.0	30.8
15	15.2	13.5	28.1
16	16.1	11.3	25.2
17	17.2	15.3	29.9
18	17.5	19.7	34.3
19	18.9	21.0	39.0
20	19.7	21.7	45.0

[10] The data in Table 4.6 were generated using SAS. The resulting values contain all digits produced by the computer but are rounded to one decimal for presentation in Table 4.6. The analyses shown in subsequent tables were performed with the original values and may differ slightly from analyses performed on the data as shown in Table 4.6. Similar differences will occur for all computer-generated data sets.

TABLE 4.7

REGRESSION RESULTS

Analysis of Variance

Source	DF	Sum of Squares	Mean Square	F Value	Prob>F
Model	2	2629.12617	1314.56309	116.775	0.0001
Error	17	191.37251	11.25721		
C Total	19	2820.49868			

Root MSE	3.35518	R-square	0.9321
Dep Mean	20.44379	Adj R-sq	0.9242
C.V.	16.41171		

Parameter Estimates

Variable	DF	Parameter Estimate	Standard Error	T for H0: Parameter=0	Prob > |T|
INTERCEP	1	1.034982	1.63888867	0.632	0.5361
X1	1	0.863462	0.38717002	2.230	0.0395
X2	1	1.035028	0.33959165	3.048	0.0073

TABLE 4.8

OUTLIER STATISTICS

OBS	RESID	STUD_R	HAT_DIAG	DFFITS	DFBETA1	DFBETA2
1	-5.401	-1.803	0.203	-0.9825	0.3220	-0.034
2	3.032	1.028	0.228	0.5600	0.1522	-0.303
3	-2.878	-0.919	0.129	-0.3515	0.1275	-0.037
4	6.113	2.142	0.277	1.5045	-1.3292	1.149
5	0.479	0.150	0.096	0.0473	-0.0081	-0.003
6	2.262	0.712	0.103	0.2381	-0.1553	0.122
7	0.106	0.033	0.085	0.0098	0.0029	-0.005
8	1.434	0.441	0.061	0.1099	-0.0227	0.007
9	-1.292	-0.396	0.053	-0.0915	-0.0047	0.012
10	-1.693	-0.518	0.051	-0.1172	0.0147	-0.014
11	1.406	0.473	0.215	0.2417	0.2017	-0.212
12	-2.392	-0.762	0.125	-0.2839	0.1812	-0.213
13	-5.280	-1.673	0.115	-0.6407	0.3716	-0.454
14	5.433	1.693	0.085	0.5508	0.3287	-0.264
15	-0.138	-0.043	0.107	-0.0146	-0.0092	0.007
16	-1.487	-0.537	0.319	-0.3592	-0.3285	0.299
17	-1.742	-0.563	0.151	-0.2326	-0.1556	0.109
18	-2.263	-0.743	0.175	-0.3375	0.0927	-0.179
19	-0.158	-0.053	0.203	-0.0258	0.0049	-0.012
20	4.461	1.504	0.219	0.8293	-0.1151	0.353

We now create three scenarios by modifying observation 10. The statistics in Table 4.8 show that this is a rather typical observation with slightly elevated leverage. The outlier scenarios are created as follows:

Scenario 1: We increase x_1 by 8 units. Although the resulting value of x_1 is not very large, this change increases the leverage of that observation because this single change decreases the correlation between the two independent variables. However, the model is used to produce the value of y; hence, this is not an outlier in the response variable.

Scenario 2: We create an outlier by increasing y by 20. Since the values of the independent variables are not changed, there is no change in leverage.

Scenario 3: We create an influential outlier by increasing y by 20 for the high-leverage observation produced in scenario 1. In other words, we have a high-leverage observation that is also an outlier, which should become an influential observation.

We now perform a regression using the data for the three scenarios. Since the resulting output is quite voluminous, we provide in Table 4.9 only the most relevant results as follows:

TABLE 4.9

ESTIMATES AND STATISTICS FOR THE SCENARIOS

	REGRESSION STATISTICS			
	Original Data	Scenario 1	Scenario 2	Scenario 3
MODEL F	116.77	119.93	44.20	53.71
ROOT MSE	3.36	3.33	5.44	5.47
β_1	0.86	0.82	0.79	2.21
STD ERROR	0.39	0.29	0.63	0.48
β_2	1.03	1.06	1.09	-0.08
STD ERROR	0.34	0.26	0.55	0.45
	OUTLIER STATISTICS			
Stud. Resid.	-0.518	-0.142	3.262	3.272
h_i	0.051	0.460	0.051	0.460
COVRATIO	1.204	2.213	0.066	0.113
DFFITS	-0.117	-0.128	1.197	4.814
SSE	191.4	188.6	503.3	508.7
PRESS	286.5	268.3	632.2	1083.5

1. The overall model statistics, F, and the residual standard deviation
2. The estimated coefficients, $\hat{\beta}_1$ and $\hat{\beta}_2$, and their standard errors
3. For observation 10 only, the studentized residual, h_i, COVRATIO, and DFFITS
4. SSE and PRESS

Scenario 1: The overall model estimates remain essentially unchanged; however, the standard errors of the coefficient estimates are smaller because the multicollinearity has been reduced, thus providing more stable parameter estimates. Note further that for observation 10 the h_i and COVRATIO may be considered "large."

Scenario 2: The outlier has decreased the overall significance of the model (smaller F) and increased the error mean square. Both coefficients have changed and their standard errors increased, primarily due to the larger error mean square. The small COVRATIO reflects the larger standard errors of the coefficients and the large DFFITS is due to the changes in the coefficients. Note, however, that the ratio of PRESS to SSE has not increased markedly, because the outlier is not influential.

Scenario 3: The overall model statistics are approximately the same as those for the noninfluential outlier. However, the estimated coefficients are now very different. The standard errors of the coefficients have decreased from those in scenario 2 because the multicollinearity has deceased and are actually not much different from those with the original data. This is why the COVRATIO is not "large." Of course, DFFITS is very large and so is the ratio of PRESS to MSE.   ◆

This very structured example should provide some insight into how the various statistics react to outliers and influential observation. Note also that in each case, the relevant statistics far exceed the guidelines for "large." Obviously, real-world applications will not be so straightforward.

EXAMPLE 4.3   Table 4.10 contains some census data on the 50 states and Washington, D.C. We want to see if the average lifespan (LIFE) is related to the following characteristics:

MALE: Ratio of males to females in percent
BIRTH: Birth rate per 1000 population
DIVO: Divorce rate per 1000 population
BEDS: Hospital beds per 100,000 population
EDUC: Percentage of population 25 years or older having completed 16 years of school
INCO: Per capita income, in dollars

The data are from Barrabba, V. P. (1979).

The first step is to perform the ordinary linear regression analysis using LIFE as the dependent and the others as independent variables. The results are shown in Table 4.11.

The regression relationship is statistically significant but not very strong, which is not unusual for this type of data and size of data set. The strongest

TABLE 4.10

DATA FOR EXAMPLE 4.3

STATE	MALE	BIRTH	DIVO	BEDS	EDUC	INCO	LIFE
AK	119.1	24.8	5.6	603.3	14.1	4638	69.31
AL	93.3	19.4	4.4	840.9	7.8	2892	69.05
AR	94.1	18.5	4.8	569.6	6.7	2791	70.66
AZ	96.8	21.2	7.2	536.0	12.6	3614	70.55
CA	96.8	18.2	5.7	649.5	13.4	4423	71.71
CO	97.5	18.8	4.7	717.7	14.9	3838	72.06
CT	94.2	16.7	1.9	791.6	13.7	4871	72.48
DC	86.8	20.1	3.0	1859.4	17.8	4644	65.71
DE	95.2	19.2	3.2	926.8	13.1	4468	70.06
FL	93.2	16.9	5.5	668.2	10.3	3698	70.66
GA	94.6	21.1	4.1	705.4	9.2	3300	68.54
HI	108.1	21.3	3.4	794.3	14.0	4599	73.60
IA	94.6	17.1	2.5	773.9	9.1	3643	72.56
ID	99.7	20.3	5.1	541.5	10.0	3243	71.87
IL	94.2	18.5	3.3	871.0	10.3	4446	70.14
IN	95.1	19.1	2.9	736.1	8.3	3709	70.88
KS	96.2	17.0	3.9	854.6	11.4	3725	72.58
KY	96.3	18.7	3.3	661.9	7.2	3076	70.10
LA	94.7	20.4	1.4	724.0	9.0	3023	68.76
MA	91.6	16.6	1.9	1103.8	12.6	4276	71.83
MD	95.5	17.5	2.4	841.3	13.9	4267	70.22
ME	94.8	17.9	3.9	919.5	8.4	3250	70.93
MI	96.1	19.4	3.4	754.7	9.4	4041	70.63
MN	96.0	18.0	2.2	905.4	11.1	3819	72.96
MO	93.2	17.3	3.8	801.6	9.0	3654	70.69
MS	94.0	22.1	3.7	763.1	8.1	2547	68.09
MT	99.9	18.2	4.4	668.7	11.0	3395	70.56
NC	95.9	19.3	2.7	658.8	8.5	3200	69.21
ND	101.8	17.6	1.6	959.9	8.4	3077	72.79
NE	95.4	17.3	2.5	866.1	9.6	3657	72.60
NH	95.7	17.9	3.3	878.2	10.9	3720	71.23
NJ	93.7	16.8	1.5	713.1	11.8	4684	70.93
NM	97.2	21.7	4.3	560.9	12.7	3045	70.32
NV	102.8	19.6	18.7	560.7	10.8	4583	69.03
NY	91.5	17.4	1.4	1056.2	11.9	4605	70.55
OH	94.1	18.7	3.7	751.0	9.3	3949	70.82
OK	94.9	17.5	6.6	664.6	10.0	3341	71.42
OR	95.9	16.8	4.6	607.1	11.8	3677	72.13
PA	92.4	16.3	1.9	948.9	8.7	3879	70.43
RI	96.2	16.5	1.8	960.5	9.4	3878	71.90
SC	96.5	20.1	2.2	739.9	9.0	2951	67.96
SD	98.4	17.6	2.0	984.7	8.6	3108	72.08
TN	93.7	18.4	4.2	831.6	7.9	3079	70.11
TX	95.9	20.6	4.6	674.0	10.9	3507	70.90
UT	97.6	25.5	3.7	470.5	14.0	3169	72.90
VA	97.7	18.6	2.6	835.8	12.3	3677	70.08
VT	95.6	18.8	2.3	1026.1	11.5	3447	71.64
WA	98.7	17.8	5.2	556.4	12.7	3997	71.72
WI	96.3	17.6	2.0	814.7	9.8	3712	72.48
WV	93.9	17.8	3.2	950.4	6.8	3038	69.48
WY	100.7	19.6	5.4	925.9	11.8	3672	70.29

TABLE 4.11

REGRESSION FOR ESTIMATING LIFE EXPECTANCY

Analysis of Variance

Source	DF	Sum of Squares	Mean Square	F Value	Prob>F
Model	6	53.59425	8.93238	6.464	0.0001
Error	44	60.80295	1.38189		
C Total	50	114.39720			

Root MSE	1.17554	R-square	0.4685	
Dep Mean	70.78804	Adj R-sq	0.3960	
C.V.	1.66064			

Parameter Estimates

Variable	DF	Parameter Estimate	Standard Error	T for H0: Parameter=0	Prob > |T|
INTERCEP	1	70.557781	4.28974713	16.448	0.0001
MALE	1	0.126102	0.04723176	2.670	0.0106
BIRTH	1	-0.516056	0.11727746	-4.400	0.0001
DIVO	1	-0.196538	0.07395330	-2.658	0.0109
BEDS	1	-0.003339	0.00097953	-3.409	0.0014
EDUC	1	0.236822	0.11102248	2.133	0.0385
INCO	1	-0.000361	0.00045979	-0.786	0.4363

coefficient shows a negative relationship with birth rate, an expected relationship since that variable may be considered a proxy for low income and other low socioeconomic factors. Note, however, that the income coefficient itself is not significant. The second strongest relationship is the negative relationship with the number of hospital beds, an unexpected result since that variable should be a proxy for the availability of medical care. The somewhat weaker ($\alpha = 0.01$) relationships with the proportion of males (positive) and divorce rate (negative) are interesting.

Table 4.12 gives the studentized residuals, hat matrix diagonals, COV-RATIO, and DFFITS statistics as well as the predicted values resulting from the regression analysis.

Although Table 4.12 provides the necessary information for finding outliers and influential observations, a plot of these statistics is more useful for the preliminary screening. Plots of the preceding statistics against the predicted values, using the first letter of the states' names as the plotting symbol, are shown in Figure 4.3.

Because some states begin with the same letter, we can use the information from Table 4.12 to augment what we see in the plots, resulting in the following conclusions:

1. Utah and, to a lesser degree, Alaska are outliers, with Utah having higher and Alaska having lower lifetimes than those estimated by the model.

TABLE 4.12

LISTING OF VARIOUS STATISTICS

OBS	STATE	PREDICT	RESIDUAL	STUDENTR	HAT	COVRATIO	DFFITS
1	AK	71.3271	-2.0171	-2.75935	0.61330	0.80331	-3.77764
2	AL	69.4415	-0.3915	-0.34865	0.08744	1.26246	-0.10683
3	AR	70.6101	0.0499	0.04474	0.10114	1.30634	0.01483
4	AZ	70.2978	0.2522	0.22878	0.12027	1.32411	0.08367
5	CA	71.6590	0.0510	0.04563	0.09499	1.29745	0.01461
6	CO	71.9730	0.0870	0.08109	0.16606	1.40702	0.03577
7	CT	72.2868	0.1932	0.18019	0.16778	1.40413	0.08001
8	DC	66.8702	-1.1602	-1.92429	0.73694	2.41318	-3.32707
9	DE	70.4192	-0.3592	-0.31749	0.07353	1.24763	-0.08852
10	FL	71.3805	-0.7205	-0.63860	0.07890	1.19475	-0.18563
11	GA	69.4238	-0.8838	-0.78960	0.09346	1.17251	-0.25243
12	HI	71.5312	2.0688	1.98759	0.21600	0.77551	1.08103
13	IA	71.4261	1.1339	0.98744	0.04583	1.05230	0.21635
14	ID	71.0405	0.8295	0.72725	0.05862	1.14647	0.18050
15	IL	70.1659	-0.0259	-0.02361	0.12984	1.34974	-0.00902
16	IN	70.2914	0.5886	0.52556	0.09229	1.23821	0.16618
17	KS	71.6500	0.9300	0.81864	0.06598	1.12947	0.21675
18	KY	70.7864	-0.6864	-0.60494	0.06827	1.18900	-0.16256
19	LA	70.3188	-1.5588	-1.38664	0.08545	0.93938	-0.42847
20	MA	70.9224	0.9076	0.80903	0.08992	1.16203	0.25337
21	MD	72.0392	-1.8192	-1.62551	0.09367	0.84010	-0.53284
22	ME	70.2533	0.6767	0.59436	0.06209	1.18364	0.15178
23	MI	70.2429	0.3871	0.34718	0.10029	1.28070	0.11474
24	MN	71.1682	1.7918	1.55052	0.03357	0.82017	0.29380
25	MO	70.7707	-0.0807	-0.07026	0.04456	1.22842	-0.01500
26	MS	68.7295	-0.6395	-0.59755	0.17127	1.33876	-0.26964
27	MT	72.0442	-1.4042	1.32327	0.08961	0.97097	-0.41884
28	NC	70.8177	-1.6077	-1.40504	0.05253	0.89903	-0.33466
29	ND	71.6705	1.1195	1.07650	0.21739	1.24510	0.56842
30	NE	71.2293	1.3707	1.18900	0.03826	0.97174	0.23830
31	NH	71.0450	0.1850	0.15973	0.02893	1.20469	0.02726
32	NJ	72.1304	-1.2004	1.13678	0.17866	1.16507	-0.52718
33	NM	70.8062	-0.4862	-0.45122	0.15985	1.35341	-0.19502
34	NV	68.7611	0.2689	0.49160	0.78351	5.22034	0.92706
35	NY	70.4696	0.0804	0.07336	0.13017	1.34921	0.02806
36	OH	70.3149	0.5051	0.44811	0.08043	1.23708	0.13131
37	OK	71.1389	0.2811	0.25138	0.09538	1.28545	0.08075
38	OR	72.5163	-0.3863	-0.35172	0.12715	1.31945	-0.13289
39	PA	70.9151	-0.4851	-0.43064	0.08158	1.24168	-0.12715
40	RI	71.4382	0.4618	0.40908	0.07772	1.24006	0.11762
41	SC	70.5163	-2.5563	-2.24879	0.06489	0.53439	-0.62248
42	SD	71.1165	0.9635	0.88105	0.13455	1.19823	0.34650
43	TN	70.0345	0.0755	0.06631	0.06244	1.25194	0.01692
44	TX	70.1801	0.7199	0.62837	0.05028	1.16116	0.14358
45	UT	69.5785	3.3215	3.58440	0.37861	0.16856	3.28714
46	VA	71.5622	-1.4822	-1.29871	0.05746	0.94790	-0.32325
47	VT	70.5113	1.1287	0.99678	0.07209	1.07882	0.27782
48	WA	72.5022	-0.7822	-0.70606	0.11177	1.22103	-0.24902
49	WI	71.4854	0.9946	0.86395	0.04086	1.08641	0.17779
50	WV	69.9235	-0.4435	-0.39886	0.10523	1.27987	-0.13546
51	WY	70.4566	-0.1666	-0.14854	0.08921	1.28513	-0.04597

Plot of STUDENTR*PREDICT=STATE.

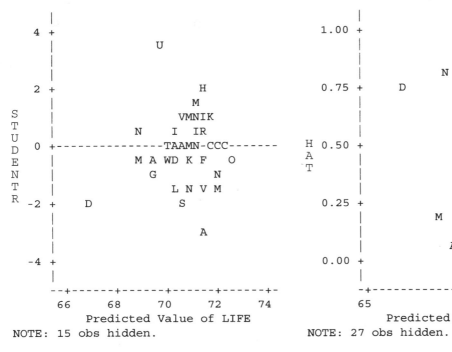

NOTE: 15 obs hidden.

Plot of HAT*PREDICT=STATE.

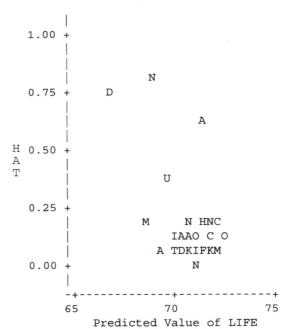

NOTE: 27 obs hidden.

Plot of COVRATIO*PREDICT=STATE.

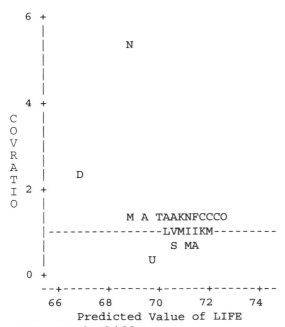

NOTE: 26 obs hidden.

Plot of DFFITS*PREDICT=STATE.

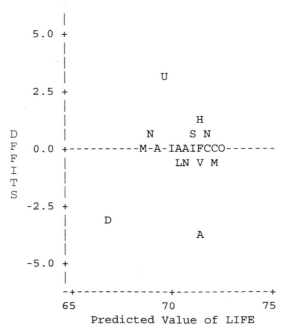

NOTE: 30 obs hidden.

FIGURE 4.3 Plots of Various Statistics

2. Alaska, the District of Columbia, and Nevada have high leverage.
3. The COVRATIO statistic shows that Nevada and, to a lesser degree, the District of Columbia cause the coefficients to be estimated with greater precision.
4. Utah, the District of Columbia, and Alaska are very influential.

Note that these conclusions were quite obvious, with no need to rely on the more formal suggestions for identifying "large" values. Obviously, we will concentrate on Alaska, the District of Columbia, Nevada, and Utah and thus reproduce in Table 4.13 the DFBETAS as well as DFFITS for these states.

We can now diagnose what is unusual about these states:

Alaska: The DFBETAS for MALE is obviously very large and negative. That state has the highest proportion of males (see Table 4.10), which is the variable that produces the high leverage. The life expectancy in Alaska is quite low; hence the large DFFITS. If Alaska is omitted from the data, the coefficient for MALE will decrease. However, the high degree of leverage causes the studentized residual to be only marginally "large."

The District of Columbia: This "state" is unusual in many respects. For our model it is important to note that it has the lowest male ratio and the highest number of hospital beds, which are the reasons for the high leverage. The high leverage causes the studentized residual not to be very large in spite of the fact that it has the lowest life expectancy. The DFFITS value is very large and appears to be caused largely by the number of hospital beds. The COVRATIO is very large, suggesting decreased precision of the estimated coefficients if that observation is deleted. We can see the meaning of these results by examining the output of the regression omitting the District of Columbia, data which is shown in Table 4.14.

First we note that the overall model has a slightly smaller R-square while the error mean square has decreased. This apparent contradiction is due to deleting the observation with the lowest value of the dependent variable, which also reduces the total sum of squares.

Looking at the coefficients we see that the major change is that the coefficient for hospital beds, whose negative sign was unexpected, is no longer statistically

TABLE 4.13

DFBETAS FOR SELECTED STATES

STATE	DFFITS	MALE	BIRTH	DIVO	BEDS	EDUC	INCO
AK	-3.778	-2.496	-1.028	0.590	-0.550	0.662	-0.917
DC	-3.327	0.741	-0.792	-0.821	-2.294	-0.735	0.244
NV	0.927	0.003	0.012	0.821	0.151	-0.239	0.281
UT	3.287	-1.289	1.967	-0.683	-1.218	0.897	-0.282

TABLE 4.14

REGRESSION RESULTS, OMITTING DC

Analysis of Variance

Source	DF	Sum of Squares	Mean Square	F Value	Prob>F
Model	6	32.40901	5.40150	4.171	0.0022
Error	43	55.68598	1.29502		
C Total	49	88.09499			

Root MSE	1.13799	R-square	0.3679	
Dep Mean	70.88960	Adj R-sq	0.2797	
C.V.	1.60530			

Parameter Estimates

Variable	DF	Parameter Estimate	Standard Error	T for H0: Parameter=0	Prob > |T|
INTERCEP	1	69.820757	4.16925640	16.747	0.0001
MALE	1	0.092205	0.04879958	1.889	0.0656
BIRTH	1	-0.426140	0.12221132	-3.487	0.0011
DIVO	1	-0.137783	0.07745305	-1.779	0.0823
BEDS	1	-0.001164	0.00144809	-0.804	0.4260
EDUC	1	0.315770	0.11458014	2.756	0.0086
INCO	1	-0.000470	0.00044845	-1.048	0.3006

significant. In other words, the highly significant negative coefficient for that variable is entirely due to that observation!

Without going into much detail, the unusual aspect of Nevada is due to the high divorce rate, while Utah has a number of unusual features with the result that the overall model fits much better when that state is deleted.   ◆

Comments

In this section we have presented methods for detecting outliers, that is, observations that have unusual features that may unduly affect the results of a regression analysis. Because outliers may occur in various ways with different effects, a number of statistics have been developed for their detection. Briefly, the types of outliers and detection statistics are as follows:

1. Outliers in the response variable, detected by residuals or preferably by studentized residuals
2. Outliers in the independent variables, called leverage, detected by the diagonals of the hat matrix

3. Observations for which a combination of leverage and outliers in the response cause a shift in the estimation of the response, measured by the DFFITS statistics for overall effect and DFBETAS for identifying individual coefficients
4. Observations for which a combination of leverage and outliers in the response cause a change in the precision of the estimated coefficients measured by the COVRATIO statistic

We have also briefly discussed the PRESS statistic, which, if substantially larger than SSE, indicates that outliers may be a problem in the data being analyzed.

Other statistics are available and may be included in some computer software packages. In addition, the rapid advances in computer hardware and software are responsible for a considerable amount of ongoing research in computer-intensive methods for outlier detection, although at the present time few of these are sufficiently developed to be routinely included in currently available computer software. Because of these possibilities, any user of a particular statistical software should scrutinize its documentation to ascertain the availability and applicability of the outlier detection methods.

We have presented various suggested statistics and guidelines for evaluating the effect of these statistics on the regression model. However, as we saw in Example 4.2, these guidelines are often somewhat "generous" in that they identify outliers that may not really be very unusual. Therefore these guidelines should be used as a preliminary screening procedure, with plots or other devices used to make a final judgment.

Further, the techniques discussed thus far are largely designed to detect single outliers. For example, if there are duplicates—or near duplicates—of influential observations, the "leave-one-out" statistics will completely fail to identify any of these. There are current research efforts that address this problem and that may result in methods that will detect duplicate influential observations.

Remedial Methods

We have presented methods for detecting outliers, but have not mentioned any statistical approach to finding remedial methods to counteract their effects. Actually, remedying the effects of outliers is not strictly a statistical problem. Obviously, we would not simply discard an outlier without further investigation. A logical, common-sense approach is desired in handling outliers, and we offer the following guidelines only as suggestions.

- Outliers may be simple recording errors—for example, incorrect recording, sloppy editing, or other sources of human error. Often the source of such errors can be found and corrections made. Unless the outlier is an obvious recording error, it should not simply be discarded.
- The observation may be an outlier because it is subject to a factor that was not included in the model. Examples include students who were sick the

day before a test, unusual weather on one or more days of an experiment, patients in a medical experiment who were exposed to another disease just before the experiment, or, as in Example 4.3, the District of Columbia, which is not really a state. A logical remedy in this case is to reanalyze that data with a revised model. Remember that this factor may exist in other observations that may not be outliers. Another option is to eliminate that observation along with any others that may be subject to that factor and restrict all inferences to observations that do not have that factor.

- Outliers may be due to unequal variances. That is, the response for some units is simply measured with less precision than others. This phenomenon is discussed in the next section.
- Finally, it may be necessary to leave the observation alone. After all, it occurred in the sample and may thus occur in the population for which inferences are being made. A quest for a better fitting model is not a reason for modifying data.

▼

Part Two: Violations of Assumptions

4.3 Unequal Variances

Another violation of the assumptions for the analysis of the linear model is the lack of constant variance of the error across all values of the data. This violation is called **heteroscedasticity**, as it is the violation of the assumption of homoscedasticity briefly discussed in the introduction to this chapter. Although heteroscedasticity may introduce bias in the estimates of the regression coefficients, its main effect is to cause incorrect estimates of variances of the estimated mean of the response variable.

Nonconstant variances are inherent when the responses follow non-normal distributions for which the variance is functionally related to the mean. The Poisson and binomial distributions are examples of such probability distributions.

General Formulation

Recall that in Section 3.3 we expressed the multiple regression model in matrix terms as

$$Y = XB + E,$$

where the E was an $n \times 1$ matrix of the error terms ϵ_i. We now define V as an $n \times n$ matrix in which the diagonal elements represent the variances of the n individual errors while the off-diagonal elements represent all their pairwise covariances. The matrix V is known as the variance covariance matrix. If the error terms are independent and have a common variance σ^2, this can be expressed as

$$V = \sigma^2 I_n,$$

where I_n is an $n \times n$ identity matrix. That is, the variances of ϵ_i are all equal to σ^2, and all their pairwise covariances are zero. This is, of course, the "usual" assumption leading to the "usual" estimates.

The least squares estimates of the coefficients were given by

$$\hat{B} = (X'X)^{-1}X'Y.$$

Suppose that the error terms are neither independent nor have the same variance. Then the variance covariance matrix, V, will not be of this simple form. The diagonal elements will still represent the variances of the individual error terms, but will not all have the same value. The off-diagonal elements will not be zero, as the individual error terms are not independent and hence have a nonzero covariance. The only restriction is that V is positive definite.

The normal equations become

$$(X'V^{-1}X)\hat{B}_g = X'V^{-1}Y,$$

where $\hat{B}_g$ is the estimated value of B. Using this notation, the **generalized least squares** estimates are obtained by

$$\hat{B}_g = (X'V^{-1}X)^{-1}X'V^{-1}Y.$$

This formula serves as the basis for partitioning of sums of squares and all other inference procedures. Although this procedure provides unbiased estimates and appropriate inference procedures with all desirable properties, it assumes that the elements of V are known, which they are normally not. In fact there is insufficient data to even estimate the $n(n + 1)/2$ distinct elements of V required for implementing this method (there are only n observations).

If the error terms are independent, but do not have constant variances, the diagonal elements of V will contain the variances of the individual errors, but the off-diagonal elements will be zero. The inverse of V contains the reciprocals of the variances on the diagonal with zeroes elsewhere. Denote these elements by $(1/\sigma_k^2)$, $k = 1, 2, \ldots, n$. Then the (i,j) element of the $X'V^{-1}X$ matrix is calculated as

$$\sum_k x_{ik}x_{kj}/\sigma_k^2,$$

where the summation is over the k subscript. The elements of $X'V^{-1}Y$ are calculated in a similar manner, and all other calculations are done in the usual way using these matrices. The effect of these formulas is that in the computation of sums of squares and cross products, the individual terms of the summation are weighted by the reciprocals of the variances of the individual observations. Therefore, the resulting method is described as **weighted regression**.[11] It is not difficult to show that these weights need only be *proportional* to the actual variance.

---

[11] Weighted regression may use weights other than the reciprocals of the variances; see Section 4.4.

The values of these variances are, of course, not normally known, but a process of assigning values to these elements is often a manageable problem. Weighted regression can be performed with all computer programs, but does require that the individual values of the weights be supplied. We will illustrate this procedure with a simple artificial example.

EXAMPLE 4.4   We have 10 observations with x-values taking the integer values from 1 to 10. The dependent variable, y, is generated using the model

$$y = x + \epsilon,$$

where ϵ is normally distributed with a known standard deviation of $0.25(x)$. In other words, the variance of the error of the first observation is $[(1)(0.25)]^2 = 0.0625$, whereas it is $[(10)(0.25)]^2 = 6.25$ for the last observation. The data are shown in Table 4.15.

The "usual" unweighted regression (using PROC REG) and requesting the 0.95 confidence intervals for the mean response provides the results in Table 4.16.

At first glance the results appear to be fine: The 0.95 confidence intervals easily include the true values of the coefficients. This type of result actually occurs quite often: The violation of the equal-variance assumption often has little effect on the estimated coefficients. However, the error mean square has no real meaning since there is no single variance to estimate. Furthermore, the standard errors and the widths of the 95% confidence intervals for the estimated conditional means are relatively constant for all observations, which is illogical since one would expect observations with smaller variances to be more precisely estimated. In other words, the unweighted analysis does not take into account the unequal variances.

We now perform a weighted regression using the (known) weights of $1/x^2$. Remember that the weights need only be proportional to the true variances, which are $(0.25x)^2$. The results are shown in Table 4.17 on page 148.

TABLE 4.15

DATA FOR REGRESSION WITH UNEQUAL VARIANCES

OBS	x	y
1	1	1.1
2	2	2.2
3	3	3.5
4	4	1.6
5	5	3.7
6	6	6.8
7	7	10.0
8	8	7.1
9	9	6.3
10	10	11.7

TABLE 4.16

RESULTS WITH UNWEIGHTED REGRESSION

Analysis of Variance

Source	DF	Sum of Squares	Mean Square	F Value	Prob>F
Model	1	88.12034	88.12034	24.679	0.0011
Error	8	28.56489	3.57061		
C Total	9	116.68523			

Root MSE	1.88961	R-square	0.7552	
Dep Mean	5.39675	Adj R-sq	0.7246	
C.V.	35.01380			

Parameter Estimates

Variable	DF	Parameter Estimate	Standard Error	T for H0: Parameter=0	Prob > |T|
INTERCEP	1	-0.287512	1.29084663	-0.223	0.8293
X	1	1.033502	0.20803877	4.968	0.0011

Obs	Dep Var Y	Predict Value	Std Err Predict	Lower95% Mean	Upper95% Mean	Residual
1	1.1355	0.7460	1.111	-1.8151	3.3071	0.3895
2	2.1830	1.7795	0.942	0.3926	3.9516	0.4035
3	3.4626	2.8130	0.792	0.9862	4.6398	0.6496
4	1.5724	3.8465	0.674	2.2919	5.4010	-2.2741
5	3.7124	4.8800	0.607	3.4813	6.2787	-1.1676
6	6.7870	5.9135	0.607	4.5148	7.3122	0.8735
7	10.0248	6.9470	0.674	5.3925	8.5015	3.0778
8	7.0825	7.9805	0.792	6.1537	9.8073	-0.8980
9	6.2852	9.0140	0.942	6.8419	11.1861	2.7288
10	11.7222	10.0475	1.111	7.4864	12.6086	1.6747

The following points are of interest:

1. All quantities relating to sums of squares of the dependent variable are affected by the weights and are not comparable to results obtained by the nonweighted regression. However, the overall F test and R-square are ratios and are therefore comparable. In this case both are somewhat larger than with the unweighted regression, but this is not always the general case.
2. The estimated coefficients are quite close to those of the unweighted regression. This is typical, although not necessarily guaranteed.
3. The standard errors and confidence intervals for the conditional means do indeed reflect the fact that the precision of estimates decreases with the larger values of x. For example, the width of the 95% confidence interval

TABLE 4.17

RESULTS WITH WEIGHTED REGRESSION

Analysis of Variance

Source	DF	Sum of Squares	Mean Square	F Value	Prob>F
Model	1	3.84787	3.84787	38.691	0.0003
Error	8	0.79562	0.09945		
C Total	9	4.64349			

Root MSE	0.31536	R-square	0.8287	
Dep Mean	1.94307	Adj R-sq	0.8072	
C.V.	16.23001			

Parameter Estimates

Variable	DF	Parameter Estimate	Standard Error	T for H0: Parameter=0	Prob > |T|
INTERCEP	1	0.188479	0.37913231	0.497	0.6325
X	1	0.928384	0.14925336	6.220	0.0003

Obs	Weight	Dep Var Y	Predict Value	Std Err Predict	Lower95% Mean	Upper95% Mean	Residual
1	1.0000	1.1355	1.1169	0.286	0.4573	1.7765	0.0186
2	0.2500	2.1830	2.0452	0.254	1.4599	2.6306	0.1377
3	0.1111	3.4626	2.9736	0.303	2.2756	3.6716	0.4890
4	0.0625	1.5724	3.9020	0.404	2.9700	4.8340	-2.3297
5	0.0400	3.7124	4.8304	0.529	3.6109	6.0498	-1.1180
6	0.0278	6.7870	5.7588	0.664	4.2283	7.2893	1.0282
7	0.0204	10.0248	6.6872	0.804	4.8339	8.5404	3.3376
8	0.0156	7.0825	7.6155	0.946	5.4329	9.7981	-0.5331
9	0.0123	6.2852	8.5439	1.091	6.0280	11.0598	-2.2587
10	0.0100	11.7222	9.4723	1.237	6.6205	12.3241	2.2499

for observation 1 is approximately 1.3, whereas the width is approximately 5.7 for observation 10. This result is the most important reason for using weighted regression. ◆

Of course, we do not normally know the true variances, and therefore must use some form of approximating the necessary weights. In general, there are two alternative methods for implementing weighted regression:

1. Estimating variances. This method can only be used if there are multiple observations at each combination level of the independent variables.
2. Using relationships. This method uses information on the relative magnitudes of the variances based on values of the observations.

We will illustrate both these methods in Example 4.5.

EXAMPLE 4.5   A block and tackle consists of a set of pulleys arranged in a manner to allow the lifting of an object with less pull than the weight of the object. Figure 4.4 illustrates such an arrangement.

The force is applied at the drum to lift the weight. The force necessary to lift the weight will be less than the actual weight; however, the friction of the pulleys will diminish this advantage, and an experiment is conducted to ascertain the loss in efficiency due to this friction.

A measurement on the load at each line is taken as the drum is repeatedly rotated to lift (UP) and release (DOWN) the weight. There are 10 independent measurements for each line for each lift and release; that is, only one line is measured at each rotation. The data are shown in Table 4.18. At this time we will only use the UP data.

If friction is present, the load on the lines should increase from line 1 to line 6, and as a first approximation, should increase uniformly, suggesting a linear regression of LOAD on LINE. However, as we shall see in Chapter 6, a straightforward linear model is not appropriate for this problem. Instead, we use a linear regression with an added indicator variable, denoted by C1, that allows line 6 to deviate from the linear regression. The model is

$$\text{LOAD} = \beta_0 + \beta_1(\text{LINE}) + \beta_2(\text{C1}) + \epsilon,$$

where C1 = 1 if LINE = 6 and C1 = 0 otherwise. This variable allows the response to deviate from the straight line for pulley number 6. The results of the regression are given in Table 4.19.

The regression is certainly significant. The loads are estimated to increase 36.88 units for each line, except that the increase from line 5 to line 6 is 38.88 + 41.92 = 78.80 units. A plot of the residuals is given in Figure 4.5. Notice the residuals seem to "fan out" as the line number increases, indicating the possibility that the variances are larger for higher line numbers.

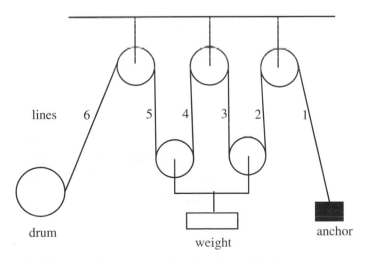

FIGURE 4.4 Illustration of a Block and Tackle

TABLE 4.18

LOADS ON LINES IN A BLOCK AND TACKLE

								LINE			
1		**2**		**3**		**4**		**5**		**6**	
UP	**DOWN**	**UP**	**DOWN**	**UP**	**DOWN**	**UP**	**DOWN**	**UP**	**DOWN**	**UP**	**DOWN**
310	478	358	411	383	410	415	380	474	349	526	303
314	482	351	414	390	418	408	373	481	360	519	292
313	484	352	410	384	423	422	375	461	362	539	291
310	479	358	410	377	414	437	381	445	356	555	300
311	471	355	413	381	404	427	392	456	350	544	313
312	475	361	409	374	412	438	387	444	362	556	305
310	477	359	411	376	423	428	387	455	359	545	295
310	478	358	410	379	404	429	405	456	337	544	313
310	473	352	409	388	395	420	408	468	341	532	321
309	471	351	409	391	401	425	406	466	341	534	318

TABLE 4.19

REGRESSION TO ESTIMATE LINE LOADS

Analysis of Variance

Source	DF	Sum of Squares	Mean Square	F Value	Prob>F
Model	2	329968.05333	164984.02667	2160.032	0.0001
Error	57	4353.68000	76.38035		
C Total	59	334321.73333			

Root MSE	8.73959	R-square	0.9870	
Dep Mean	412.26667	Adj R-sq	0.9865	
C.V.	2.11989			

Parameter Estimates

Variable	DF	Parameter Estimate	Standard Error	T for H0: Parameter=0	Prob > |T|
INTERCEP	1	276.200000	2.89859252	95.288	0.0001
LINE	1	36.880000	0.87395853	42.199	0.0001
C1	1	41.920000	4.00498111	10.467	0.0001

Weighted Regression by Estimating the Variances In this example we have multiple observations for each line number, so we can estimate these variances by calculating the sample variance for each value of the variable LINE. These are shown in Table 4.20 as an optional output from PROC ANOVA of the

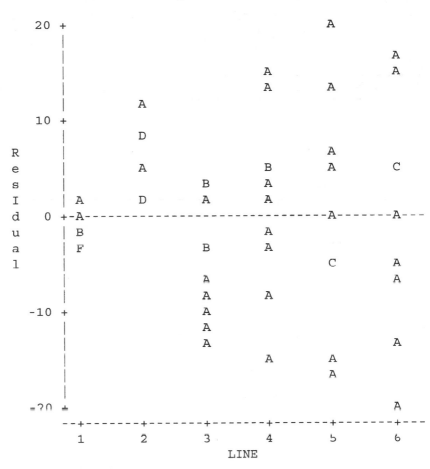

FIGURE 4.5 Residuals from Regression

SAS System. The increase in variability with line numbers is quite evident.[12] We need only square the standard deviations to get the estimates of the variances to construct the appropriate weights. That is, $w_i = 1/(\text{st dev})^2$.

The reciprocals of these variances are now used in a weighted regression using PROC REG with a WEIGHT statement. The results of the regression, as well as the estimated values and standard errors for the estimated means, are shown in Table 4.21.[13]

Notice that the predicted means for each value of LINE for the unweighted and the weighted regressions are very similar. This is expected, as the estimation

---

[12] The standard deviations for lines 5 and 6 are identical, a result that is certainly suspicious. However, there is no obviously suspicious feature in the numbers for these two lines.

[13] The results presented at the bottom of the table are not directly available with PROC REG.

TABLE 4.20

MEANS AND VARIANCES FOR DIFFERENT LINES

Level of LINE	N	Mean	SD
		------------LOAD------------	
1	10	310.900000	1.5951315
2	10	355.500000	3.7490740
3	10	382.300000	5.9637796
4	10	424.900000	9.2189419
5	10	460.600000	11.8902388
6	10	539.400000	11.8902388

TABLE 4.21

WEIGHTED REGRESSION ON LOAD

Analysis of Variance

Source	DF	Sum of Squares	Mean Square	F Value	Prob>F
Model	2	7991.86646	3995.93323	2544.453	0.0001
Error	57	89.51558	1.57045		
C Total	59	8081.38204			

Root MSE	1.25318	R-square	0.9889	
Dep Mean	328.67531	Adj R-sq	0.9885	
C.V.	0.38128			

Parameter Estimates

Variable	DF	Parameter Estimate	Standard Error	T for H0: Parameter=0	Prob > |T|
INTERCEP	1	273.428692	1.09612387	249.451	0.0001
LINE	1	38.081361	0.68824350	55.331	0.0001
C1	1	37.483143	5.71382732	6.560	0.0001

	PREDICTED MEAN		S.D. ERROR OF MEAN	
LINE	UNWEIGHTED	WEIGHTED	UNWEIGHTED	WEIGHTED
1	313.08	311.506	2.14075	0.61012
2	349.96	349.581	1.51374	0.70123
3	386.84	387.656	1.23596	1.24751
4	423.72	425.731	1.51374	1.88837
5	460.60	463.805	2.14075	2.55357
6	539.40	539.400	2.76370	4.70476

is not affected by the different variances. The standard errors of the predicted means for the weighted regression reflect the decrease in precision as line number increases. This result is, of course, more in line with the larger dispersion of weights.

The use of estimated variances has one drawback: Even if there are multiple observations at each value of the independent variables, there usually are not many. Therefore, the estimates of the variances are likely to be unstable. For example, the estimated standard deviation for line 3 (above) has a 0.95 confidence interval from 6.723 to 15.17, which overlaps the estimated standard deviations of lines 5 and 6. Care should be taken before using weights based on estimated variances. In this case, the consistency of the increases in variances provides additional evidence that the estimated variances are reasonable.   ◆

Weights Based on Relationships

In Example 4.4, the variance was generated to be proportional to x; hence, using the reciprocal of x did provide correct weights. If in Example 4.5 we make the assumption that the pulleys have essentially equivalent friction, the readings are independent, and the variances increase uniformly across line numbers, then the reciprocal of the line number may be used instead of the estimated variances. The reader may want to verify that using these weights does indeed provide almost identical results to those obtained above.

Knowledge about the underlying distribution of the error often provides a theoretical relationship between the mean and variance. For example, for many applications ranging from biological organisms to economic data, variability is proportional to the mean, which is equivalent to the standard deviation being proportional to the mean. In this case, a weighted regression with weights of $1/y^2$ or $1/\hat{\mu}^2$ would be appropriate. Alternatively, if the response is a frequency, the underlying distribution may be related to the Poisson distribution, for which the variance is identical to the mean, and appropriate weights would be $1/y$ or $1/\hat{\mu}$. And finally, if the response is in the form of a proportion, the underlying distribution is likely to be the binomial where the variance is $p(1 - p)/n$, where p is the proportion of successes.[14] Although these are the most common distributions with a relationship between the mean and the variance, others do exist.

Obviously, if the error term of a regression model has a distribution in which the mean and variance are related, it does not have a normal distribution. One of the assumptions necessary to carry out inferences on the parameters of the model was that the error terms were approximately normally distributed. Fortunately, the lack of constant variance and the lack of normality can usually be corrected by the same remedial action. That is, when we attempt to correct one problem, we also correct the other. A standard approach for a regression analysis where the error terms do not have constant variance is to perform a transformation on the dependent variable as follows:

---

[14] Regression with a binomial response is presented in detail in Chapter 8.

> • When the standard deviation is proportional to the mean, use the logarithm (either base e or base 10).
> • If the distribution is related to the Poisson, use the square root.
> • If the distribution is related to the binomial, use the arcsine of the square root of the proportion.

Unfortunately, the use of such transformations often makes interpretation of the results difficult. Transformations on the dependent variable may also change the character of the model, and in particular the error term. For example, the logarithmic transformation would require that the error term be multiplicative rather than additive. These transformations and others are discussed at length in Chapter 8.

Knowledge of the distribution of the error term can, however, be very beneficial in doing weighted regression. That is, the theoretical distribution might give a form for the variance of the error term that can be used in constructing weights. We will examine an example of the use of weighted regression in which the error can be considered to have the Poisson distribution.

EXAMPLE 4.6   There is considerable variation among individuals in their perception of what specific acts constitute a crime. To get an idea of factors that influence this perception, a sample of 45 college students were given the following list of acts and asked how many of these they perceived as constituting a crime:

Aggravated assault	Armed robbery	Arson
Atheism	Auto theft	Burglary
Civil disobedience	Communism	Drug addiction
Embezzlement	Forcible rape	Gambling
Homosexuality	Land fraud	Nazism
Payola	Price fixing	Prostitution
Sexual abuse of child	Sex discrimination	Shoplifting
Striking	Strip mining	Treason
Vandalism		

The response variable, CRIMES, is the number of these activities perceived by the individual students as a crime. Variables describing personal information that may influence perceptions are:

AGE: Age of interviewee
SEX: Coded 0: female, 1: male
INCOME: Income of parents ($1000)

The data are given in Table 4.22, and the results of a regression using number of items considered a crime (CRIMES) as the dependent variable are shown in Table 4.23.

The regression is quite significant and indicates that, within this group of students, the number of activities considered a crime is higher for males and also

TABLE 4.22

DATA ON PERCEPTIONS OF CRIME

AGE	SEX	INCOME	CRIMES
19	0	56	13
19	1	59	16
20	0	55	13
21	0	60	13
20	0	52	14
24	0	54	14
25	0	55	13
25	0	59	16
27	1	56	16
28	1	52	14
38	0	59	20
29	1	63	25
30	1	55	19
21	1	29	8
21	1	35	11
20	0	33	10
19	0	27	6
21	0	24	7
21	1	53	15
16	1	63	23
18	1	72	25
18	1	75	22
18	0	61	16
19	1	65	19
19	1	70	19
20	1	78	18
19	0	76	16
18	0	53	12
31	0	59	23
32	1	62	25
32	1	55	22
31	0	57	25
30	1	46	17
29	0	35	14
29	0	32	12
28	0	30	10
27	0	29	8
26	0	28	7
25	0	25	5
24	0	33	9
23	0	26	7
23	1	28	9
22	0	38	10
22	0	24	4
22	0	28	6

TABLE 4.23

REGRESSION FOR PERCEPTIONS OF CRIME

Analysis of Variance

Source	DF	Sum of Squares	Mean Square	F Value	Prob>F
Model	3	1319.63143	439.87714	71.375	0.0001
Error	41	252.67968	6.16292		
C Total	44	1572.31111			

Root MSE	2.48252	R-square	0.8393	
Dep Mean	14.35556	Adj R-sq	0.8275	
C.V.	17.29311			

Parameter Estimates

Variable	DF	Parameter Estimate	Standard Error	T for H0: Parameter=0	Prob > |T|
INTERCEP	1	-10.319628	2.24824563	-4.590	0.0001
AGE	1	0.405292	0.07517626	5.391	0.0001
SEX	1	2.318084	0.82435218	2.812	0.0075
INCOME	1	0.290934	0.02516492	11.561	0.0001

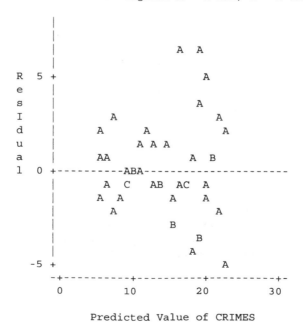

Plot of RCRIME*PCRIME.   Legend: A = 1 obs, B = 2 obs, etc.

Predicted Value of CRIMES

FIGURE 4.6 Residual Plot

increases with age and income. The residual plot shown in Figure 4.6 indicates larger residuals for the higher predicted values. This is because the response is a frequency (number of activities) whose distribution is related to the Poisson, where the variance is identical to the mean.

The relationship between the mean and variance implies that we should use the response means as weights. Of course, we do not know the weights, but we can use the reciprocals of the estimated responses as weights. To do this, we first perform the unweighted regression, compute the estimated (predicted) values, and use the reciprocals of these values as weights in the second weighted regression. This can be done with virtually any computer package, although it will be easier with some than with others. The results of this method, using PROC REG and a listing of estimated means and their standard errors for some selected observations, are shown in Table 4.24.

As before, weighting has little effect on the overall fit of the model and the estimated regression parameters. Also, the predicted values are quite similar for both methods, but the standard errors are larger for larger values of the predicted response when weighting is used. The differences are not large, but do more nearly represent the nature of the underlying distribution of the response variable.   ◆

4.4 Robust Estimation

We have seen how violations of assumptions, such as outliers and non-normal distributions of errors, can adversely affect the results of a regression analysis based on the least squares method. We have also presented some methods for detecting such violations and have suggested some remedial methods, not all of which are universally useful. Another approach is to use methods whose validity is assured in the presence of such violations. Two types of analyses are used in this context:

1. Nonparametric methods, whose results are not given in terms of parameters such as means and variances
2. Robust methods that do make inferences on the usual parameters, but whose results are not seriously affected by violations of assumptions

Because the usual objective of a regression analysis is to estimate the parameters of the regression model and to do statistical inferences on these estimates, the use of nonparametric methods is usually not effective.[15] Instead, we introduce a principle of robust estimation called the iteratively reweighted least square (IWLS) procedure, which yields something called the **M-estimator**. The IWLS procedure is an adaptation of least squares and is, in fact, implemented as a weighted regression.

As we have seen, parameters estimated by the least squares method may be extensively influenced by observations with large residuals. This can be demonstrated

---

[15] There is a body of methodology called nonparametric regression that is used to fit curves to data where the curves are not based on models containing the usual regression parameters. These methods do, however, generally use least squares and do require the usual assumptions on the error.

TABLE 4.24

RESULTS OF WEIGHTED REGRESSION

Analysis of Variance

Source	DF	Sum of Squares	Mean Square	F Value	Prob>F
Model	3	108.65190	36.21730	94.144	0.0001
Error	41	15.77279	0.38470		
C Total	44	124.42469			

Root MSE	0.62024	R-square	0.8732	
Dep Mean	11.92761	Adj R-sq	0.8640	
C.V.	5.20007			

Parameter Estimates

Variable	DF	Parameter Estimate	Standard Error	T for H0: Parameter=0	Prob > |T|
INTERCEP	1	-9.753598	2.00466905	-4.865	0.0001
AGE	1	0.379488	0.07515332	5.050	0.0001
SEX	1	2.178195	0.77587317	2.807	0.0076
INCOME	1	0.293055	0.02187276	13.398	0.0001

OBS	CRIMES	PREDICTED VALUES UNWEIGHTED	WEIGHTED	STANDARD ERRORS UNWEIGHTED	WEIGHTED
1	4	5.5792	5.6285	0.70616	0.49228
2	5	7.0860	7.0600	0.66198	0.48031
3	6	5.2361	5.3692	0.75115	0.55408
5	7	5.1739	5.2490	0.73051	0.51454
8	8	8.9467	8.8924	0.93894	0.80415
10	9	9.0082	9.0249	0.54281	0.39358
12	10	7.3870	7.5070	0.63073	0.46508
15	11	10.6923	10.6508	0.83064	0.72668
16	12	12.3951	12.6091	0.68129	0.63510
18	13	13.6732	13.8677	0.66600	0.63118
22	14	12.9148	13.0750	0.59185	0.54530
26	15	15.9291	15.9258	0.62338	0.61558
27	16	16.8641	16.9251	0.67561	0.68251
32	17	17.5402	17.2898	0.78969	0.79759
33	18	22.7972	22.8726	0.82235	0.84340
36	19	20.0644	20.1487	0.74112	0.76615
37	20	22.2466	21.9572	1.26156	1.27875
38	22	21.1138	21.2345	0.83147	0.85358
40	23	16.8120	16.9589	0.81744	0.82514
42	25	22.0808	21.8922	0.74478	0.78496

by the **influence function**, which is illustrated in Figure 4.7. In this plot the vertical axis represents a standardized measure of influence and the horizontal axis a standardized measure of the residual, $(y - \hat{\mu}_{y|x})$. The left-hand graph illustrates the influence function for least squares and shows that with this method, the influence increases linearly with the magnitude of the residuals.

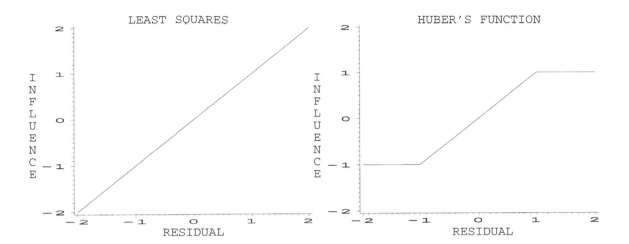

FIGURE 4.7 Influence Functions

Since violations of assumptions often result in large residuals, a robust method should attempt to reduce the influence of observations with large residuals. One method for doing this is to use Huber's influence function, which is illustrated in the right-hand portion of Fig. 4.7. It can be seen that this function gives the same influence as least squares up to some arbitrarily chosen value of the residual (usually denoted by r, which is arbitrarily set to 1 in this plot), beyond which all observations will have the same influence. In other words, the procedure ignores information about magnitudes of residuals greater than r. The larger the chosen value of r, the closer the results are to those of least squares and the less robust are the results. On the other hand, smaller values of r reduce the influence of extreme observations, thus resulting in greater robustness. However, because more information from the data is down-weighted, there is greater bias of estimates and less power for inferences.

Commonly used values of r are one to one and one-half times some measure of variability. The standard deviation is such a measure of variability, but because its value is also influenced by the extreme observations we may wish to downplay, an alternative estimator such as the median of the absolute values of the residuals is often used.

Other measures of variability and methods for determining r may be used. Additionally, other influence functions may be useful. For example, an influence function that specifies zero influence for observations with residuals greater than r is equivalent to eliminating those observations. Other robust estimators are given in Montgomery and Peck (1982), Section 9.3.2. Another consideration is that we know the residuals of outliers for observations having high leverage may not be large, and thus it may be worthwhile to use an influence function based on the magnitudes of the DFFITS statistic.

It turns out that the use of influence functions is equivalent to weighted regression where the weights are determined by the residuals. Since the residuals depend on the estimated regression, the method is implemented by performing a sequence of weighted least squares regressions in a method known as **iteratively reweighted least squares**. The method proceeds as follows:

1. Perform an ordinary (unweighted) least squares regression and obtain an initial set of residuals, $e = (y - \hat{\mu}_{y|x})$. Compute the desired measure of variation of residuals and determine a value of r.
2. Generate weights according to the rule
    If $e > r$, then $w = r/e$
    If $e < -r$, then $w = -r/e$
    Else $w = 1$
3. Perform a weighted regression using the weights obtained in step 2, and obtain a new set of residuals.
4. Compute the desired measure of variation of residuals obtained by the weighted regression and compare with those obtained from the unweighted regression or a previous iteration. If the differences in estimates are large, say in the second digit, use the new measure of variation to compute a new value of r and go to step 2; else stop.

A few comments on the results of this procedure are in order:

- The estimated coefficients are biased estimates of the "true" least squares coefficients. However, their robustness may make them more useful.
- The error mean square from the partitioning of sums of squares is a function of the weights, which are on the average less than unity; hence, that value will be smaller than that for ordinary least squares. In other words, the "true" residual mean square will be larger.
- The same iterated procedure can be used for down-weighting observations with large values of some other statistic, such as DFFITS.

EXAMPLE 4.3
REVISITED

In Example 4.3 we investigated the factors influencing the expected lifetimes of individuals in the various states of the United States. We noted that there were a number of states that could be considered unusual, and it may be of interest to examine the results of a robust regression that down-weights the data from some of these states. The data are shown in Table 4.10 and the residuals are shown in Table 4.12. One- and one-half times the median of the absolute values of the residuals is 1.029 (compare with the residual standard deviation of 1.175 in Table 4.11). We generate a set of weights using $r = 1.029$ and the rule in step 2 above, and then perform a weighted regression.

This regression produces an estimated residual standard deviation of 0.913, which is sufficiently different from the original 1.175 that we perform another iteration. This second iteration produces an estimated standard deviation of 0.902, and although performing another iteration may be worthwhile, we will not do so here.

The estimated coefficients from the three regressions (OLS is the unweighted regression) and the computed residual standard deviation (RMSE, which is affected by the weights) are shown in Table 4.25. The changes in the estimated coefficients are not large, although the changes for MALE and BIRTH may be considered of interest.

TABLE 4.25

REGRESSION COEFFICIENTS FOR M-ESTIMATOR

ITER	INTERCEP	MALE	BIRTH	DIVO	BEDS	EDUC	INCO	RMSE
OLS	70.557	0.12610	-0.51606	-0.19654	-0.00334	0.23682	-0.00036	1.17554
1	68.708	0.16262	-0.60073	-0.19025	-0.00292	0.20064	-0.00037	0.91263
2	68.567	0.16563	-0.60499	-0.19271	-0.00296	0.19820	-0.00037	0.90198

An examination of residuals and weights provides an indication of how the M-estimator works. The residuals for observations whose weights were decreased, shown in order of the magnitudes of the weights obtained in the second iteration, are shown in Table 4.26. Here we see that the largest change is the weight for Utah, and we can also see that the results of the M-estimator are somewhat similar to what the DFBETAS (Table 4.13) indicate would occur if that state were omitted. Note, however, that the weight for Nevada, a state with high leverage, was not reduced, and the second largest weight change was for South Carolina, which was not identified as an influential or outlying observation. ◆

In summary, then, there is no magic bullet! It is fair to say that all diagnostic statistics and remedial methods, including those not presented here, should be considered exploratory in nature, intended to diagnose data problems and suggest how statistical results may be affected by these data problems.

TABLE 4.26

SELECTED RESIDUAL FROM SECOND ITERATION

STATE	OLS RESID	FIRST WEIGHT	ITER 1 RESID	SECOND WEIGHT	ITER 2 RESID
UT	3.32152	0.30980	4.08335	0.29836	4.10023
SC	-2.55633	0.40253	-2.49943	0.38767	-2.50562
HW	2.06880	0.49739	1.97114	0.47902	1.95080
AK	-2.01712	0.51013	-2.14899	0.49129	-2.19056
MD	-1.81916	0.56565	-1.79916	0.54476	-1.79797
MN	1.79184	0.57427	1.70406	0.55306	1.70201
NC	-1.60770	0.64004	-1.58095	0.61641	-1.59269
LA	-1.55884	0.66010	-1.39831	0.63573	-1.40058
MT	-1.48422	0.69329	-1.61906	0.66769	-1.63704
VA	-1.48217	0.69425	-1.51230	0.66861	-1.51599
NE	1.37071	0.75070	1.20437	0.72298	1.19665
NJ	-1.20043	0.85719	-1.18570	0.82553	-1.19569
IA	1.13386	0.90752	1.00061	0.87400	0.98907
VT	1.12873	0.91165	1.08216	0.87798	1.09185
DC	-1.16019	0.88692	-0.89235	1.00000	-0.79724
ND	1.11950	0.91916	0.66137	1.00000	0.63547

4.5 Correlated Errors

In all of the regression models considered so far we have assumed that the random error terms are uncorrelated random variables. In some instances this assumption may be violated for reasons beyond the control of the investigator, which occurs primarily when the selection of sample units is not strictly random. The most frequent violation of this assumption occurs in time series data where observations are collected in successive time periods. Examples of time series include monthly observations of economic variables such as unemployment or gross national product, or weather data collected daily at a weather station. Such time dependencies may be in the form of seasonal trends or cycles, as well as dependence on what has occurred in previous time periods. Less frequent examples of dependent errors occur when sample observations are "neighbors" in some physical sense, such as adjacent plants in a seeded plot.

Models of time-dependent errors are usually linear; hence, the lack of independence among errors is measured by correlations, and nonindependent errors are usually referred to as correlated errors. Such time-dependent errors are said to be **autocorrelated** or **serially correlated**.

As indicated in Section 4.3, the effects of correlated errors can best be examined in terms of the variance matrix used for generalized least squares. If the errors have the same variance but are correlated, we can write the variances as

$$V = \text{Variance}(E) = R\sigma^2,$$

where R has ones on the diagonal and there are at least some nonzero off-diagonal elements. These nonzero elements reflect the correlations of the error terms. Because there may be as many as $(n)(n-1)$ correlations in R, we cannot estimate them using sample data. Instead we formulate alternative models that incorporate certain types of correlated error structure.

Autoregressive Models

In time series, the most popular model for correlated errors is the **autoregressive model**, in which the error of any period t is linearly dependent on previous errors. That is, the error at time t can be expressed as

$$\epsilon_t = \rho_1 \epsilon_{t-1} + \rho_2 \epsilon_{t-2} + \ldots + \delta_t,$$

where the ϵ_{t-i} refer to the errors for the ith previous period, and ρ_i is the correlation between the tth and $(t-i)$th error. We assume the δ_t, called the disturbances, are independent normal random variables with mean zero and variance σ^2. We will concentrate here on the **first-order autoregressive model**, which describes the special case where the error in period t is directly correlated only with the error in period $(t-1)$, the error of the previous period; that is,

$$\epsilon_t = \rho \epsilon_{t-1} + \delta_t.$$

It can be shown that this model results in a simplified form of the V matrix,

$$V = \frac{\sigma^2}{1 - \rho^2} \begin{bmatrix} 1 & \rho & \rho^2 & \cdots & \rho^{n-1} \\ \rho & 1 & \rho & \cdots & \rho^{n-2} \\ \rho^2 & \rho & 1 & \cdots & \rho^{n-3} \\ . & . & . & \cdots & . \\ . & . & . & \cdots & . \\ . & . & . & \cdots & . \\ \rho^{n-1} & \rho^{n-2} & \rho^{n-3} & \cdots & 1 \end{bmatrix},$$

where ρ is the first-order autocorrelation; that is, the correlation of the error in period t to the error of period $(t - 1)$. The matrix V now contains only two parameters (σ and ρ), and although the generalized least squares procedure requires that ρ be known, it can be estimated from the data.

EXAMPLE 4.7   To illustrate the effects of a first-order autoregressive model, we generate data for two "time series." In both models the mean response increases one unit for each 10 units of time t, starting with 1. This means that a regression model of the form

$$y = 0.1t + \epsilon$$

is used to generate the data. Table 4.27 presents 30 simulated observations where the two models have response variables Y and YT, respectively. For Y the error terms are specified as

$$\epsilon_t,$$

which are $N(\mu = 0, \sigma = 0.5)$, and independent, and for YT,

$$c_t = \rho c_{t-1} + \delta_t,$$

where $\rho = 0.9$, and the δ_t are independent and normally distributed with mean 0, standard deviation 0.5. In other words, the model for Y assumes independent errors, while YT assumes a first-order autocorrelation model with a correlation of 0.9. The effect of the autocorrelation can be seen in the two plots of response against T in Figure 4.8, where the dots are the observed values of the response and the line shows the population regression line. The differences or deviations from the line to the data points are thus the actual errors produced by the simulation.

Now if the errors are independent and normally distributed, the probability of a positive residual should be 0.5; the probability of two successive positive residuals should be $(0.5)(0.5) = 0.25$; for three, the probability should be 0.0625, and so forth, and equivalently for successive negative residuals. In other words, observations should switch from above to below the line quite frequently, and there should not be long runs of observations above or below the line. We can see that this is the case for the uncorrelated errors response, but definitely not for the correlated errors response.

We now perform regressions of Y and YT on T using ordinary least squares; that is, we ignore the possible effects due to correlated errors. The results are shown in Table 4.28, giving only the residual standard deviation (ROOT MSE),

TABLE 4.27

ARTIFICIAL TIME-SERIES DATA

T	YT	Y
1	−1.64647	−0.64018
2	−0.86142	0.71041
3	−1.15231	−0.19704
4	−1.63068	−0.32360
5	−1.22167	0.60594
6	−1.46028	0.08923
7	−0.90551	0.94874
8	−0.78536	0.65960
9	−0.57114	0.85568
10	−0.50542	0.81861
11	0.67700	2.03188
12	1.50125	1.88194
13	1.73901	1.46789
14	2.65335	2.25824
15	2.29300	1.16498
16	1.60861	0.89491
17	1.34306	1.33531
18	1.84158	2.16283
19	2.58981	2.55239
20	2.00224	1.38141
21	2.55263	2.55061
22	3.23378	2.82641
23	3.43901	2.50862
24	3.76315	2.73804
25	3.69364	2.46681
26	3.26156	2.18728
27	2.70972	2.11432
28	2.71701	2.70826
29	2.81698	2.89168
30	2.02513	2.09984

the parameter estimates, and the test for H_0: $\beta_1 = 0.1$ (the true value) from PROC REG of the SAS System. The result of the test for H_0: $\beta_1 = 0.1$ for Y shows that the estimate is indeed right on the mark; the null hypothesis is definitely not rejected (p-value $= 0.9691$). However, for the correlated errors response, the hypothesis H_0: $\beta = 0.1$ for YT is rejected (p-value $= 0.0001$). Remember also that the true error standard deviation is 0.5. For the uncorrelated errors response, the residual standard deviation (Root MSE) of 0.528 is quite close to that value, while that statistic from the correlated errors response is 0.813, which differs significantly from 0.5. ◆

   This example illustrates the problems associated with using least squares procedures when we have correlated errors, particularly positive autocorrelated errors. If the data result from a time series that has positively autocorrelated errors and a regular regression analysis is performed, the following important consequences must be addressed:

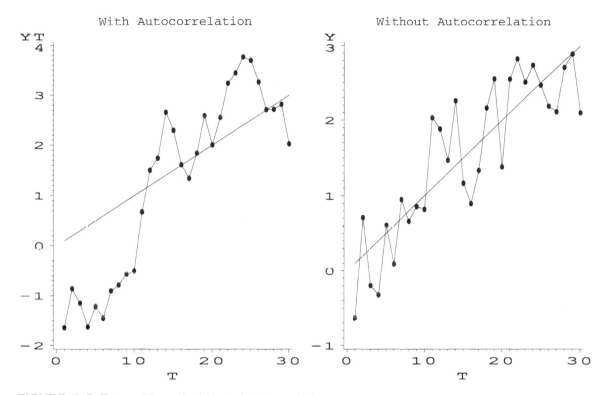

FIGURE 4.8 Data with and without Autocorrelation

1. Confidence intervals and hypothesis tests using the t and F distributions are not strictly applicable. This is illustrated by the hypothesis test on β_1 in the preceding example.
2. The estimated regression coefficients are still unbiased estimators of the regression parameters, but they no longer have the property of being the best linear unbiased estimators.
3. The value MSE may seriously underestimate the variance of the error terms. The fact that our example showed a value higher is likely an anomaly due to sampling fluctuations. The fact that normally MSE will underestimate the variance can be shown by

$$\text{Var}(\epsilon_t) = \frac{\sigma^2}{1 - \rho^2}.$$

For a nonzero correlation the actual variance of the error terms will be higher than σ^2. Since MSE estimates σ^2, it follows that it will underestimate the variance of the error terms in proportion to the value of ρ.
4. As a result of 3, the standard error of the coefficients computed from the least squares analysis may also seriously underestimate the true standard error of these coefficients.

TABLE 4.28

REGRESSIONS FOR TIME SERIES DATA

Correlated Errors

Root MSE   0.81291   R-square   0.8076

Parameter Estimates

Variable	DF	Parameter Estimate	Standard Error	T for H0: Parameter=0	Prob > |T|
INTERCEP	1	-1.624377	0.30441155	-5.336	0.0001
T	1	0.185919	0.01714711	10.843	0.0001

Test: TEST1   Numerator: 16.5914 DF: 1   F value: 25.1074
               Denominator: 0.660818 DF: 28   Prob>F: 0.0001

Uncorrelated Errors

Root MSE   0.52832   R-square   0.7403

Parameter Estimates

Variable	DF	Parameter Estimate	Standard Error	T for H0: Parameter=0	Prob > |T|
INTERCEP	1	-0.018211	0.19784182	-0.092	0.9273
T	1	0.099564	0.01114418	8.934	0.0001

Test: TEST 1   Numerator: 0.0004   DF: 1   F value: 0.0015
               Denominator: 0.279123 DF: 28   Prob>F: 0.9691

This example has shown that the use of ordinary least squares estimation on data having autocorrelated errors can give misleading results. For this reason it is important to ascertain the degree of autocorrelation and, if it is found to exist, to employ alternative methodology.

Diagnostics for Autocorrelation

We noted earlier that positively correlated errors[16] tend to create long series of residuals with the same sign. This pattern is usually evident in a residual plot where the residuals are plotted against the time variable. Such a plot for the simulated time-series data is shown for both the correlated and uncorrelated errors data in Figure 4.9.

[16] Although the signs of autocorrelations can be of any sign, positive autocorrelations are most common.

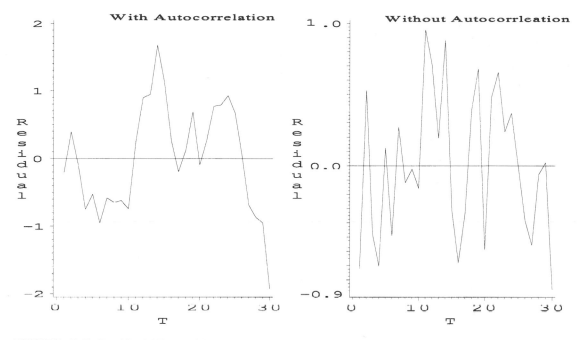

FIGURE 4.9 Residual Plots with and without Autocorrelation

The larger magnitudes of errors in the correlated errors case is quite evident. Also, we have previously noted that autocorrelated errors produce a tendency for longer series of residuals above or below zero, and this phenomenon is quite obvious in this example. A test for trends of this type is furnished by the runs test (Ostle and Mensing, 1975, Section 14.4), which can be used for any nonrandom pattern.

Because at this time we are only interested in nonrandomness caused by a first-order autocorrelation, we use the **Durbin–Watson test**, which is more specifically designed to test for this type of correlated errors. This test uses the residuals from the ordinary least squares regression. Let e_t be the residual from the tth observation. Then the Durbin–Watson test statistic, denoted by D, is calculated as follows:

$$D = \frac{\sum_{t>2}(e_t - e_{t-1})}{\sum_{t>2}e_t^2}.$$

Because e_{t-1} cannot be computed for the first observation, both sums start with observation 2.

The sampling distribution of this statistic is somewhat unusual. The range of the distribution is from 0 to 4, and under the null hypothesis of no autocorrelation, the mean of that distribution is close to 2. Positive autocorrelation makes adjacent differences small, hence tending to reduce the numerator. Therefore, the rejection region for positive correlations is in the lower tail of the distribution. Additionally, the computations of critical values for the distribution depend not only on sample size, but also on the number of independent variables and the pattern of the independent variables. Therefore, the critical values are not exact, but are only good approximations.

Critical values for the Durbin–Watson statistic for selected sample sizes and numbers of independent variables are given in Appendix Table A.5. Two values, labeled D_L and D_U, are given for each significance level for various combinations of sample size and number of independent variables. For testing the null hypothesis of no first-order autocorrelation, the computed statistic D is compared to the values in the table. If the calculated value is less than D_L, reject the hypothesis of no first-order autocorrelation; if it is greater than D_U, accept; otherwise, defer judgment.[17]

When the Durbin–Watson test gives indeterminate results, a reasonable procedure would be to treat the inconclusive results as suggesting the presence of autocorrelation and employ one of the remedial measures suggested in the following section. If such an action does not lead to substantially different results, the assumption of no autocorrelation would appear to be valid, and the ordinary least squares procedure should be valid. If the remedial action does lead to substantially different results, the analysis using the remedial method should be used.

For Example 4.7, the value of the Durbin–Watson statistic is 1.680 for Y and 0.433 for YT. From the table of critical values, for $n = 30$ and one independent variable, DL = 1.13 and DU = 1.26 for $\alpha = 0.01$; hence, we accept the hypothesis of no first-order autocorrelation for Y, while we readily reject that hypothesis for YT.

A preliminary estimate of the first-order autocorrelation coefficient, ρ, can be obtained by calculating the simple correlation between e_t and e_{t-1}. In Example 4.7, the estimates of the autocorrelation coefficients are 0.079 and 0.682 for Y and YT, respectively.

Remedial Methods

The autoregressive model is only one of a large number of statistical models used for analyzing data with time-dependent error structures. Data with time-ordered effects are commonly called **time series**. The analyses used on time series may be used to do the following:

- Study the nature of the time-dependent error structure
- Study the nature of the underlying regression model
- Provide for prediction of the response variable

We present here two methods that generally work well with the first-order autoregressive model:

- A method for analyzing the underlying structure and providing for limited prediction for the autoregressive model using a transformation incorporating estimates of the autocorrelation coefficients

---

[17] For a test of negative autocorrelation, subtract tabled critical values from 4.

- A method for studying the underlying model for a first-order autoregressive model that uses a simple redefinition of the model (also a
  transformation)

Details on these as well as additional methods for a wide spectrum of applications are described in books and references on the general topic of time series
(e.g., Fuller, 1996).

Alternative Estimation Technique

We present here the results of the Yule–Walker procedure as implemented by
PROC AUTOREG of the SAS/ETS software. This method first estimates the
model using ordinary least squares methods. It then computes the autocorrelations using the residuals from the least squares regression. It then solves the
Yule–Walker equations (see Gallant and Goebel, 1976) to obtain initial estimates
of the regression parameters, which are referred to as the autoregressive parameters. The procedure then transforms the observations using the solutions to the
Yule–Walker equations and re-estimates the regression coefficients with the
transformed data. This is equivalent to a generalized least squares with the ap
propriate weights.

EXAMPLE 4.7
REVISITED

PROC AUTOREG can be instructed to either select the order of the autocorrelation or use an order specified by the user. Because the data for YT in Example
4.7 were generated with a first order process, we will specify a first-order model.
The results are shown in Table 4.29.

The top portion of the output reproduces the results of the ordinary least
squares estimation as shown in Table 4.28 (SBC and AIC are two measures of
the effectiveness of a regression that we have not discussed).

The next portion contains the estimated autocorrelations and autoregressive
parameters. For a higher-order process, the autoregressive parameters are equivalent to partial regression estimates. The "T-ratio" is equivalent to the t test for
regression coefficients, but since the sampling distribution is only approximately
a t distribution, p-values are not given.

The last portion provides the statistics for the estimated regression using the
estimated autoregressive parameters. Note that the coefficient for T is not very
different from the ordinary least squares one, but the stated standard error is
larger and now approaches the value suggested by the simulation results. In fact,
the 0.95 confidence interval (using the stated standard error and t distribution
with 27 degrees of freedom) does include the true value.

One feature of this estimation process is that it provides two estimates of the
response variable:

1. The estimated response based only on the regression, called the **structural**
   portion of the model:

$$\hat{\mu}_{y|x} = -1.410 + 0.1600T$$

TABLE 4.29

RESULTS OF ESTIMATION WITH AUTOREGRESSION

ORDINARY LEAST SQUARES ESTIMATES

SSE	18.5029	DFE	28
MSE	0.6608177	ROOT MSE	0.8129069
SBC	77.4406	AIC	74.63821
REG RSQ	0.8076	TOTAL RSQ	0.8076
DURBIN-WATSON	0.4325		

VARIABLE	DF	B VALUE	S.D. ERROR	T RATIO	APPROX PROB
INTERCPT	1	-1.62437668	0.304411545	-5.336	0.0001
T	1	0.18591945	0.017147112	10.843	0.0001

ESTIMATES OF AUTOCORRELATIONS

LAG	COVARIANCE	CORRELATION	-1 9 8 7 6 5 4 3 2 1 0 1 2 3 4 5 6 7 8 9 1
0	0.616763	1.000000	| |******************|
1	0.420697	0.682104	| |*************

PRELIMINARY MSE=   0.3298041

ESTIMATES OF THE AUTOREGRESSIVE PARAMETERS

LAG	COEFFICIENT	S.D. ERROR	T RATIO
1	-0.68210432	0.14073007	-4.846898

YULE-WALKER ESTIMATES

SSE	7.597834	DFE	27
MSE	0.2814013	ROOT MSE	0.5304727
SBC	55.01729	AIC	50.81369
REG RSQ	0.4283	TOTAL RSQ	0.9210

VARIABLE	DF	B VALUE	S.D. ERROR	T RATIO	APPROX PROB
INTERCPT	1	-1.41015149	0.665042005	-2.120	0.0433
T	1	0.15985075	0.035540088	4.498	0.0001

2. The estimated responses based on both the structural and autoregressive portions of the model. These estimates are those obtained by the structural portion plus an estimate of the effect of the autocorrelation.

These two estimates are shown in Figure 4.10, where the dots show the actual values of the response, the solid line indicates the estimated structural model, and the dashed line represents the estimated model including both portions.

The relative effectiveness of the two models are shown in the computer output, where "Reg RSQ" and "Total RSQ" give the coefficients of determination

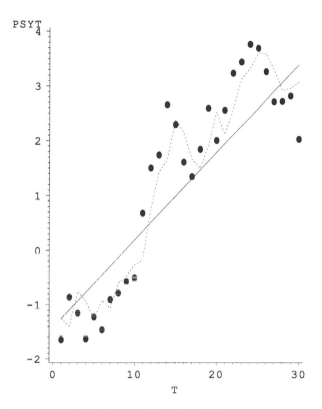

FIGURE 4.10 Plots of Predicted Values

for the structural and full models, respectively. In this example, the values of 0.428 for the structural model and 0.921 for the full model clearly show the importance of the autocorrelation. ◆

The relevance of the two models depends on the purpose of the analysis. If the purpose is to estimate the structural relationship, the structural model provides the appropriate estimates and the autocorrelation is simply a nuisance that must be accounted for. On the other hand, if the purpose is to provide the best predictor of the response, the full model, which uses all available parameter estimates, is the one to use, and the regression coefficient may be considered to be of secondary importance.

Model Modification

Another method of analysis is to redefine the model to account for the autocorrelation. The first-order autoregressive model essentially states that the observed response in any period t depends on the response at period $(t - 1)$. If we now assume that the first-order autocorrelation is close to unity, it is logical to model the change in the period-to-period responses, called the **first differences**. In other words, the dependent variable is the difference

$$d_t = y_t - y_{t-1}.$$

The use of this model may also suggest the use of first differences for some or all of the independent variables to provide a more interpretable model, especially when these variables are also subject to autocorrelation. In any case, the model using first differences provides a different model than that using the actually observed variables, and often the two do not provide comparable results.

E X A M P L E 4 . 8   In recent years there has been some argument as to whether the inflation rates in the 1970s and 1980s were due to changes in energy prices or to federal government deficits. Table 4.30 shows the data on the consumer price index (CPI), the

TABLE 4.30

INFLATION DATA

OBS	YEAR	CPI	ENERGY	PDEF
1	1960	29.6	22.4	0.0
2	1961	29.9	22.5	0.2
3	1962	30.2	22.6	0.4
4	1963	30.6	22.6	0.3
5	1964	31.0	22.5	0.4
6	1965	31.5	22.9	0.1
7	1966	32.4	23.3	0.2
8	1967	33.4	23.8	0.5
9	1968	34.8	24.2	1.4
10	1969	36.7	24.8	−0.2
11	1970	38.8	25.5	0.1
12	1971	40.5	26.5	1.1
13	1972	41.8	27.2	1.1
14	1973	44.4	29.4	0.6
15	1974	49.3	38.1	0.2
16	1975	53.8	42.1	2.1
17	1976	56.9	45.1	2.6
18	1977	60.6	49.4	1.6
19	1978	65.2	52.5	1.6
20	1979	72.6	65.7	1.0
21	1980	82.4	86.0	1.8
22	1981	90.9	97.7	1.8
23	1982	96.5	99.2	2.7
24	1983	99.6	99.9	3.7
25	1984	103.9	100.9	2.7
26	1985	107.6	101.6	2.8
27	1986	109.6	88.2	2.5
28	1987	113.6	88.6	1.5

index of energy prices (ENERGY), and the federal deficit in percentage of gross national product (PDEF) for the years 1960 through 1984 (*Statistical Abstract of the United States,* 1988).

The least squares regression analysis, using CPI as the dependent and ENERGY and PDEF as independent variables, as produced by PROC REG of the SAS System is given in Table 4.31. The highly significant ($p < 0.0001$) coefficient for ENERGY and the nonsignificant coefficient for PDEF ($p = 0.6752$) appear to strongly support the contention that inflation rates are primarily affected by energy prices. However, the Durbin–Watson statistic clearly indicates a first-order autocorrelation (DL = 1.84, for $\alpha = 0.01$); hence, the use of an alternative analysis is indicated.

A logical model for this analysis consists of using the first differences in CPI as the dependent variable. It is also logical to use the first differences of energy prices as one independent variable. It is not logical to use the first differences of the deficit rates, since it is the actual deficit that creates additional money supply. Denoting DCPI and DENERGY as the first differences in CPI and ENERGY and using PDEF as is, the results of the regression, using output

TABLE 4.31

REGRESSION FOR INFLATION DATA

```
                          Analysis of Variance

                          Sum of          Mean
Source          DF        Squares         Square      F Value      Prob>F

Model           2     22400.35045      11200.17522    258.680      0.0001
Error          25      1082.43634         43.29745
C Total        27     23482.78679


        Root MSE        6.58008      R-square     0.9539
        Dep Mean       58.86071      Adj R-sq     0.9502
        C.V.           11.17907


                      Parameter Estimates

                  Parameter      Standard    T for H0:
Variable   DF     Estimate       Error       Parameter=0    Prob > |T|

INTERCEP    1     13.169788      2.39023711      5.510       0.0001
ENERGY      1      0.894806      0.07134979     12.541       0.0001
PDEF        1      0.888278      2.09485486      0.424       0.6752


Durbin-Watson D                 0.360
(For Number of Obs.)               28
1st Order Autocorrelation       0.633
```

from PROC REG, are shown in Table 4.32. For this model, coefficients for both DENERGY and PDEF are positive and significant (P < 0.0001); hence, the use of this model indicates that both factors appear to contribute positively to inflation.[18] Additionally, the Durbin–Watson statistic is barely in the "no decision" region, indicating only limited evidence of the existence of a first-order autocorrelation.

Some additional comments on this model:

- There are only 27 observations because the first difference cannot be computed for the first observation.
- The coefficient of determination is only 0.7998, compared to 0.9539 for the observed variables model. On the other hand, the residual standard deviation is 1.182, compared to 6.580 for the observed variables model. These results illustrate the fact that the two models are not comparable because the two response variables are measured in entirely different scales.
- The estimated coefficient for DENERGY estimates the annual change in CPI associated with a one-unit increase in annual change in the energy

---

TABLE 4.32

FIRST-DIFFERENCE MODEL

Analysis of Variance

Source	DF	Sum of Squares	Mean Square	F Value	Prob>F
Model	2	133.93363	66.96682	47.929	0.0001
Error	24	33.53303	1.39721		
C Total	26	167.46667			

Root MSE	1.18204	R-square	0.7998	
Dep Mean	3.11111	Adj R-sq	0.7831	
C.V.	37.99402			

Parameter Estimates

Variable	DF	Parameter Estimate	Standard Error	T for H0: Parameter=0	Prob > |T|
INTERCEP	1	0.866422	0.37658229	2.301	0.0304
DENERGY	1	0.337497	0.03981449	8.477	0.0001
PDEF	1	1.099549	0.21911776	5.018	0.0001

Durbin-Watson D	1.511
(For Number of Obs.)	27
1st Order Autocorrelation	0.207

---

[18] Before we can claim to have discovered an important result, it must be noted that this model is quite incomplete and subject to valid criticism. Its primary use here is to illustrate the method.

price index, whereas the coefficient for PDEF estimates the annual change in CPI associated with a unit change in the deficit percentage (holding the other variable constant). These coefficients are also not necessarily comparable to those of the observed variables model. This is especially true for this example because the first-differences model uses first-differences for one independent variable and observed values for the other. ◆

4.6 Chapter Summary

In this chapter we have discussed problems that arise when the observed values of the response variable fail to fulfill the assumptions underlying regression analyses based on least squares estimation. We have presented three major areas of concern:

- The existence of outliers that may bias estimates of both coefficients and error variance
- The existence of nonconstant variance of the residuals
- The existence of correlated errors

In each case we have used simulations to illustrate some possible effects of these violations of assumptions, provided tools to help diagnose the nature of possible violations, and suggested some remedial methods. As is often the case with such exploratory analyses, the diagnostic tools are not always useful, and the remedies may not provide the desired results. In any case, the analyst's awareness of these types of problems should help to provide more valid and useful analyses.

CHAPTER EXERCISES

Exercises 1 through 4 can be worked with only material from the first part of this chapter. The remaining exercises deal with topics from the second part.

1. The data in Table 4.33 resulted from a small sample taken in an old urban neighborhood. The variable INCOME represents the monthly income of the head of the household, and the variable AGE represents the age of that person.

    (a) Perform the regression of income on age and plot the residuals. Do the residuals indicate the presence of an outlier?

    (b) Calculate the studentized residuals for the regression. Do these indicate any problems?

    (c) Calculate the DFFITS and DFBETAS for the data. Discuss the results.

2. It is a well-known fact that differences between summer and winter temperatures, called the temperature ranges, increase with latitude; that is, as you go

TABLE 4.33

INCOME DATA

AGE	INCOME	AGE	INCOME
25	1200	33	1340
32	1290	22	1000
43	1400	44	1330
26	1000	25	1390
33	1370	39	1400
48	1500	55	2000
39	6500	34	1600
59	1900	58	1680
62	1500	61	2100
51	2100	55	2000

north. In Table 4.34 are temperature ranges (RANGE) and latitudes (LAT) of some selected U.S. cities. The data are available in File REG04P02. Investigate the relationship between range and latitude and check for outliers. Determine a reason for outliers (an atlas may help).

3. The Galapagos Islands off the coast of Ecuador are a well-known source of data for various types of biological studies. Such a study, reported by Hamilton and Rubinoff in *Science* in 1963, attempts to relate the number of plant species on an island to various characteristics of the island. The variables used are:

AREA: Area in square miles
HEIGHT: Maximum elevation in feet above mean sea level
DSNEAR: Distance to nearest island in miles
DCENT: Distance to the center of the archipelago
ARNEAR: Area of nearest island in square miles
SPECIES: Number of plant species found

The data are shown in Table 4.35 on page 178 and are available in File REG04P03. Perform a regression to estimate the number of species. Look for outliers and influential observation(s). Again, a map will be helpful.

4. Data were obtained from the 1988 *Statistical Abstract of the United States* to determine factors related to state expenditures on criminal activities (courts, police, etc). The variables are:

STATE: The standard two-letter abbreviation (DC is included)
EXPEND: State expenditures on criminal activities ($1000)
BAD: The number of persons under criminal supervision
CRIME: Crime rate per 100,000
LAWYERS: The number of lawyers in the state
EMPLOY: The number of persons employed in the state
POP: The population of the state (1000)

The data are available in File REG04P04.

TABLE 4.34

TEMPERATURES AND LATITUDE

City	State	LAT	RANGE
Montgome	AL	32.3	18.6
Tuscon	AZ	32.1	19.7
Bishop	CA	37.4	21.9
Eureka	CA	40.8	5.4
San_Dieg	CA	32.7	9.0
San_Fran	CA	37.6	8.7
Denver	CO	39.8	24.0
Washingt	DC	39.0	24.0
Miami	FL	25.8	8.7
Talahass	FL	30.4	15.9
Tampa	FL	28.0	12.1
Atlanta	GA	33.6	19.8
Boise	ID	43.6	25.3
Moline	IL	41.4	29.4
Ft_Wayne	IN	41.0	26.5
Topeka	KS	39.1	27.9
Louisv	KY	38.2	24.2
New_Orl	LA	30.0	16.1
Caribou	ME	46.9	30.1
Portland	ME	43.6	25.8
Alpena	MI	45.1	26.5
St_Cloud	MN	45.6	34.0
Jackson	MS	32.3	19.2
St_Louis	MO	38.8	26.3
Billings	MT	45.8	27.7
N_Platte	ND	41.1	28.3
L_Vegas	NV	36.1	25.2
Albuquer	NM	35.0	24.1
Buffalo	NY	42.9	25.8
NYC	NY	40.6	24.2
C_Hatter	NC	35.3	18.2
Bismark	ND	46.8	34.8
Eugene	OR	44.1	15.3
Charestn	SC	32.9	17.6
Huron	SD	44.4	34.0
Knoxvlle	TN	35.8	22.9
Memphis	TN	35.0	22.9
Amarillo	TX	35.2	23.7
Brownsvl	TX	25.9	13.4
Dallas	TX	32.8	22.3
SLCity	UT	40.8	27.0
Roanoke	VA	37.3	21.6
Seattle	WA	47.4	14.7
Grn_Bay	WI	44.5	29.9
Casper	WY	42.9	26.6

TABLE 4.35

PLANT SPECIES

OBS	Island	AREA	HEIGHT	DSNEAR	DCENT	ARNEAR	SPECIES
1	Culpepper	0.9	650	21.7	162	1.8	7
2	Wenman	1.8	830	21.7	139	0.9	14
3	Tower	4.4	210	31.1	58	45.0	22
4	Jervis	1.9	700	4.4	15	203.9	42
5	Bindloe	45.0	1125	14.3	54	20.0	47
6	Barrington	7.5	899	10.9	10	389.0	48
7	Gardiner	0.2	300	1.0	55	18.0	48
8	Seymour	1.0	500	0.5	1	389.0	52
9	Hood	18.0	650	30.1	55	0.2	79
10	Narborough	245.0	4902	3.0	59	2249.0	80
11	Duncan	7.1	1502	6.4	6	389.0	103
12	Abingdon	20.0	2500	14.1	75	45.0	119
13	Indefatigable	389.0	2835	0.5	1	1.0	193
14	James	203.0	2900	4.4	12	1.9	224
15	Chatham	195.0	2490	28.6	42	7.5	306
16	Charles	64.0	2100	31.1	31	389.0	319
17	Albemarle	2249.0	5600	3.0	17	245.0	325

Perform the regression of EXPEND on the other variables. Interpret and comment on usefulness of the results. Look for outliers and influential observations. Can omitting some of these be justified? If so, redo the regression without deleted observations. Are the results more reasonable?

5. File REG04P05 contains monthly sorghum grain prices received by farmers in a South Texas county from 1980 through 1987. The variables are:

N: Sequential month number from 1 through 96
Year: Actual year
Month: Labeled 1 through 12
Price: Dollars per bushel

Fit a linear trend using Price as the dependent variable and N as the independent variable. Check for autocorrelation. Redo the analysis if necessary.

6. In Chapter 2, Exercise 5, CPI data for the years 1960 to 1994 were given with the instructions to perform simple linear regression using years as the independent variable. Repeat part (b) of Exercise 5 assuming the first-order autoregressive model. Compare the results with those obtained in Chapter 2.

7. Pecan production in the United States is believed to be subject to a biennial fluctuation; that is, a "good" year is followed by a "bad" year. In Table 4.36 are production data for native (USQN) and improved (USQI) pecan production, in million pounds, for the years 1970–1991 as provided by the USDA *Fruit and Tree-nut Situation Yearbook*. The data are available in File REG04P07. Estimate the trend in production over the years and perform an analysis to determine if the biennial production cycle exists.

TABLE 4.36

PECAN PRODUCTION

Year	USQN	USQI
1970	73.08	81.52
1971	104.10	143.10
1972	94.11	88.99
1973	131.70	144.00
1974	51.50	85.60
1975	136.70	110.10
1976	25.80	77.30
1977	98.70	137.90
1978	86.20	164.50
1979	109.50	101.10
1980	55.00	128.50
1981	164.55	174.55
1982	46.90	168.20
1983	102.75	167.25
1984	63.17	169.23
1985	91.90	152.50
1986	90.05	182.65
1987	82.55	179.65
1988	122.70	185.30
1989	73.20	161.00
1990	41.20	143.50
1991	82.80	145.00

5

Multicollinearity

5.1 Introduction

So far all of our examples of regression analyses have been based on a specified model for describing the behavior of the response variable, and the regression analyses were performed to confirm the validity of that model. Such analyses are referred to as **confirmatory** analyses. However, in many statistical analyses, and especially in regression analyses, the specification of the model is somewhat nebulous and a large part of the statistical analysis is devoted to a search for an appropriate model. Such analyses are referred to as **exploratory** analyses.

In a regression setting, an exploratory analysis often consists of specifying an initial model that contains a large number of variables. Because computing considerations are no longer a major consideration, the number of variables is often limited only by data availability, and it is hoped that the statistical analysis will magically reveal the correct model. One result of using a large number of variables is that many of the variables in such a model will be correlated because they may be measuring similar factors. The existence of high correlations among the independent variables in a regression model is known as **multicollinearity**, and although it is not a violation of the assumptions underlying a regression analysis, it often impairs the usefulness of a regression analysis, particularly with respect to the interpretation of the regression coefficients.

Because multicollinearity is so often encountered, we will devote this entire chapter to this topic. In Section 5.2 we use some artificially generated data to show how multicollinearity affects the results of a regression analysis. Then, in Section 5.3, we provide some tools to study the existence and nature of the multicollinearity, and in Section 5.4 we present some remedial methods that may help to provide useful results.

One very popular method used to combat multicollinearity is the use of variable selection, a statistically based method for selecting a subset of the initially chosen set of independent variables that, ideally, will produce a model that has lost very little in precision while being subject to lesser multicollinearity. Because this methodology is so often used (and misused) and because it is also used when multicollinearity is not a serious problem, it is presented in Chapter 6.

5.2 The Effects of Multicollinearity

In this section we present artificially generated data sets of 50 observations, all using the model

$$y = 4.0x_1 + 3.5x_2 + 3.0x_3 + 2.5x_4 + 2.0x_5 + 1.5x_6 + 1.0x_7 + \epsilon,$$

where ϵ is the random error. Each of the independent variables has values from -0.5 to 0.5 generated from the uniform distribution, standardized to have mean of 0 and range of 1. Because all the independent variables have approximately equal dispersion, the magnitudes of the coefficients provide an easy visual guide to the relative importance (degree of statistical significance) of each of the independent variables, with x_1 being the most important, x_2 the next most important, and so forth.

Three data sets are generated according to these specifications and differ only in the degree and nature of the multicollinearity among independent variables. In order to provide for easier comparisons among the three sets, the magnitude of the random error is generated to provide an R-square of approximately 0.85 for each case.

EXAMPLE 5.1   No Multicollinearity   The independent variables in this example are generated with zero population correlations. The purpose of this example is to provide a basis for comparison with results using data sets having multicollinearity. The results of the regression analysis, using PROC REG of the SAS System, are shown in Table 5.1.

The results conform to expectations: The regression is significant ($p < 0.0001$) and the R-square value is 0.86. The 0.95 confidence intervals for all coefficients include the true parameter values, and the magnitude of the t statistics for testing that the coefficients are 0 decrease consistently from x_1 to x_7. For six of the seven coefficients the p-values suggest a high degree of statistical significance, whereas β_7 appears to contribute little to the predictive ability of the model.

TABLE 5.1

RESULTS OF REGRESSION WITH NO MULTICOLLINEARITY (EXAMPLE 5.1)

Analysis of Variance

Source	DF	Sum of Squares	Mean Square	F Value	Prob>F
Model	7	184.89000	26.41286	37.067	0.0001
Error	42	29.92779	0.71257		
C Total	49	214.81779			

Root MSE	0.84414	R-square	0.8607	
Dep Mean	-0.40533	Adj R-sq	0.8375	
C.V.	-208.25958			

Parameter Estimates

Variable	DF	Parameter Estimate	Standard Error	T for H0: Parameter=0	Prob > |T|
INTERCEP	1	-0.121615	0.13162405	-0.924	0.3608
X1	1	4.664889	0.43608835	10.697	0.0001
X2	1	2.791484	0.49951208	5.588	0.0001
X3	1	2.975358	0.45938390	6.477	0.0001
X4	1	2.348426	0.38722774	6.065	0.0001
X5	1	1.582507	0.43453080	3.642	0.0007
X6	1	1.313007	0.43038059	3.051	0.0039
X7	1	0.664433	0.47345660	1.403	0.1679

Another feature of regression when there is no multicollinearity is that the total and partial regression coefficients are nearly the same. This is shown in Table 5.2, where the first line shows the partial coefficients and other lines show the estimated coefficients for the individual simple linear regressions.

TABLE 5.2

PARTIAL AND TOTAL REGRESSION COEFFICIENTS WITH NO MULTICOLLINEARITY

MODEL	RMSE	INTERCEPT	X1	X2	X3	X4	X5	X6	X7
			Partial Coefficients						
ALL	0.84414	-0.12162	4.66	2.79	2.98	2.35	1.58	1.31	0.66
			Total Coefficients						
X1	1.52750	-0.59354	4.98	.	.	.	.	.	.
X2	2.05315	-0.33729	.	2.02	.	.	.	.	.
X3	1.92956	-0.20467	.	.	3.11	.	.	.	.
X4	2.09845	-0.41001	.	.	.	0.80	.	.	.
X5	2.06229	-0.26412	.	.	.	.	1.63	.	.
X6	1.96728	-0.32502	.	.	.	.	.	2.60	.
X7	2.10530	-0.39360	.	.	.	.	.	.	0.79

Furthermore, the residual standard deviations (RMSE) of all the one-variable models are much larger than that of the multiple regression, indicating that at least several variables are needed.   ◆

EXAMPLE 5.2   "Uniform" Multicollinearity   For this example, the independent variables are generated so that correlations between all "adjacent" variables, that is, between x_j and x_{j+1}, are 0.98. The matrix of sample correlations for the generated sample is shown in Table 5.3. From this table it can be seen that the correlations among adjacent variables are all close to 0.98; the other correlations are, by definition,[1] somewhat lower, but still quite strong.

The results of the regression using these data are shown in Table 5.4. The R-square value is very close to that of the no-multicollinearity case.[2] There is, however, a big change in the estimated coefficients, and only β_5 may be considered statistically significant ($P < 0.05$), a surprising result because this certainly does not correspond to the most "important" variable. Furthermore, three coefficients, including that for x_1, have negative signs that certainly do not agree with the true model. Actually, this result is a direct consequence of the large standard errors of the estimates of the coefficients: In general they are six to eight times larger than those resulting from the no-multicollinearity data. Thus, all confidence intervals do include the true population coefficients. We will see later that a multicollinearity assessment statistic is based on the difference in standard errors.

What we have here is the main feature of regression analyses when multicollinearity is present:

> Although the model may fit the data very well, the individual coefficients may not be very useful.

TABLE 5.3

CORRELATIONS AMONG INDEPENDENT VARIABLES, EXAMPLE 5.2

Variable	X1	X2	X3	X4	X5	X6	X7
X1	1.0000	0.9834	0.9622	0.9550	0.9254	0.9169	0.8963
X2	0.9834	1.0000	0.9806	0.9654	0.9413	0.9290	0.9079
X3	0.9622	0.9806	1.0000	0.9813	0.9608	0.9545	0.9347
X4	0.9550	0.9654	0.9813	1.0000	0.9737	0.9629	0.9346
X5	0.9254	0.9413	0.9608	0.9737	1.0000	0.9789	0.9513
X6	0.9169	0.9290	0.9545	0.9629	0.9789	1.0000	0.9807
X7	0.8963	0.9079	0.9347	0.9346	0.9513	0.9807	1.0000

[1] Population correlations between x_j and x_{j+2} are 0.98^2, between x_j and x_{j+3} are 0.98^3, and so forth. Also, variables x_2 to x_7 do not have uniform distributions; however, the variances of all variables are still 1/12.

[2] Because all coefficients are positive, multicollinearity causes the response variable to span a larger range; hence, with a constant R-square, the residual mean square is also larger.

TABLE 5.4

RESULTS OF REGRESSION WITH MULTICOLLINEARITY (EXAMPLE 5.2)

Analysis of Variance

Source	DF	Sum of Squares	Mean Square	F Value	Prob>F
Model	7	1302.35790	186.05113	44.992	0.0001
Error	42	173.67978	4.13523		
C Total	49	1476.03768			

Root MSE	2.03353	R-square	0.8823	
Dep Mean	0.97697	Adj R-sq	0.8627	
C.V.	208.14732			

Parameter Estimates

Variable	DF	Parameter Estimate	Standard Error	T for H0: Parameter=0	Prob > |T|
INTERCEP	1	0.063117	0.32376278	-0.195	0.8464
X1	1	-2.018039	5.51355286	-0.366	0.7162
X2	1	8.083684	7.93787184	1.018	0.3143
X3	1	3.832642	7.82826473	0.490	0.6270
X4	1	-3.251330	7.23563916	-0.449	0.6555
X5	1	14.184539	6.67995255	2.123	0.0397
X6	1	3.675018	8.74613529	0.420	0.6765
X7	1	-5.286719	5.61215537	-0.942	0.3516

In other words, multicollinearity does not affect the overall fit of the model, and therefore does not affect the model's ability to obtain point estimates of the response variable or to estimate the residual variation. However, multicollinearity does reduce the effectiveness of a regression analysis if its purpose is to determine the effects of the various independent factor variables. Furthermore, the large standard errors of the coefficients also increase the standard errors of the estimated conditional means and predicted values.

When multicollinearity is present, the values of the partial regression coefficients may be quite different from the total regression coefficients as shown in Table 5.5. Here we can see, for example, that the partial coefficient for X1 is −2.02, whereas the total coefficient is 16.27! The comparisons of the other coefficients show similar results. This is the direct consequence of the definitions of partial and total regression coefficients: The partial coefficient is the change in $\hat{\mu}_{y|x}$ associated with a unit change in the respective x *holding constant* all other independent variables, whereas the total coefficient is that change *ignoring* all other variables.

Another feature of multicollinearity is that the model with all variables may not fit the data better than models with fewer variables. Again this result is a direct consequence of multicollinearity: Correlated variables may be considered as

TABLE 5.5

TOTAL AND PARTIAL REGRESSION COEFFICIENTS WITH MULTICOLLINEARITY (EXAMPLE 5.2)

MODEL	RMSE	INTERCEPT	X1	X2	X3	X4	X5	X6	X7
					Partial Coefficients				
ALL	2.03353	-0.06312	-2.02	8.08	3.83	-3.25	14.18	3.68	-5.29
					Total Coefficients				
X1	2.47698	-0.18011	16.27	.	.	.	.	.	.
X2	2.23214	-0.06115	.	17.45	.	.	.	.	.
X3	2.15505	-0.14525	.	.	18.43	.	.	.	.
X4	2.17881	-0.39417	.	.	.	18.64	.	.	.
X5	2.04840	-0.15513	.	.	.	.	19.33	.	.
X6	2.33456	0.02735	.	.	.	.	.	18.47	.
X7	2.69789	-0.04245	.	.	.	.	.	.	17.31

largely measuring the same phenomenon, therefore one variable may do almost as well as combinations of several variables. In this example, the model with all seven variables has a residual standard deviation (RMSE) of 2.03, while the residual standard deviation of the model with only x_5 is almost as small and the largest one-variable RMSE is only 2.70. Actually, this makes sense, because highly correlated variables tend to provide almost identical information. ◆

In Section 3.4 we presented a method of obtaining partial regression coefficients from residuals, where the coefficients are obtained by performing simple linear regressions using the residuals from the regression on all other independent variables. Multicollinearity implies strong relationships among the independent variables that will cause the variances of the residuals from these regressions to have small magnitudes. Now the precision of the estimated regression coefficient in a simple linear regression (Section 2.3) is inversely related to the dispersion of the independent variable. Since multicollinearity reduces the dispersion of the residuals, which are the independent variables in these regressions, the precision of a partial regression coefficient will be poor in the presence of multicollinearity. We will see later that a multicollinearity assessment statistic is based on this argument.

Actually, the poor performance of partial regression coefficients in the presence of multicollinearity can be explained in a more practical manner. If a set of variables are strongly related, it means that changing one variable while holding the other variables constant does not really occur in the data. In other words, the partial regression coefficient is trying to estimate a phenomenon that is not occurring in the data, and data cannot be made to show something that is not there.

EXAMPLE 5.3   Several Multicollinearities   In this example, the population correlation patterns are more complicated, as follows:

Correlations among x_1, x_2, and x_3 are as in Example 5.2; that is, correlations among adjacent variables are 0.98.

Correlations among x_4, x_5, and x_6 are similar to those of Example 5.2, except correlations among adjacent variables are 0.95, but they are uncorrelated with x_1, x_2, and x_3.

Correlations of x_7 with all other variables are zero.

The sample correlation matrix of the independent variables for the 50 sample observations is shown in Table 5.6 and is seen to agree with the specifications.

The results of the regression using these data are shown in Table 5.7. The results are somewhat similar to those of Example 5.2. The statistics for the model are virtually unchanged: The entire regression is highly significant[3] and the standard errors for the coefficients of x_1, x_2, and x_3 are somewhat similar to those of Example 5.2. The standard errors for the coefficients x_4, x_5, and x_6 are somewhat smaller that those of Example 5.2 because the correlations among these variables are not quite as high. Note, however, that the standard error for the coefficient of x_7 is about the same as for the no-multicollinearity case. This result shows that, regardless of the degree of multicollinearity among variables in the model, the standard error of the coefficient for a variable not correlated with other variables is not affected by the multicollinearity among the other variables.   ◆

Now that we have seen what multicollinearity can do in a "controlled" situation, we will see what it does in a "real" example.

EXAMPLE 5.4   **Basketball Statistics**   The data are NBA team statistics published by the *World Almanac and Book of Facts* for the 1976/77 through 1978/79 seasons. The following variables are used here:

FGAT   Attempted field goals
FGM   Field goals made
FTAT   Attempted free throws
FTM   Free throws made
OFGAT   Attempted field goals by opponents
OFGAL   Opponent field goals allowed
OFTAT   Attempted free throws by opponents

TABLE 5.6

CORRELATIONS AMONG INDEPENDENT VARIABLES, EXAMPLE 5.3

	X1	X2	X3	X4	X5	X6	X7
X1	1.0000	0.9794	0.9653	0.1075	0.2000	0.1895	0.1954
X2	0.9794	1.0000	0.9816	0.0918	0.1803	0.1718	0.2018
X3	0.9653	0.9816	1.0000	0.1424	0.2364	0.2346	0.1957
X4	0.1075	0.0918	0.1424	1.0000	0.9417	0.9028	-.0295
X5	0.2000	0.1803	0.2364	0.9417	1.0000	0.9573	0.0755
X6	0.1895	0.1718	0.2346	0.9028	0.9573	1.0000	0.1028
X7	0.1954	0.2018	0.1957	-.0295	0.0755	0.1028	1.0000

[3] Again the difference in the magnitudes of the error mean square is due to the larger variation in the values of the response variable.

TABLE 5.7

REGRESSION RESULTS FOR EXAMPLE 5.3

Analysis of Variance

Source	DF	Sum of Squares	Mean Square	F Value	Prob>F
Model	7	790.72444	112.96063	44.035	0.0001
Error	42	107.74148	2.56527		
C Total	49	898.46592			

Root MSE	1.60165	R-square	0.8801	
Dep Mean	0.46520	Adj R-sq	0.8601	
C.V.	344.29440			

Parameter Estimates

Variable	DF	Parameter Estimate	Standard Error	T for H0: Parameter=0	Prob > |T|
INTERCEP	1	0.717614	0.24052877	2.983	0.0047
X1	1	12.174225	3.58069346	3.400	0.0015
X2	1	-5.395881	5.24621191	-1.029	0.3096
X3	1	3.363046	4.13109139	0.814	0.4202
X4	1	4.373243	2.55233121	1.713	0.0940
X5	1	0.691321	3.67341938	0.188	0.8516
X6	1	0.684917	2.62334085	0.261	0.7953
X7	1	1.971058	0.79394132	2.483	0.0171

OFTAL	Opponent free throws allowed
DR	Defensive rebounds
DRA	Defensive rebounds allowed
OR	Offensive rebounds
ORA	Offensive rebounds allowed
WINS	Season wins

The data are shown in Table 5.8.

We perform a regression of WINS on the other variables for the purpose of determining what aspects of team performance lead to increased wins.

One would expect some high correlations among the independent variables in this data set. For example, it is logical that the more goals are attempted, the more are made. The correlations are shown in Table 5.9. At first glance, there do not appear to be many very highly correlated variables. Only the correlations between attempted and made free throws for both sides have correlations above 0.9. However, as we will see, there are indeed other sources of multicollinearity, indicating that absence of high pairwise correlations does not necessarily imply that multicollinearity is not present.

TABLE 5.8

NBA DATA

Obs	Region	FGAT	FGM	FTAT	FTM	OFGAT	OFGAL	OFTAT	OFTAL	DR	DRA	OR	ORA	WINS
1	1	7322	3511	2732	2012	7920	3575	2074	1561	2752	2448	1293	1416	50
2	1	7530	3659	2078	1587	7610	3577	2327	1752	2680	2716	974	1163	40
3	1	7475	3366	2492	1880	7917	3786	1859	1404	2623	2721	1213	1268	30
4	1	7775	3462	2181	1648	7904	3559	2180	1616	2966	2753	1241	1110	44
5	1	7222	3096	2274	1673	7074	3279	2488	1863	2547	2937	1157	1149	22
6	1	7471	3628	2863	2153	7788	3592	2435	1803	2694	2473	1299	1363	55
7	1	7822	3815	2225	1670	7742	3658	2785	2029	2689	2623	1180	1254	43
8	1	8004	3547	2304	1652	7620	3544	2830	2135	2595	2996	1306	1312	24
9	1	7635	3494	2159	1682	7761	3539	2278	1752	2850	2575	1235	1142	32
10	1	7323	3413	2314	1808	7609	3623	2250	1695	2538	2587	1083	1178	27
11	1	7873	3819	2428	1785	8011	3804	1897	1406	2768	2541	1309	1178	54
12	1	7338	3584	2411	1815	7626	3542	2331	1747	2712	2506	1149	1252	47
13	1	7347	3527	2321	1820	7593	3855	2079	1578	2396	2453	1119	1122	29
14	1	7523	3464	2613	1904	7306	3507	2861	2160	2370	2667	1241	1234	37
15	1	7554	3676	2111	1478	7457	3600	2506	1907	2430	2489	1200	1225	31
16	2	7657	3711	2522	2010	8075	3935	2059	1512	2550	2687	1110	1329	44
17	2	7325	3535	2103	1656	7356	3424	2252	1746	2632	2232	1254	1121	49
18	2	7479	3514	2264	1622	7751	3552	1943	1462	2758	2565	1185	1167	48
19	2	7602	3443	2183	1688	7712	3486	2448	1833	2828	2781	1249	1318	35
20	2	7176	3279	2451	1836	7137	3409	2527	1909	2512	2533	1244	1121	31
21	2	7688	3451	1993	1468	7268	3265	2325	1748	2563	2711	1312	1202	43
22	2	7594	3794	2234	1797	8063	3808	1996	1494	2594	2576	1030	1345	52
23	2	7772	3580	2655	1887	8065	3767	1895	1437	2815	2683	1349	1166	44
24	2	7717	3568	2331	1690	7938	3659	2213	1661	2907	2747	1309	1273	39
25	2	7707	3496	2116	1569	7620	3474	2113	1574	2676	2779	1187	1214	43
26	2	7691	3523	1896	1467	7404	3571	2238	1699	2421	2525	1301	1195	28
27	2	7253	3335	2316	1836	6671	3162	2930	2193	2359	2606	1160	1160	41
28	2	7760	3927	2423	1926	7970	3798	2343	1759	2619	2531	1096	1297	48
29	2	7498	3726	2330	1845	7625	3795	2211	1627	2504	2315	1256	1186	47
30	2	7802	3708	2242	1607	7623	3755	2295	1732	2380	2628	1303	1301	30
31	2	7410	3505	2534	1904	6886	3367	2727	2045	2341	2440	1381	1176	46
32	2	7511	3517	2409	1848	8039	3864	2246	1666	2676	2664	1234	1486	26
33	2	7602	3556	2103	1620	7150	3600	2423	1837	2256	2587	1229	1123	30
34	3	7471	3590	2783	2053	7743	3585	2231	1635	2700	2481	1288	1269	50
35	3	7792	3764	1960	1442	7539	3561	2543	1933	2495	2637	1169	1317	44
36	3	7840	3668	2072	1553	7753	3712	2330	1721	2519	2613	1220	1265	30
37	3	7733	3561	2140	1706	7244	3422	2513	1912	2593	2739	1222	1097	40
38	3	7840	3522	2297	1714	7629	3599	2252	1705	2584	2770	1409	1378	36
39	3	7186	3249	2159	1613	7095	3306	1907	1425	2705	2559	1292	1055	44
40	3	7883	3801	2220	1612	7728	3715	2404	1832	2480	2617	1239	1234	44
41	3	7441	3548	2705	2068	7799	3678	2365	1740	2736	2546	1177	1267	48
42	3	7731	3601	2262	1775	7521	3564	2635	2004	2632	2684	1208	1232	31
43	3	7424	3552	2490	1832	7706	3688	2177	1662	2601	2494	1229	1244	38
44	3	7783	3500	2564	1904	7663	3634	2455	1841	2624	2793	1386	1350	31
45	3	7041	3330	2471	1863	7273	3565	1980	1466	2577	2367	1248	1065	40
46	3	7773	3906	2021	1541	7505	3676	2415	1819	2370	2437	1157	1229	38
47	3	7644	3764	2392	1746	7061	3434	2897	2170	2404	2547	1191	1156	48

(Continued)

TABLE 5.8 (*Continued*)

NBA DATA

Obs	Region	FGAT	FGM	FTAT	FTM	OFGAT	OFGAL	OFTAT	OFTAL	DR	DRA	OR	ORA	WINS
48	3	7311	3517	2841	2046	7616	3631	2277	1713	2596	2429	1307	1218	47
49	3	7525	3575	2317	1759	7499	3586	2416	1868	2530	2605	1225	1299	38
50	3	7108	3478	2184	1632	7408	3682	2029	1549	2544	2377	1224	1095	31
51	4	7537	3623	2515	1917	7404	3408	2514	1889	2703	2510	1260	1197	49
52	4	7832	3724	2172	1649	7584	3567	2282	1699	2639	2640	1300	1256	46
53	4	7657	3663	1941	1437	7781	3515	1990	1510	2628	2625	1177	1348	53
54	4	7249	3406	2345	1791	7192	3320	2525	1903	2493	2594	1059	1180	34
55	4	7639	3439	2386	1646	7339	3394	2474	1863	2433	2651	1355	1257	40
56	4	7836	3731	2329	1749	7622	3578	2319	1749	2579	2743	1166	1202	49
57	4	7672	3734	2095	1576	7880	3648	2050	1529	2647	2599	1136	1365	45
58	4	7367	3556	2259	1717	7318	3289	2282	1747	2686	2523	1187	1187	58
59	4	7654	3574	2081	1550	7368	3425	2408	1820	2629	2794	1183	1185	43
60	4	7715	3445	2352	1675	7377	3384	2203	1670	2601	2600	1456	1121	47
61	4	7516	3847	2299	1765	7626	3775	2127	1606	2379	2424	1083	1238	50
62	4	7706	3721	2471	1836	7801	3832	2295	1760	2413	2322	1392	1294	43
63	4	7397	3827	2088	1606	7848	3797	1931	1415	2557	2486	949	1288	47
64	4	7338	3541	2362	1806	7059	3448	2501	1889	2435	2350	1256	1080	45
65	4	7484	3504	2298	1732	7509	3475	2108	1567	2591	2453	1310	1156	52
66	4	7453	3627	1872	1367	7255	3493	2155	1604	2513	2533	1169	1147	38

TABLE 5.9

CORRELATIONS FOR NBA DATA

VAR	FGAT	FGM	FTAT	FTM	OFGAT	OFGAL	OFTAT	OFTAL	DR	DRA	OR	ORA
FGAT	1.00	0.56	-.24	-.30	0.42	0.29	0.14	0.13	0.11	0.48	0.23	0.37
FGM	0.56	1.00	-.15	-.11	0.47	0.57	-.04	-.05	-.10	-.22	-.28	0.36
FTAT	-.24	-.15	1.00	0.94	0.19	0.17	0.06	0.03	0.15	-.17	0.31	0.20
FTM	-.30	-.11	0.94	1.00	0.20	0.22	0.05	0.02	0.14	-.23	0.14	0.19
OFGAT	0.42	0.47	0.19	0.20	1.00	0.79	-.54	-.55	0.54	0.11	-.10	0.60
OFGAL	0.29	0.57	0.17	0.22	0.79	1.00	-.44	-.46	0.03	-.16	-.17	0.44
OFTAT	0.14	-.04	0.06	0.05	-.54	-.44	1.00	0.99	-.37	0.21	0.08	-.01
OFTAL	0.13	-.05	0.03	0.02	-.55	-.46	0.99	1.00	-.39	0.20	0.09	-.03
DR	0.11	-.10	0.15	0.14	0.54	0.03	-.37	-.39	1.00	0.26	0.08	0.12
DRA	0.48	-.22	-.17	-.23	0.11	-.16	0.21	0.20	0.26	1.00	-.00	0.23
OR	0.23	-.28	0.31	0.14	-.10	-.17	0.08	0.09	0.08	-.00	1.00	0.00
ORA	0.37	0.36	0.20	0.19	0.60	0.44	-.01	-.03	0.12	0.23	0.00	1.00

The results of the regression using WINS as the dependent variable and all 12 performance statistics are shown in Table 5.10. The *F*-value for the test for the model as well as the coefficient of determination suggest a rather well-fitting model. However, only two coefficients (FGM and OFGAL) have *p*-values less

TABLE 5.10

RESULTS OF REGRESSION FOR NBA DATA

Analysis of Variance

Source	DF	Sum of Squares	Mean Square	F Value	Prob>F
Model	12	3968.07768	330.67314	26.638	0.0001
Error	53	657.92232	12.41363		
C Total	65	4626.00000			

Root MSE	3.52330	R-square	0.8578	
Dep Mean	41.00000	Adj R-sq	0.8256	
C.V.	0.59341			

Parameter Estimates

Variable	DF	Parameter Estimate	Standard Error	T for H0: Parameter=0	Prob > |T|
INTERCEP	1	36.173788	22.39378508	1.615	0.1122
FGAT	1	-0.017783	0.01386721	-1.282	0.2053
FGM	1	0.069801	0.01315666	5.305	0.0001
FTAT	1	-0.003571	0.01044877	-0.342	0.7339
FTM	1	0.027675	0.01141602	2.424	0.0188
OFGAT	1	0.022016	0.01217051	1.809	0.0761
OFGAL	1	-0.075084	0.01139379	-6.590	0.0001
OFTAT	1	0.016453	0.01756946	0.936	0.3533
OFTAL	1	-0.043636	0.02207290	-1.977	0.0533
DR	1	-0.013764	0.01137661	-1.210	0.2317
DRA	1	0.007346	0.01274171	0.576	0.5667
OR	1	0.026541	0.01853874	1.432	0.1581
ORA	1	-0.021398	0.01393127	-1.536	0.1305

than 0.01, and one other (FTM) has a *p*-value less than 0.05. These results would suggest that only field goals made by the team and opposition are important and the coefficients do have the expected sign, while the number of free throws made by the team has a marginal positive effect.

Results of this nature are, of course, what we expect if multicollinearity is present, although these apparent contradictions between model and coefficient statistics are not as severe as those of the artificial examples. Also, we have seen that another result of multicollinearity is that the partial and total coefficients tend to be different. The partial and total coefficients for the NBA data are shown in Table 5.11. The differences are indeed quite marked, although again not as much as in the artificial data. Also, in this example, single-variable regressions do not fit as well as the multiple regression. However, note that many single-variable regressions have nearly equal residual mean squares. ◆

TABLE 5.11

PARTIAL AND TOTAL COEFFICIENTS, NBA DATA

Model	RMSE	FGAT	FGM	FTAT	FTM	OFGAT	OFGAL	OFTAT	OFTAL	DR	DRA	OR	ORA
					Partial Regression Coefficients								
ALL	3.52330	-.02	.070	-.00	.028	.022	-.08	.016	-.04	-.01	.007	.027	-.02
					Total Regression Coefficients								
FGAT	8.50124	.000	.	.	.	.	.	.	.	.	.	.	.
FGM	7.75699	.	.022	.	.	.	.	.	.	.	.	.	.
FTAT	8.30182	.	.	.008	.	.	.	.	.	.	.	.	.
FTM	8.29139	.	.	.	.011	.	.	.	.	.	.	.	.
OFGAT	8.39625	.	.	.	.	.004	.	.	.	.	.	.	.
OFGAL	8.48802	.	.	.	.	.	-.00	.	.	.	.	.	.
OFTAT	8.28360	.	.	.	.	.	.	-.01	.	.	.	.	.
OFTAL	8.24302	.	.	.	.	.	.	.	-.01	.	.	.	.
DR	8.21042	.	.	.	.	.	.	.	.	.015	.	.	.
DRA	7.84466	.	.	.	.	.	.	.	.	.	-.02	.	.
OR	8.50124	.	.	.	.	.	.	.	.	.	.	-.00	.
ORA	8.49853	.	.	.	.	.	.	.	.	.	.	.	.003

5.3   Diagnosing Multicollinearity

We have seen the effects of multicollinearity, and if these effects are seen in an analysis, we may conclude that multicollinearity exists. It is, however, useful to have additional tools that can indicate the magnitude of (and assist in identifying) the variables that are involved in the multicollinearity. Two frequently used tools are the **variance inflation factor** and **variance proportions**.

Variance Inflation Factors

In Section 3.4 we noted that the variance of an estimated partial regression coefficient,

$$\text{Var}(\hat{\beta}_j) = \text{MSE } c_{jj},$$

where MSE is the error mean square and c_{jj} is the jth diagonal element of $(X'X)^{-1}$. We have already seen that multicollinearity has no effect on the residual mean square; hence, the large variances of the coefficients must be associated with large values of the c_{jj}. It can be shown that

$$c_{jj} = \frac{1}{(1 - R_j^2)\Sigma(x_j - \bar{x}_j)^2},$$

where R_j^2 is the coefficient of determination of the "regression" of x_j on all other independent variables in the model. In Chapter 2 we saw that $\Sigma(x_j - \bar{x}_j)^2$ is the denominator of the formula for the variance of the regression coefficient in a simple linear regression. If there is no multicollinearity $R_j^2 = 0$; then the variance as well as the estimated coefficient is the same for the total and partial regression coefficients. However, correlations among any independent variables cause R_j^2 to increase, effectively increasing the magnitude of c_{jj} and consequently increasing the variance of the estimated coefficient. In other words, the variance of $\hat{\beta}_j$ is increased or *inflated* by the quantity $[1/(1 - R_j^2)]$. This statistic is computed for each coefficient, and the statistics $[1/(1 - R_j^2)], j = 1, 2, \ldots, m$ are known as the **variance inflation factors**, often simply denoted by **VIF**.

We have previously noted that when multicollinearity exists, it is difficult to vary one variable while holding the others constant, thus providing little information on a partial regression coefficient. The variance inflation factor quantifies this effect by stating that the effective dispersion of that independent variable is reduced by $(1 - R_j^2)$, which then increases the variance of that estimated coefficient.

Table 5.12 shows the variance inflation factors for the three artificial data sets (Examples 5.1, 5.2, and 5.3). From this table we can see the following:

Example 5.1: Because there is no multicollinearity, we would expect variance inflation factors to be unity. Actually, the VIF are all slightly larger than 1, because although the population correlations are zero, the sample coefficients are not exactly zero.

Example 5.2: Variance inflation factors range from 28 to 66, indicating that the variances of the partial coefficients are vastly inflated by the existence of multicollinearity. No wonder that none of the estimated coefficients was statistically significant. Note also that all of the VIF are of the same order of magnitude because all variables are uniformly correlated.

Example 5.3: The VIF values for X1, X2, and X3 have similar magnitudes as those of Example 5.2, because the correlations among these are the same as those in Example 5.2. The VIF values for X4, X5, and X6 are smaller than those for the first three coefficients, because

TABLE 5.12

VARIANCE INFLATION FACTORS FOR EXAMPLES 5.1, 5.2, AND 5.3

	Example	5.1	5.2	5.3
	INTERCEP	0.00000	0.0000	0.0000
	X1	1.10507	32.8267	25.0931
	X2	1.07279	61.9082	51.3006
COEFFICIENT	X3	1.10424	54.6922	31.9486
	X4	1.12248	45.4948	10.2479
	X5	1.06575	36.8241	20.7060
	X6	1.11844	65.8494	12.5494
	X7	1.03638	28.6475	1.1465

the correlations among these are smaller. The VIF value for X7 is close to unity because X7 was generated to have zero (population) correlation with any of the other variables. Thus, the variance inflation factors illustrate the fact that although extreme multicollinearity may be present, variances of coefficients of uncorrelated variables are not affected by correlations among the other variables.

Before continuing we will want to know how large a VIF must be before the degree of multicollinearity is considered to seriously affect the estimation of the corresponding coefficient. Because we are working with exploratory analyses there is no "significance" test; hence, any cutoff value must be based on practical considerations. A popular cutoff value is 10. This value has no theoretical basis, but is convenient as it is easy to spot in a listing of VIF values. For example, we can easily see that all variables in Example 5.2 and all but one in Example 5.3 are involved in "serious" multicollinearity.

However, VIF values must also be evaluated relative to the overall fit of the model under study. For example, if the model R^2 is 0.9999, then VIF values of 10 will not be large enough to seriously affect the estimates of the coefficients, whereas if the model R^2 is 0.25, then VIF values of only 7 may cause poor estimates. It may therefore be useful to compare the VIF values with the equivalent statistic for the regression model: $1/(1 - R^2_{model})$. Any VIF values larger than this quantity imply stronger relationships among the independent variables than their relationship to the response.

Finally, the effect of sample size is not affected by multicollinearity; hence, variances of regression coefficients based on very large samples may still be quite reliable in spite of multicollinearity.

Table 5.13 shows the variance inflation factors for Example 5.4. All but one of the VIF values exceeds 10, indicating that multicollinearity definitely exists.

TABLE 5.13

VIF VALUES FOR NBA DATA

Variable	DF	Variance Inflation
INTERCEP	1	0.00000000
FGAT	1	46.16283766
FGM	1	23.02705631
FTAT	1	27.83995306
FTM	1	18.98467286
OFGAT	1	71.80453440
OFGAL	1	18.54001592
OFTAT	1	102.80792238
OFTAL	1	94.38613969
DR	1	13.88371365
DRA	1	17.65046315
OR	1	17.17856425
ORA	1	8.14903188

The regression model R^2 is 0.8578; hence, $1/(1 - R^2_{model}) = 7.03$, indicating that many of the correlations among the independent variables are indeed stronger than the regression relationship.

Looking closer, it can be seen that the largest VIF values occur with the scoring variables; that is, for the field goal and free throws. The source of the multicollinearity can be diagnosed as coming from high correlations among the total number of attempts and the number of attempts made. That is, the more attempts made, the more scores are achieved. Similar but weaker correlations apparently also exist between rebounds and rebounds allowed.

Variance Proportions

Linear dependencies are defined as the existence of exact linear relationships among variables or equivalently the existence of linear functions of variables being equal to zero. Multicollinearity can be thought of as "near" linear dependencies among the independent variables.

In a regression setting, the existence of linear dependencies among the independent variables makes the $X'X$ matrix singular and prevents finding unique estimates of the regression coefficients. The practical interpretation of this fact is that if an *exact* linear relationship exists among a set of variables, it is impossible to change one while holding the others constant; hence, a partial regression coefficient cannot be defined.

In a similar manner, then, multicollinearity is defined by the existence of linear functions of variables being "near" zero. Variance proportions may provide information on the existence of such linear functions.

Variance proportions are a by-product of **principal components** analysis, which is the simplest of many multivariate procedures for analyzing the structure of a set of correlated variables that are commonly grouped under the heading of **factor analysis**. We will see later that principal components are also useful as a possible remedial method for overcoming the effects of multicollinearity.

Principal Components

Because principal components are intended to study correlation patterns, the analysis is based on standardized variables to avoid confusion introduced by differing variances among the variables.[4] Thus, if X is an $n \times m$ matrix of observed standardized variables, then $X'X$ is the correlation matrix.

Principal component analysis is a procedure that creates a set of new variables, z_i, $i = 1, 2, \ldots, m$, which are linearly related to the original set of standardized variables, x_i, $i = 1, 2, \ldots, m$. The equations relating the z_i to the x_i are of the form

$$z_i = v_{i1} x_1 + v_{i2} x_2 \ldots v_{im} x_m, i = 1, 2, \ldots, m,$$

---

[4] In this application, we standardize by subtracting the mean and dividing by the standard deviation. In some applications, variables need not be standardized. For simplicity we will not discuss that option.

which can be represented by the matrix equation

$$Z = XV,$$

where V is an $m \times m$ matrix of coefficients (v_{ij}) that describe the relationships between the two sets of variables. The Z variables are called a *linear transformation* of the X variables.

There exists an infinite number of such transformations. However, the principal components transformation creates a unique set of variables, z_i, to have the following properties:

1. The variables are uncorrelated, that is, $Z'Z$ is a diagonal matrix with diagonal elements, λ_i.
2. z_1 has the largest possible variance, z_2 the second largest, and so forth.

The principal components transformation is obtained by finding the *eigenvalues* and *eigenvectors*[5] (sometimes called characteristic values and vectors) of the correlation matrix, where the eigenvalues, denoted by $\lambda_1, \lambda_2, \ldots, \lambda_m$, are the variances of the corresponding z_i, and the matrix of eigenvectors are the columns of V, the so-called transformation matrix that relates the z variables to the x variables. That is, the first column of V provides the coefficients for the equation

$$z_1 = v_{11}x_1 + v_{21}x_2 \ldots v_{m1}x_m,$$

and so forth.

We illustrate with the principal components for a sample of 100 observations from a bivariate population with a correlation of 0.9 between the two variables. The sample correlation matrix is

$$X'X = \begin{bmatrix} 1 & 0.922 \\ 0.922 & 1 \end{bmatrix}.$$

The principal component analysis provides the transformation

$$z_1 = 0.707x_1 + 0.707x_2$$

$$z_2 = 0.707x_1 - 0.707x_2,$$

and the eigenvalues provide the estimated variances of the components:

$$\hat{\sigma}_{z_1}^2 = 1.922$$

$$\hat{\sigma}_{z_2}^2 = 0.078.$$

The left portion of Figure 5.1 shows a scatterplot of the original variables (labeled X1 and X2), and the right portion shows the scatterplot of the principal component variables (labeled PRIN1 and PRIN2).

The plot of the original variables is typical for a pair of highly correlated standardized variables. The plot of the principal components is typical of a pair of uncorrelated variables, where one variable (PRIN1 in this case) has a much

---

[5] The theoretical derivation and computation of eigenvalues and eigenvectors are beyond the scope of this book. However, procedures for these computations are readily available in many statistical software packages.

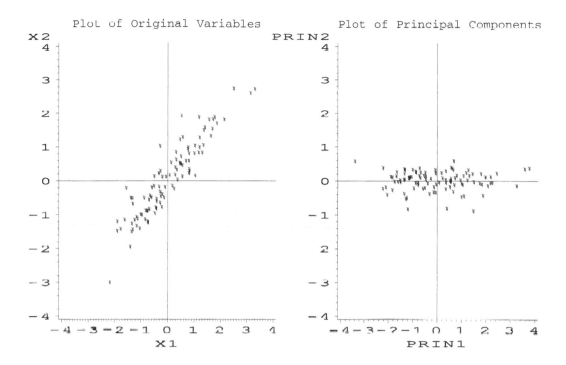

FIGURE 5.1 Two-Variable Principal Components

larger variance. A close inspection shows that the pattern of the data points is identical in both plots, illustrating the fact that the principal component transformation is simply a rigid rotation of axes such that the resulting variables have zero correlation and, further, that this first component variable has the largest variance, and so forth.

Note further that the sum of the variances for both sets of variables is 2.0. This result illustrates the fact that the principal component transformation has not altered the total variability of the set of variables (as measured by the sum of variances) but has simply apportioned them differently among the principal component variables.

The eigenvectors show that the first principal component variable consists of the sum of the two original variables, whereas the second consists of the difference. This is an expected result: For any two variables, regardless of their correlation, the sum provides the most information on their variability, and the difference explains the rest.

In fact, this structure of the principal component variables shows what happens for different correlation patterns. If the correlation is unity, the sum explains all of the variability; that is, the variance of the first component is 2.0. The variance of the second component is zero because there are no differences between the two variables. If there is little or no correlation, both components will have variances near unity.

For larger number of variables, the results are more complex but will have the following features:

> 1. The more severe the multicollinearity, the larger the differences in magnitudes among the variances (eigenvalues) of the components. However, the sum of eigenvalues is always equal to the number of variables.
> 2. The coefficients for the transformation (the eigenvectors) show how the principal component variables relate to the original variables.

We will see later that the principal component variables with large variances may help to interpret results of a regression where multicollinearity exists. However, when trying to diagnose the reasons for multicollinearity, the focus is on the principal components with very small variances.

Remember that linear dependencies are defined by the existence of a linear function of variables being equal to zero, which will result in one or more principal components having zero variance. Multicollinearity is a result of "near" linear dependencies among the independent variables that will consequently result in principal components with near zero variances. The coefficients of the transformation for those components provide some information on the nature of that multicollinearity. However, a set of related statistics called the **variance proportions** are more useful.

Without going into the mathematical details (which are not very instructive), variance proportions indicate the relative contribution from each principal component to the variance of each regression coefficient. Consequently, the existence of a relatively large contribution to the variances of several coefficients by a component with a small eigenvalue (a "near" collinearity) may indicate which variables contribute to the overall multicollinearity. For easier comparison among coefficients, the variance proportions are standardized to sum to 1, so that the individual elements represent the **proportions** of the variance of the coefficient attributable to each component.

The variance proportions for a hypothetical regression using the two correlated variables we have used to illustrate principal components are shown in Table 5.14 as provided by PROC REG of the SAS System.[6]

The first column reproduces the eigenvalues and the second the condition indices, which are an indicator of possible roundoff error in computing the inverse that may occur when there are many variables and multicollinearity is extreme. Condition numbers must be very large (usually at least in the hundreds) for roundoff error to be considered a problem.

---

[6] In some references (e.g., Belsley *et al.*, 1980), the variance proportions are obtained by computing eigenvalues and eigenvectors from the standardized matrix of uncorrected (raw) sums of squares and cross products of the m independent variables and the dummy variable representing the intercept. This procedure implies that the intercept is simply another coefficient that may be subject to multicollinearity. In most applications, especially where the intercept is beyond the range of the data, the results will be misleading. The subheading "(Intercept adjusted)" in this output shows that the intercept is not included in these statistics.

TABLE 5.14

VARIANCE PROPORTIONS

Collinearity Diagnostics(intercept adjusted)

Number	Eigenvalue	Condition Index	Var Prop X1	Var Prop X2
1	1.90119	1.00000	0.0494	0.0494
2	0.09881	4.38647	0.9506	0.9506

The columns headed by the names of the independent variables (X1 and X2 in this example) are the variance proportions. We examine these proportions looking for components with small eigenvalues, which in this case is component 2 (the variance proportions are 0.9506 for both coefficients). The second principal component consists of the difference between the two variables. The small variance of this component shows that differences between these variables cannot be very large; hence, it is logical that the second component contributes to the instability of the regression coefficients. In other words, the sum of the two variables (the first component) provides almost all the information needed for the regression.

EXAMPLE 5.3 REVISITED

Variance Proportions   Table 5.15 shows the eigenvalues and variance proportions for Example 5.3, which was constructed with two separate groups of correlated variables. We can immediately see that there are three large and four relatively small eigenvalues, indicating rather severe multicollinearity.

Remember that variables involved in multicollinearities are identified by relatively large variance proportions in principal components with small eigenvalues (variances). In this example, these are components 5 through 7. We can see

TABLE 5.15

VARIANCE PROPORTIONS FOR EXAMPLE 5.3

Collinearity Diagnostics(intercept adjusted)

Number	Eigenvalue	Condition Index	Var Prop X1	Var Prop X2	Var Prop X3	Var Prop X4	Var Prop X5	Var Prop X6	Var Prop X7
1	3.47296	1.00000	0.0021	0.0010	0.0017	0.0035	0.0022	0.0035	0.0040
2	2.41274	1.19976	0.0023	0.0012	0.0016	0.0087	0.0036	0.0058	0.0046
3	0.94862	1.91340	0.0004	0.0002	0.0003	0.0005	0.0000	0.0002	0.8844
4	0.08698	6.31898	0.0013	0.0011	0.0021	0.5923	0.0022	0.4084	0.0876
5	0.03580	9.84930	0.3877	0.0049	0.2368	0.1787	0.3413	0.1181	0.0069
6	0.03020	10.72409	0.2798	0.0018	0.1559	0.2004	0.6487	0.4411	0.0030
7	0.01271	16.53153	0.3263	0.9899	0.6015	0.0160	0.0020	0.0229	0.0095

that variables x_1, x_2, and x_3 have relatively large variance proportions in component 7, revealing the strong correlations among these three variables. Further, variables x_5 and x_6 in component 6 and x_1, x_3, and x_5 in component 5 show somewhat large proportions, and although they do not mirror the built-in correlation pattern, they do show that these variables are involved in multicollinearities. Variable x_7 shows very small proportions in all of these components because it is not correlated with any other variables. ◆

EXAMPLE 5.4
REVISITED

Variance Proportions   Table 5.16 shows the eigenvalues and variance proportions for Example 5.4, the NBA data.

There are two very small eigenvalues, indicating two sets of almost linear dependencies. However, the variance proportions associated with these eigen-

---

TABLE 5.16

VARIANCE PROPORTIONS FOR NBA DATA

Collinearity Diagnostics (intercept adjusted)

Number	Eigenvalue	Condition Index	Var Prop FGAT	Var Prop FGM	Var Prop FTAT	Var Prop FTM	Var Prop OFGAT
1	3.54673	1.00000	0.0002	0.0009	0.0001	0.0002	0.0010
2	2.39231	1.21760	0.0021	0.0018	0.0035	0.0053	0.0000
3	2.09991	1.29961	0.0007	0.0002	0.0030	0.0037	0.0001
4	1.63382	1.47337	0.0003	0.0052	0.0000	0.0002	0.0001
5	0.97823	1.90412	0.0021	0.0005	0.0000	0.0011	0.0001
6	0.59592	2.43962	0.0010	0.0120	0.0001	0.0008	0.0001
7	0.44689	2.81719	0.0030	0.0010	0.0021	0.0038	0.0000
8	0.20082	4.20249	0.0054	0.0338	0.0113	0.0095	0.0051
9	0.05138	8.30854	0.0727	0.1004	0.1756	0.3290	0.0041
10	0.04365	9.01395	0.0204	0.0677	0.1253	0.1563	0.1055
11	0.00632	23.68160	0.3279	0.3543	0.2128	0.2052	0.2157
12	0.00403	29.67069	0.5642	0.4221	0.4663	0.2849	0.6682

Number	Var Prop OFGAL	Var Prop OFTAT	Var Prop OFTAL	Var Prop DR	Var Prop DRA	Var Prop OR	Var Prop ORA
1	0.0030	0.0004	0.0004	0.0012	0.0000	0.0001	0.0030
2	0.0000	0.0001	0.0001	0.0003	0.0023	0.0004	0.0009
3	0.0001	0.0009	0.0009	0.0000	0.0009	0.0016	0.0076
4	0.0026	0.0001	0.0001	0.0126	0.0096	0.0035	0.0000
5	0.0001	0.0002	0.0001	0.0023	0.0065	0.0393	0.0035
6	0.0026	0.0005	0.0004	0.0438	0.0064	0.0013	0.0546
7	0.0197	0.0001	0.0001	0.0134	0.0235	0.0014	0.1157
8	0.0877	0.0040	0.0054	0.0158	0.0116	0.0049	0.0044
9	0.0285	0.0004	0.0000	0.0122	0.0733	0.0177	0.0032
10	0.1370	0.0000	0.0022	0.1040	0.0636	0.0448	0.0358
11	0.1796	0.3960	0.5324	0.1723	0.3571	0.2832	0.2041
12	0.5390	0.5974	0.4578	0.6220	0.4452	0.6019	0.5672

values appear to involve almost all variables, and hence, no sets of correlated variables can be determined. Eigenvectors 9 and 10 may also be considered small, but there are also no large variance proportions for these components. In other words, the multicollinearities in this data set appear to involve almost all variables.   ◆

Inconclusive results from the analysis of variance proportions are quite common. However, this analysis is generally available with most computer programs for regression at small cost in computer resources. Therefore, if multicollinearity is suspected, the analysis is worth doing.

5.4 Remedial Methods

We have shown two sets of statistics that may be useful in diagnosing the extent and nature of multicollinearity, and we will now explore various remedial methods for lessening the effects of this multicollinearity. The choice of remedial methods to be employed depends to a large degree on the primary purpose of the regression analysis. In this context we distinguish between two related but different purposes:

1. *Estimation.* The purpose of the analysis is to obtain the best estimate of the mean of the response variable for a given set of values of the independent variables without being very concerned with the contribution of the individual independent variables. That is, we are not particularly interested in the partial regression coefficients.
2. *Analysis of structure.* The purpose of the analysis is to determine the effects of the individual independent variables; that is, the magnitudes and significances of the individual partial regression coefficients. We are, of course, also interested in good estimates of the response, because if the overall estimation is poor, the coefficients may be useless.

There are three major approaches for remedial methods:

1. Variable selection
2. Redefining variables
3. Biased estimation

As we have already indicated, variable selection will be presented in Chapter 6.

Redefining Variables

We have already shown that one remedy for multicollinearity is to redefine the independent variables. For example, it is well known that if two variables, say, x_1 and x_2, are correlated, then the redefined variables

$$z_1 = x_1 + x_2 \text{ and } z_2 = x_1 - x_2$$

are uncorrelated. In fact, this redefinition (with a change in scale) was obtained by a principal component analysis for two correlated variables. Now, if these new variables have some useful meaning in the context of the data, their use in the regression provides a model with no multicollinearity. Analysis of this model will, however, yield the same overall statistics because the linear transformation does not affect the overall model. This is easily shown as follows. Given a model with two independent variables,

$$y = \beta_0 + \beta_1 x_1 + \beta_2 x_2 + \epsilon,$$

and a model using the z variables

$$y = \alpha_0 + \alpha_1 z_1 + \alpha_2 z_2 + \epsilon,$$

then, using the definitions of the z variables:

$$y = \alpha_0 + \alpha_1(x_1 + x_2) + \alpha_2(x_1 - x_2) + \epsilon$$
$$y = \alpha_0 + (\alpha_1 + \alpha_2)x_1 + (\alpha_1 - \alpha_2)x_2 + \epsilon.$$

Thus,

$$\beta_0 = \alpha_0,$$
$$\beta_1 = \alpha_1 + \alpha_2, \text{ and}$$
$$\beta_2 = \alpha_1 - \alpha_2.$$

Of course, things are not as simple when there are more than two variables. There are two distinct procedures for creating variable redefinitions:

1. Methods based on knowledge of the variables
2. Method based on a statistical analysis

Methods Based on Knowledge of the Variables

The use of linear functions and/or ratios involving the independent variables can often be employed in providing useful models with reduced multicollinearity. For example, the independent variables may be measures of different characteristics of, say, a biological organism. In this case, as the total size of the organism increases, so do the other measurements. Now, if x_1 is a measure of overall size and the other variables are measurements of width, height, girth, etc., then using x_1 as is and redefining all others as x_j/x_1 or $x_i - x_1$, the resulting variables will exhibit much less multicollinearity. Of course, using ratios will cause some change in the fit of the model, whereas the use of differences will not.

In other applications the variables may be data from economic time series, where all variables are subject to inflation and increased population and therefore are correlated. Converting these variables to a deflated and/or per capita basis will reduce multicollinearity.

EXAMPLE 5.2
REVISITED

In this example all adjacent variables are equally correlated. Because the variables in this example may take negative values, ratios are not useful. Instead we will use x_1 as is and define $x_{jD} = x_j - x_1, j = 2, 3, \ldots, 7$. The results of the regression using these redefined variables are shown in Table 5.17.

A number of features of this analysis are of interest:

1. The overall model statistics are the same because the redefinitions are linear transformations.
2. The variance inflation factors have been dramatically decreased; the maximum VIF is now 1.36.
3. The coefficients for x_1 and x_{2D} are highly significant ($p < 0.0001$) and positive, whereas x_{3D} and x_{5D} have p-values somewhat less than 0.05 and exhibit positive coefficients. This means that only two variables are really necessary for estimating the behavior of y. In other words, because the variables are so highly correlated, almost all of the information on the independent variables is conveyed by two variables.

Of course, in this example, as well as in Example 5.3, we know how the variables were constructed and therefore we have information that allows us to specify appropriate redefinitions needed to reduce the multicollinearity. In most practical applications we must use our knowledge of expected relationships among the variables to specify redefinitions. ◆

TABLE 5.17

REGRESSION FOR EXAMPLE 5.2 WITH REDEFINED VARIABLES

Analysis of Variance

Source	DF	Sum of Squares	Mean Square	F Value	Prob>F
Model	7	1302.35790	186.05113	44.992	0.0001
Error	42	173.67978	4.13523		
C Total	49	1476.03768			

Root MSE	2.03353	R-square	0.8823
Dep Mean	0.97697	Adj R-sq	0.8627
C.V.	208.14732		

Parameter Estimates

Variable	DF	Parameter Estimate	Standard Error	T for H0: Parameter=0	Prob > |T|	Variance Inflation
INTERCEP	1	-0.063117	0.32376278	-0.195	0.8464	0.00000000
X1	1	19.219795	1.12235545	17.125	0.0001	1.36026904
X2D	1	21.237834	5.78294732	3.672	0.0007	1.22128356
X3D	1	13.154150	5.51866497	2.384	0.0217	1.17395677
X4D	1	9.321508	5.95608651	1.565	0.1251	1.17213414
X5D	1	12.572838	5.05049170	2.489	0.0168	1.14922038
X6D	1	-1.611701	5.67873577	-0.284	0.7779	1.15960538
X7D	1	-5.286719	5.61215537	-0.942	0.3516	1.09947949

EXAMPLE 5.4
REVISITED
The 12 independent variables in this example contain four pairs of variables that appear to contribute to the multicollinearity: the attempts and successes (or failures) for the two ways to make points (field goals and free throws) for the team and the opponent. We have noted that one reason for multicollinearity is that the more attempts at making points, the more are made. Of course, this is not an exact relationship, because teams have different proportions of successes. We will replace the number of field goals made (FGM) with the percentage of field goals made (100 $\times$ FGM/FGAT). We then do the same for FTM and OFGAT. The regression will then be done using the following variables:

1. The four attempt variables, used as is:
   FGAT = Attempted field goals
   FTAT = Attempted free throws
   OFGAT = Opponents attempted field goals
   OFTAT = Opponents attempted free throws
2. The four percentages of successes:
   FGPC = 100 $\times$ FGM/FGAT, percentage field goals made
   FTPC = 100 $\times$ FTM/FTAT, percentage free throws made
   OFGPC = 100 $\times$ OFGAL/OFGAT, percentage field goals allowed
   OFTPC = 100 $\times$ OFTAL/OFTAT, percentage free throws allowed
3. The two pairs of rebound variables:
   DR = Defensive rebounds
   DRA = Defensive rebounds allowed
   OR = Offensive rebounds
   ORA = Offensive rebounds allowed

The results of the regression of WINS on these variables are shown in Table 5.18.

Because we are using ratios rather than linear functions, the overall model statistics are not exactly the same as those of the original model; however, the fit of the model is essentially unchanged. The variance inflation factors of most variables have decreased markedly, although some are still large enough to suggest that additional multicollinearity exists. However, the decreased multicollinearity has increased the number of statistically significant coefficients, with four (instead of two) having p-values less than 0.0001 and four others (rather than one) having p-values less than 0.05.

An interesting result is that for both teams, the *percentage* of field goals and the *attempted* free throws are the most important factors affecting the score. Of more marginal importance are the number of attempted field goals of both teams and the percentage of attempted free throws by the opposing team. Also, all significant coefficients have the expected signs.   ◆

Methods Based on Statistical Analyses

When practical or intuitive redefinitions are not readily available, statistical analyses may reveal some useful redefinitions. Statistical analyses of relationships among a set of variables are a subset of the field of **multivariate analysis**. One of the simplest multivariate methods is **principal component** analysis,

TABLE 5.18

NBA REGRESSION WITH REDEFINED VARIABLES

Dependent Variable: WINS

Analysis of Variance

Source	DF	Sum of Squares	Mean Square	F Value	Prob>F
Model	12	3949.87658	329.15638	25.802	0.0001
Error	53	676.12342	12.75705		
C Total	65	4626.00000			

Root MSE	3.57170	R-square	0.8538	
Dep Mean	41.00000	Adj R-sq	0.8208	
C.V.	8.71147			

Parameter Estimates

Variable	DF	Parameter Estimate	Standard Error	T for H0: Parameter=0	Prob > |T|	Variance Inflation
INTERCEP	1	91.577172	74.63143145	1.227	0.2252	0.00000000
FGAT	1	0.016395	0.00843670	1.943	0.0573	16.62678256
FGPC	1	5.106327	0.99900002	5.111	0.0001	15.57357502
FTAT	1	0.017431	0.00402202	4.334	0.0001	4.01398023
FTPC	1	0.576731	0.26381734	2.186	0.0332	2.16568727
OFGAT	1	-0.014879	0.00765608	-1.943	0.0573	27.65005104
OFGPC	1	-5.475758	0.85624464	-6.395	0.0001	6.98285853
OFTAT	1	-0.016348	0.00397200	4.116	0.0001	5.11301768
OFTPC	1	-1.043793	0.51818342	-2.014	0.0491	1.26447241
DR	1	-0.011238	0.01135633	0.990	0.3269	13.46183963
DRA	1	0.004860	0.01279597	0.380	0.7056	17.32191542
OR	1	0.022881	0.01853993	1.234	0.2226	16.71826418
ORA	1	-0.019162	0.01398778	-1.370	0.1765	7.99412376

which has already been presented as a basis for variance proportions. In our study of variance proportions, we focused on the components having small variances. However, in principal component analysis per se, we focus on those principal components having large eigenvalues.

Remember that principal components consist of a set of uncorrelated variables produced by a linear transformation of the original standardized variables, that is,

$$Z = XV,$$

where Z is the matrix of principal component variables, X is the matrix of standardized original variables, and V, the matrix of eigenvectors, is the matrix of coefficients for the transformation. Because the original variables have been standardized, each has variance 1 and therefore each variable contributes equally to the total variability of the set of variables. The principal components, however, do not have equal variances. In fact, by construction, the first component has the maximum possible variance, the second has the second largest variance, and so forth. Therefore, principal components with large variances contribute more to the total variability of the model than those with smaller variances.

The columns of V are coefficients that show how the principal component variables are related to the original variables. These coefficients may allow useful interpretations, and if they do, a regression using these uncorrelated variables may provide a useful regression with independent variables that have no multicollinearity.

Remember that the bivariate sample with correlation of 0.9 produced principal components with sample variances of 1.922 and 0.078. This means that the first component, z_i, accounts for 1.922/2.0 = 0.961 or 96.1% of the total variability of the two variables. This is interpreted as saying that virtually all of the variability is contained in one dimension. The variable z_i is often called a **factor**. In situations where there are several variables, it is of interest to do the following:

1. See how many factors account for most of the variability. This generally (but not always) consists of principal components having variances of 1 or greater.
2. Examine the coefficients of the transformations (the eigenvectors) to see if the components with large variances have any interpretations relative to the definitions of the original variables.

EXAMPLE 5.3 REVISITED

Principal Components   The results of performing a principal component analysis on the data for Example 5.3 are shown in Table 5.19 as produced by PROC PRINCOMP of the SAS System. The "factors" are labeled PRIN1 through PRIN7 in decreasing order of the eigenvalues. The column labeled "Difference" is the difference between the current and next largest eigenvalue; the column labeled "Proportion" is the proportion of total variation (which is $m = 7$, because each variable has unit variance) accounted for by the current component, and the column labeled "Cumulative" is the sum of proportions up to the current one.

The results show that the first three eigenvalues are much larger than the rest. In fact, the cumulative proportions (last column) show that these three components account for over 97% of the total variation, implying that this set of seven variables essentially has only three dimensions or factors. This result confirms that the data were generated to have three uncorrelated sets of variables.

The eigenvectors are the coefficients of the linear equations relating the components to the original variables. These show the following:

1. The first component is an almost equally weighted function of the first six variables, with slightly larger coefficients for the first three.
2. The second component consists of the differences between the first and second set of three variables.
3. The third component is almost entirely a function of variable seven.

These results do indeed identify three factors:

1. Factor 1 is an overall score that implies that the first six variables are correlated.

TABLE 5.19

PRINCIPAL COMPONENTS FOR EXAMPLE 5.3

Eigenvalues of the Correlation Matrix

	Eigenvalue	Difference	Proportion	Cumulative
PRIN1	3.47296	1.06023	0.496138	0.49614
PRIN2	2.41274	1.46412	0.344677	0.84081
PRIN3	0.94862	0.86164	0.135517	0.97633
PRIN4	0.08698	0.05118	0.012425	0.98876
PRIN5	0.03580	0.00560	0.005114	0.99387
PRIN6	0.03020	0.01749	0.004314	0.99818
PRIN7	0.01271	.	0.001815	1.00000

Eigenvectors

NO	PRIN1	PRIN2	PRIN3	PRIN4	PRIN5	PRIN6	PRIN7
X1	0.425602	-.375712	-.102102	0.054059	-.590171	0.460425	-.322589
X2	0.422062	-.388053	-.097393	0.069024	0.094833	-.052110	0.803318
X3	0.438769	-.354859	-.101187	-.076926	0.520381	-.387786	-.494192
X4	0.351059	0.463049	-.068805	0.726602	0.256052	0.249027	-.045612
X5	0.395749	0.424155	0.019161	-.062730	-.503004	-.636901	0.022746
X6	0.390176	0.418592	0.053939	-.667663	0.230328	0.408846	0.060446
X7	0.126929	-.113086	0.980735	0.093468	0.016789	0.010178	-.011767

2. Factor 2 is the difference between the two sets of correlated variables.
3. Factor 3 is variable 7.

Among these factors, factor 1 by itself does not correspond to the pattern that generated the variables, although it may be argued that in combination with factor 2 it does. This result illustrates the fact that principal components are not guaranteed to have useful interpretation. ◆

Because principal components do not always present results that are easily interpreted, additional methods have been developed to provide more useful results. These methods fall under the topic generally called **factor analysis**. Most of these methods start with principal components and use various geometric rotations to provide for better interpretation. Presentation of these methods is beyond the scope of this book. A good discussion of factor analysis can be found in Johnson and Wichern (1988).

Principal Component Regression

If a set of principal components has some useful interpretation, it may be possible to use the component variables as independent variables in a regression. That is, we use the model

$$Y = Z\gamma + \epsilon,$$

where γ is the vector of regression coefficients.[7] The coefficients are estimated by least squares:

$$\hat{\gamma} = (Z'Z)^{-1} Z'Y.$$

Since the principal components are uncorrelated, $Z'Z$ is a diagonal matrix, and the variances of the regression coefficients are not affected by multicollinearity.[8] Table 5.20 shows the results of using the principal components for Example 5.3 in such a regression.

The results have the following features:

1. The model statistics are identical to those of the original regression because the principal components are simply a linear transformation using all of the information from the original variables.
2. The only clearly significant coefficients are for components 1 and 2, which together correspond to the structure of the variables. Notice that the

TABLE 5.20

PRINCIPAL COMPONENT REGRESSION, EXAMPLE 5.3

Analysis of Variance

Source	DF	Sum of Squares	Mean Square	F Value	Prob>F
Model	7	790.72444	112.96063	44.035	0.0001
Error	42	107.74148	2.56527		
C Total	49	898.46592			

Root MSE	1.60165	R-square	0.8801	
Dep Mean	0.46520	Adj R-sq	0.8601	
C.V.	344.29440			

Parameter Estimates

Variable	DF	Parameter Estimate	Standard Error	T for H0: Parameter=0	Prob > |T|
INTERCEP	1	0.465197	0.22650710	2.054	0.0463
PRIN1	1	2.086977	0.12277745	16.998	0.0001
PRIN2	1	-0.499585	0.14730379	-3.392	0.0015
PRIN3	1	0.185078	0.23492198	0.788	0.4352
PRIN4	1	0.828500	0.77582791	1.068	0.2917
PRIN5	1	-1.630063	1.20927168	-1.348	0.1849
PRIN6	1	1.754242	1.31667683	1.332	0.1899
PRIN7	1	-3.178606	2.02969950	-1.566	0.1248

[7] The principal component variables have zero mean; hence, the intercept is μ and is separately estimated by $\bar{y}$. If the principal component regression also uses the standardized dependent variable, then the intercept is zero.

[8] Alternatively, we can compute $\hat{\gamma} = V\hat{B}$, where $\hat{B}$ is the vector of regression coefficients using the standardized independent variables.

coefficient for component 3 (which corresponds to the "lone" variable X7) is not significant.

EXAMPLE 5.4
REVISITED

NBA Data, Principal Component Regression   Table 5.21 shows the results of the principal component analysis of the NBA data, again provided by PROC PRINCOMP.

Because this is a "real" data set, the results are not as obvious as those for the artificially generated Example 5.3. It appears that the first six components are of importance, as they account for almost 94% of the variability. The coefficients of these components do not allow very clear interpretation, but the following tendencies are of interest:

1. The first component is largely a positive function of opponents' field goals and a negative function of opponents' free throws. This may be considered as a factor describing opponent teams' prowess on the court rather than on the free throw line.
2. The second component is similarly related to the team's activities on the court as opposed to on the free-throw line.
3. The third component stresses team and opponent free throws, with some additional influence of offensive rebounds allowed. This could describe the variation in total penalties in games.
4. The fourth component is a function of defensive rebounds and a negative function of field goals made, and may describe the quality of the defense.
5. The fifth component is almost entirely a function of offensive rebounds.

If we are willing to accept that these components have some useful interpretation, we can perform a regression using them. The results are shown in Table 5.22. We first examine the coefficients for the components with large variances.

An interesting feature of this regression is that the three most important coefficients (having the smallest *p*-values) relate to components 6, 7, and 8, which have quite small variances and would normally be regarded as relatively "unimportant" components. This type of result is not overly common, as usually the most important components tend to produce the most important regression coefficients. However, because of this result, we will need to examine these components and interpret the corresponding coefficients.

1. Component 6 is a positive function of field goals made and defensive rebounds, and a negative function of offensive rebounds allowed. Its positive and strong contribution to the number of wins appears to make sense.
2. Component 7 is a positive function of opponents' field goals allowed and defensive rebounds allowed, and a negative function of offensive rebounds allowed. The negative regression coefficient does make sense.
3. Component 8, which results in the most significant and negative coefficient, is a negative function of all home-team scoring efforts and a positive function of all opponent scoring efforts. Remembering that a double negative is positive, this one seems obvious.

TABLE 5.21

PRINCIPAL COMPONENT ANALYSIS

Principal Component Analysis
Eigenvalues of the Correlation Matrix

	Eigenvalue	Difference	Proportion	Cumulative
PRIN1	3.54673	1.15442	0.295561	0.29556
PRIN2	2.39231	0.29241	0.199359	0.49492
PRIN3	2.09991	0.46609	0.174992	0.66991
PRIN4	1.63382	0.65559	0.136151	0.80606
PRIN5	0.97823	0.38231	0.081519	0.88758
PRIN6	0.59592	0.14903	0.049660	0.93724
PRIN7	0.44689	0.24606	0.037240	0.97448
PRIN8	0.20082	0.14945	0.016735	0.99122
PRIN9	0.05138	0.00773	0.004282	0.99550
PRIN10	0.04365	0.03733	0.003638	0.99914
PRIN11	0.00632	0.00230	0.000527	0.99966
PRIN12	0.00403	.	0.000336	1.00000

Eigenvectors

	PRIN1	PRIN2	PRIN3	PRIN4	PRIN5	PRIN6
FGAT	0.180742	0.476854	0.266317	0.139942	0.307784	0.169881
FGM	0.277100	0.310574	0.106548	-.440931	0.109505	0.405798
FTAT	0.090892	-.483628	0.417400	-.003158	-.013379	0.039545
FTM	0.101335	-.489356	0.382850	-.087789	-.144502	0.098000
OFGAT	0.510884	0.040398	0.090635	0.089731	-.079677	0.062453
OFGAL	0.443168	0.020078	0.066332	-.281097	0.045099	-.170165
OFTAT	-.358807	0.169955	0.434395	-.116152	-.135725	0.171628
OFTAL	-.369109	0.175132	0.419559	-.116617	-.109270	0.157648
DR	0.242131	-.102177	-.029127	0.535598	-.178294	0.602160
DRA	-.018575	0.309761	0.180227	0.526906	-.335944	-.258506
OR	-.063989	-.129132	0.236684	0.315635	0.812732	-.116822
ORA	0.294223	0.134651	0.360877	0.001297	-.168208	-.514845

	PRIN7	PRIN8	PRIN9	PRIN10	PRIN11	PRIN12
FGAT	0.248934	-.222913	-.415273	-.202644	-.309409	-.323937
FGM	-.101451	-.395466	0.344674	0.260927	0.227146	0.197888
FTAT	0.160719	-.250823	0.501105	-.390179	-.193585	-.228698
FTM	0.180189	-.190206	-.566482	0.359895	0.156944	0.147620
OFGAT	-.008903	0.271516	-.123055	-.575011	0.312988	0.439669
OFGAL	0.403500	0.571521	0.164850	0.332933	-.145131	-.200655
OFTAT	-.049914	0.287093	0.048670	0.011021	-.507408	0.497431
OFTAL	-.054739	0.320340	0.001115	-.095745	0.563739	-.417231
DR	-.288428	0.209786	0.093180	0.251060	-.123006	-.186518
DRA	0.430931	-.203078	0.257780	0.221353	0.199654	0.177923
OR	-.104360	0.130140	0.124897	0.183196	0.175397	0.204092
ORA	-.649060	-.085094	-.036331	0.112823	-.102560	-.136459

TABLE 5.22

PRINCIPAL COMPONENT RERGRESSION, NBA DATA

Analysis of Variance

Source	DF	Sum of Squares	Mean Square	F Value	Prob>F
Model	12	3968.07768	330.67314	26.638	0.0001
Error	53	657.92232	12.41363		
C Total	65	4626.00000			

Root MSE	3.52330	R-square	0.8578	
Dep Mean	41.00000	Adj R-sq	0.8256	
C.V.	8.59341			

Parameter Estimates

Variable	DF	Parameter Estimate	Standard Error	T for H0: Parameter=0	Prob > |T|
INTERCEP	1	41.000000	0.43368800	94.538	0.0001
PRIN1	1	1.100603	0.23204835	4.743	0.0001
PRIN2	1	-1.044853	0.28254255	-3.698	0.0005
PRIN3	1	-0.154270	0.30157323	-0.512	0.6111
PRIN4	1	-0.939146	0.34189366	-2.747	0.0082
PRIN5	1	1.076344	0.44184815	2.436	0.0182
PRIN6	1	5.430269	0.56610946	9.592	0.0001
PRIN7	1	-4.184893	0.65372432	-6.402	0.0001
PRIN8	1	-11.124680	0.97517988	-11.408	0.0001
PRIN9	1	0.213121	1.92798218	0.111	0.9124
PRIN10	1	-1.489054	2.09167212	-0.712	0.4797
PRIN11	1	2.746444	5.49527617	0.500	0.6193
PRIN12	1	16.642533	6.88503526	2.417	0.0191

We now continue with the most important components.

4. The first component, which measures the opponent team's field activity as against its free-throw activity, has a significant and positive coefficient ($P < 0.005$). When considered in light of the effects of component 8, it may indicate that when opponents have greater on-court productivity as opposed to free-throw productivity, it helps the "home" team.

5. The second component, which is similar to the first component as applied to the "home" team, has a significant negative coefficient. It appears that these two components mirror each other.

6. Component 3, which relates to free throws by both teams, does not produce a significant coefficient.

7. Component 4 indicates that defensive rebounds, both made and allowed, as well as fewer field goals made, contribute positively to the number of

wins. Fortunately, the small value of the coefficient indicates this puzzler is not too important.

8. Component 5, associated with offensive rebounds, is positive.

It is fair to say that although the results do make sense, especially in light of apparent interplay of component 8 with 1 and 2, the results certainly are not clear-cut. This type of result occurs often with principal component analyses and is the reason for the existence of other factor analysis methods. However, it is recommended that these methods be used with caution as many of them involve subjective choices of transformations (or rotations), and therefore p-values must be used carefully. ◆

EXAMPLE 5.5   **Mesquite Data**   Mesquite is a thorny bush that grows in the Southwestern U.S. Great Plains. Although the use of mesquite chips enhances the flavor of barbecue, its presence is very detrimental to livestock pastures. Eliminating mesquite is very expensive, so it is of interest to have a method to estimate the total biomass of mesquite in a pasture. One way to do this is to obtain certain easily measured characteristics of mesquite in a sample of bushes and use these to estimate biomass:

DIAM1: The wider diameter
DIAM2: The narrower diameter
TOTHT: The total height
CANHT: The height of the canopy
DENS: A measure of the density of the bush

The response is:

LEAFWT: A measure of biomass

Table 5.23 contains data on these measures from a sample of 19 mesquite bushes.

The linear regression of leaf weight on the five measurements produces the results shown in Table 5.24. The regression is certainly significant, with a p-value of less than 0.0001, and the coefficient of determination is reasonably large. However, the residual standard deviation of 180 is quite large compared with the mean leaf weight of 548; hence, there is considerable variability about the estimated conditional means. When we turn to the coefficients, we see that the variance inflation factors are not extremely large, yet the smallest p-value for any coefficient is 0.014, indicating that the multicollinearity is affecting the precision of the estimated coefficients. This result is due to the fact that the largest variance inflation factor of 6.0 implies a coefficient of determination of 0.83 for the relationship of TOTHT to the other variables, which is about the same as the coefficient of determination for the regression model. In other words, the relationships among the independent variables are essentially as strong as the regression relationship. As we have noted, the effect of multicollinearity is to some degree relative to the strength of the regression; hence, it does have some effect in this example.

If the purpose of the analysis is to simply assess the feasibility of estimating leaf weight, the lack of useful coefficient estimates is not of great concern.

TABLE 5.23

MESQUITE DATA

OBS	DIAM1	DIAM2	TOTHT	CANHT	DENS	LEAFWT
1	2.50	2.3	1.70	1.40	5	723.0
2	2.00	1.6	1.70	1.40	1	345.0
3	1.60	1.6	1.60	1.30	1	330.9
4	1.40	1.0	1.40	1.10	1	163.5
5	3.20	1.9	1.90	1.50	3	1160.0
6	1.90	1.8	1.10	0.80	1	386.6
7	2.40	2.4	1.60	1.10	3	693.5
8	2.50	1.8	2.00	1.30	7	674.4
9	2.10	1.5	1.25	0.85	1	217.5
10	2.40	2.2	2.00	1.50	2	771.3
11	2.40	1.7	1.30	1.20	2	341.7
12	1.90	1.2	1.45	1.15	2	125.7
13	2.70	2.5	2.20	1.50	3	462.5
14	1.30	1.1	0.70	0.70	1	64.5
15	2.90	2.7	1.90	1.90	1	850.6
16	2.10	1.0	1.80	1.50	2	226.0
17	4.10	3.8	2.00	1.50	2	1745.1
18	2.80	2.5	2.20	1.50	1	908.0
19	1.27	1.0	0.92	0.62	1	213.5

TABLE 5.24

REGRESSION FOR MESQUITE DATA

```
                        Analysis of Variance

                        Sum of          Mean
        Source      DF  Squares         Square      F Value   Prob>F

        Model        5 2774582.7673 554916.55347    17.142    0.0001
        Error       13 420823.11897  32371.00915
        C Total     18 3195405.8863

           Root MSE    179.91945   R-square   0.8683
           Dep Mean    547.54211   Adj R-sq   0.8177
           C.V.         32.85947

                        Parameter Estimates

                    Parameter    Standard    T for H0:                 Variance
    Variable   DF   Estimate     Error       Parameter=0   Prob > |T|  Inflation

    INTERCEP   1   -633.944717  174.89405467   -3.625       0.0031     0.00000000
    DIAM1      1    421.214437  147.45417084    2.857        0.0135     5.89602328
    DIAM2      1    179.019942  125.43117278    1.427        0.1771     4.57677978
    TOTHT      1     13.116884  245.54250956    0.053        0.9582     6.01963153
    CANHT      1   -110.427969  287.77343582   -0.384        0.7074     5.03118745
    DENS       1     -0.190209   31.07269108   -0.006        0.9952     1.36573723
```

However, if we also wish to study how the various measurements affect leaf weight, the results of the regression are not particularly useful, and we will want to try some remedial methods.

We start with a principal component regression. The results of the principal component analysis is shown in Table 5.25, which also includes the correlation matrix for the independent variables.

The correlation matrix shows high correlations among the two diameter and two height variables and somewhat lower correlations between the diameter and height variables. Density appears to be uncorrelated with all other variables. The eigenvalues show one very large eigenvalue and two others that may have some importance. The eigenvectors allow the following interpretations of the principal components:

1. The first component is a function of all size variables. This is the size factor and simply shows that larger bushes have larger dimensions. This very obvious factor also accounts for about two-thirds of the total variability.

TABLE 5.25

PRINCIPAL COMPONENTS FOR THE MESQUITE DATA

Principal Component Analysis
Correlation Matrix

	DIAM1	DIAM2	TOTHT	CANHT	DENS
DIAM1	1.0000	0.8767	0.7255	0.6835	0.3219
DIAM2	0.8767	1.0000	0.6354	0.5795	0.2000
TOTHT	0.7255	0.6354	1.0000	0.8794	0.3943
CANHT	0.6835	0.5795	0.8794	1.0000	0.2237
DENS	0.3219	0.2000	0.3943	0.2237	1.0000

Eigenvalues of the Correlation Matrix

	Eigenvalue	Difference	Proportion	Cumulative
PRIN1	3.33309	2.44433	0.666619	0.66662
PRIN2	0.88877	0.31655	0.177753	0.84437
PRIN3	0.57221	0.45672	0.114443	0.95881
PRIN4	0.11550	0.02507	0.023100	0.98191
PRIN5	0.09043	.	0.018086	1.00000

Eigenvectors

	PRIN1	PRIN2	PRIN3	PRIN4	PRIN5
DIAM1	0.501921	-.124201	0.382765	-.624045	-.443517
DIAM2	0.463408	-.261479	0.563632	0.471797	0.420247
TOTHT	0.501393	0.045730	-.422215	0.521831	-.544004
CANHT	0.474064	-.138510	-.580993	-.339074	0.550957
DENS	0.239157	0.946006	0.141387	-.026396	0.164894

2. The second component represents the density, which the correlations showed to be an independent factor.
3. The third component has positive coefficients for the diameters and negative coefficients for the heights. Values of this component variable will increase with diameters and decrease with height; thus, it may be called a "fatness" component.

The results of the principal component regression are shown in Table 5.26. The regression statistics show that the size component is by far the strongest, but the fatness component is also important and indicates that short, fat bushes have more leaf weight. Density has no effect, as was seen in the original regression.

The principal component regression suggests that we may have limited success with a set of redefined variables. Redefined variables based on knowledge are usually preferred, as they are more easily interpretable functions of specific variables, in contrast to the principal components, which involve all variables, albeit sometimes with small coefficients. We will try the following variables:

SIZE = DIAM1 + DIAM2 + TOTHT + CANHT, a measure of overall size

FAT = DIAM1 + DIAM2 − TOTHT − CANHT, a measure of fatness

TABLE 5.26

PRINCIPAL COMPONENT REGRESSION FOR MESQUITE DATA

Analysis of Variance

Source	DF	Sum of Squares	Mean Square	F Value	Prob>F
Model	5	2774582.7673	554916.55347	17.142	0.0001
Error	13	420823.11897	32371.00915		
C Total	18	3195405.8863			

Root MSE	179.91945	R-square	0.8683
Dep Mean	547.54211	Adj R-sq	0.8177
C.V.	32.85947		

Parameter Estimates

Variable	DF	Parameter Estimate	Standard Error	T for H0: Parameter=0	Prob > |T|
INTERCEP	1	547.542105	41.27635308	13.265	0.0001
PRIN1	1	193.054335	23.22834006	8.311	0.0001
PRIN2	1	-65.368191	44.98293892	-1.453	0.1699
PRIN3	1	204.389045	56.06126485	3.646	0.0030
PRIN4	1	-107.186966	124.78283609	-0.859	0.4059
PRIN5	1	-99.228970	141.02247638	-0.704	0.4941

> OBLONG = DIAM1 − DIAM2, a measure of how oblong, as opposed to circular, is the shape of the bush
>
> HIGH = TOTHT − CANHT, a measure of the size of the stem portion of the bush[9]

The results of the regression using these four variables and DENS are shown in Table 5.27.

Because we have used a complete set of linear transformations, the model statistics are identical. The variance inflations show that there is still some multicollinearity, but it is weaker. The results agree with those of the principal component regression, although the coefficients are not as strong in terms of statistical significance. In other words, the principal component regression produced stronger coefficients, which, however, are not as readily interpretable.   ◆

Biased Estimation

The sampling distribution of a statistic is used to assess the usefulness of a statistic as an estimate of a parameter. The mean of the sampling distribution of the statistic, called the *expected value,* indicates how well, on the average, the sta-

TABLE 5.27

REGRESSION WITH REDEFINED VARIABLES FOR MESQUITE DATA

Analysis of Variance

Source	DF	Sum of Squares	Mean Square	F Value	Prob>F
Model	5	2774582.7673	554916.55347	17.142	0.0001
Error	13	420823.11897	32371.00915		
C Total	18	3195405.8863			

Root MSE	179.91945	R-square	0.8683	
Dep Mean	547.54211	Adj R-sq	0.8177	
C.V.	32.85947			

Parameter Estimates

Variable	DF	Parameter Estimate	Standard Error	T for H0: Parameter=0	Prob > |T|	Variance Inflation
INTERCEP	1	-633.944717	174.89405467	-3.625	0.0031	0.00000000
SIZE	1	125.730823	34.06428655	3.691	0.0027	2.47971084
FAT	1	174.386366	61.35208837	2.842	0.0139	2.13528111
OBLONG	1	121.097247	129.28960471	0.937	0.3660	1.16344094
HIGH	1	61.772427	252.43906664	0.245	0.8105	1.50442281
DENS	1	-0.190209	31.07269108	-0.006	0.9952	1.36573723

[9] Products and ratios could easily have been used and may be more easily justified.

tistic approximates the parameter. The standard deviation of that distribution, called the standard error of the estimate, indicates the precision of the estimate. If the expected value is the same as the value of the parameter, the estimate is said to be *unbiased*, and the smaller the standard error, the more precise it is.

In many cases there exist several alternative procedures, called *estimators*, for obtaining an estimate of a parameter. The effectiveness of different estimators is evaluated on the basis of the expected values and standard errors of their sampling distributions.

We have made almost exclusive use of the least squares estimator, which is known to produce unbiased estimates, and among unbiased estimators, it can be shown to produce the smallest standard errors. Thus, least squares is known as the best unbiased estimator.[10] However, other estimators may be used. For example, the median can be used as an alternative to the mean as a measure of central tendency. In Section 4.4 we introduced the *M* estimator as an alternative to the least squares estimator to reduce the effect of outliers. The reason for using alternative estimators is that their sampling distributions may have some attractive features.

For example, suppose that we want to estimate a parameter whose value is 100. Two estimators are proposed; one is least squares and the other is something else. Their respective sampling distributions appear in Figure 5.2.

The sampling distribution of the least squares estimator is shown by the solid line. It is seen to be unbiased because its mean is indeed 100, and the

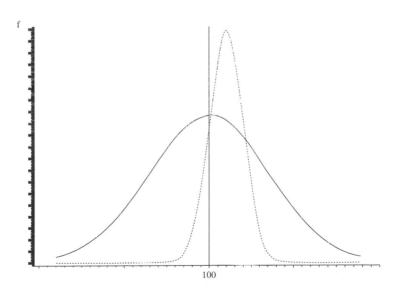

100

FIGURE 5.2 Unbiased and Biased Estimators

---

[10] Actually, it is known as the best linear unbiased estimator (BLUE) because it is a linear function of the response variable. The distinction is of no importance here.

standard error is 10. The alternative estimator is seen to be biased; its mean is 104, but its standard error is smaller: it is 4. The question is Which of these estimates is "better"?

A widely used statistic for comparing estimators is their **mean squared error**, which is defined as the variance plus the square of the bias. In the example illustrated in Figure 5.2, the least squares estimator has a variance of $10^2 = 100$ and there is no bias; hence,

$$\text{Mean squared error} = 100 + 0 = 100.$$

For the alternative estimator the variance is $4^2 = 16$, and the bias is 4; hence,

$$\text{Mean squared error} = 16 + 16 = 32.$$

Therefore, we may conclude that the biased estimator is "better," although the usefulness of this result may depend on different penalties for incorrect answers.

Recent research in remedial methods for combating the effects of multicollinearity has resulted in developing biased estimators for partial regression coefficients that have smaller standard errors. The reduction is obtained by artificially reducing multicollinearity. Such a procedure, which effectively alters the data, results in estimates that are biased. However, this research has also found that some of these estimators do have smaller mean squared errors than does the least squares estimator. Unfortunately, the magnitudes of the mean squared errors of these estimators are functions of the unknown parameters; hence, it is not possible to know if a particular biased estimator does indeed provide that smaller mean squared error. Therefore, results obtained by these biased estimation methods should be viewed as exploratory rather than confirmatory.

Two popular biased estimation methods for multiple regression are **ridge regression** and **incomplete principal component regression**. We present the basic formulation for both of these, but since they tend to give similar results, we present an example of only the incomplete principal component estimator.

Ridge Regression

Define X and Y as the matrices of *standardized* independent and dependent variables. The least squares estimator of the regression coefficients is

$$\hat{B} = (X'X)^{-1}X'Y.$$

Note that we do not need the dummy variable for the intercept, which is 0 for the model with standardized variables. The ridge regression estimator is

$$\hat{B}_k = (X'X + kI)^{-1}X'Y,$$

where kI is a diagonal matrix with all elements consisting of an arbitrary small constant k. All other statistics are calculated as usual, with the exception that the inverse $(X'X + kI)^{-1}$ is used in place of $(X'X)^{-1}$.

Remember that when all variables are standardized, $X'X$ is the correlation matrix with the diagonal elements 1 and the off-diagonal elements the simple correlations, r_{ij}. However, for the ridge regression estimator, the diagonal elements of $X'X$ are $(1 + k)$, and hence the "effective" correlation between x_i and x_j now becomes

$$\frac{r_{ij}}{1 + k},$$

where r_{ij} is the sample correlation between x_i and x_j. In other words, all correlations are artificially reduced by the factor, $1/(1 + k)$, thereby reducing the multicollinearity. Larger values of k reduce multicollinearity, but increase the bias, whereas a k of zero reproduces the least squares estimates. The question becomes one of determining what value of k should be used.

The formulas for calculating the value of k that minimizes the total mean squared error of the set of regression coefficients are functions of the unknown values of the population coefficients. Using the least squares coefficients in these formulas is not advised because multicollinearity makes these estimates unreliable. The most commonly used procedure is to calculate the ridge regression coefficients for a set of values of k and plot the resulting regression coefficients against k. These plots, called ridge plots, often show large changes in the estimated coefficients for smaller values of k, which, as k increases, ultimately "settle down" to a steady progression toward zero. An optimum value for k is said to occur when these estimates appear to "settle down." This procedure is obviously quite subjective, but has been used to obtain useful estimates.

Incomplete Principal Component Regression

We have already seen that we can use principal components to provide a regression that may furnish insight into the structure of the regression model. When the regression uses all of the principal components, which we have done, the overall model is the same as that using the original variables; that is, it produces the same estimates of the response.

We already know that the principal components corresponding to very small eigenvalues describe linear combinations of the independent variables that are "almost" linearly dependent. In many, but not all, cases, these components do not contribute to the overall fit of the model. Hence, a model without these components could be useful. However, since the components do not correspond to the original variables, the resulting regression is not useful for inferences about the role of the various independent variables. It would, however, be useful to see what this reduced model implies about the regression coefficients for the original variables.

Principal components are obtained through a linear transformation of the standardized independent variables,

$$Z = XV,$$

where Z is the matrix of principal components and V is the matrix of eigenvectors that are the coefficients of the linear transformation. In principal component regression, we obtain the estimated model equation

$$\hat{Y} = Z\hat{\gamma},$$

where $\hat{\gamma}$ is the set of principal component regression coefficients. Because this regression is the same as when the original variables have been used, it can be seen that

$$\hat{Y} = Z\hat{\gamma} = X\hat{B},$$

and

$$\hat{B} = V\hat{\gamma}.$$

Incomplete principal component regression uses the above relationship, omitting columns of V corresponding to principal components with very small variances that also do not contribute to the regression. That is, if these criteria suggest keeping only the first p coefficients, then the estimated coefficients for the original variables can be calculated as

$$\hat{B}_p = V_p\hat{\gamma}_p,$$

where V_p is the matrix containing the first p columns of V and $\hat{\gamma}_p$ is a matrix containing the first p principal component regression coefficients. The variances of the estimated coefficients are the diagonal elements of

$$\mathrm{Var}(\hat{B}_p) = \sigma^2[V_p(Z_p'Z_p)^{-1}V_p'],$$

where Z_p is the matrix containing the first p principal component variables. Normally the mean square from the full regression is used as the estimate of σ^2, and the square root of the resulting variance is used as a standard error to provide t statistics. For comparison and interpretive purposes, the coefficients can be converted back to reflect the original units by multiplying the coefficients by the ratios of the respective standard deviations.[11]

EXAMPLE 5.3 **A Sampling Experiment** In order to provide some ideas on the results to be
REVISITED expected from an incomplete principal component regression, we have computed both the ordinary least squares and the incomplete principal component regressions deleting the last four principal components for 250 samples from the population specified for Example 5.3. We delete four (keeping three) because in this example the eigenvalues reflect the known fact that there are three distinct sets of variables. Table 5.28 gives population coefficients (BETA) and the means, standard deviations, and standard errors of the mean of the resulting empirical sampling distributions of both estimates.

Using the standard errors, we can see that the means of the least squares coefficients are all within the 0.95 confidence intervals of the population values, and the standard deviations clearly show the effects of the high correlations among X1, X2, and X3, the more moderate correlations among X4, X5, and X6,

---

[11] Procedures for performing incomplete principal component regression as well as other biased estimation methods are currently not directly available in the SAS System, or in many other widely used statistical analysis software packages. Therefore, these methods are most readily performed with the aid of matrix manipulation programs, such as PROC IML of the SAS System, which is the method used to obtain the results in this chapter.

TABLE 5.28

SAMPLING DISTRIBUTIONS OF REGRESSION COEFFICIENTS, EXAMPLE 5.3

VARIABLE	BETA	LEAST SQUARES MEAN	STD	STD ERR	INCOMPLETE PRINC. COMP. MEAN	STD	STD ERR
X1	4.0	4.01	5.10	0.32	3.46	0.32	0.020
X2	3.5	3.21	7.47	0.47	3.49	0.33	0.021
X3	3.0	3.20	5.36	0.34	3.46	0.34	0.021
X4	2.5	2.43	2.85	0.18	1.99	0.40	0.025
X5	2.0	2.22	4.23	0.26	2.02	0.40	0.025
X6	1.5	1.31	3.56	0.23	1.98	0.41	0.026
X7	1.0	1.12	1.10	0.07	1.13	1.02	0.064

and the lack of correlations with X7. This confirms the unbiasedness and large variances of the least squares estimates.

The incomplete principal components coefficients show an interesting pattern. There are now three distinct groups: the first three, the second three, and X7. The coefficients within the two groups of correlated variables are equivalent in all respects, and that for X7 is essentially the least squares estimate. In other words, this method of estimation recognizes the three groups of distinct variables, but cannot distinguish among those within a group of correlated variables.

These results are, of course, due to our knowledge of the structure of the data helping us to choose the number of components to be deleted. In fact, deleting only three does not provide nearly as informative results. Because the number of "effective" components is rarely known in practical cases, the question of how many components to delete is not as easily answered. In fact, some authorities (e.g., Rawlings, 1988, p. 348) caution against eliminating too many components.[12] For example, there are no very small eigenvalues in the principal components obtained for Example 5.5, and therefore an incomplete principal component regression may not be suitable for this data set. ◆

EXAMPLE 5.4
REVISITED

Incomplete Principal Component Regression on the NBA Data The principal component analysis of the independent variables in the NBA data showed two very small eigenvalues and two others that were moderately small. Looking at the principal component regression, we see that of these four, component 12 does appear to be of some importance. It is generally suggested that components should be deleted only if the eigenvalues are very small *and* the components do not contribute to the regression. These results would seem to suggest that component 12 should be kept, and one should omit only 11, or possibly 9, 10, and 11. Because we are in an exploratory mode, we can actually perform all of these, as well as the elimination of 11 and 12. The reader may wish

---

[12] Considering the relatively low cost of computing, it may not be unreasonable to examine results of different numbers of deleted components.

TABLE 5.29

INCOMPLETE PRINCIPAL COMPONENT REGRESSION FOR NBA DATA

COEFF	LEAST SQUARES ESTIMATES			INCOMPLETE PRINCIPAL COMPONENT		
	BETAHAT	STD ERR	T	BETAHAT	STD ERR	T
FGAT	-0.0178	0.0139	-1.2824	0.0114	0.0046	2.4954
FGM	0.0698	0.0132	5.3054	0.0452	0.0062	7.2695
FTAT	-0.0036	0.0104	-0.3417	0.0161	0.0059	2.7179
FTM	0.0277	0.0114	2.4242	0.0104	0.0082	1.2709
OFGAT	0.0220	0.0122	1.8089	-0.0049	0.0041	-1.1717
OFGAL	-0.0751	0.0114	-6.5899	-0.0524	0.0060	-8.6792
OFTAT	0.0165	0.0176	0.9364	-0.0108	0.0014	-7.5928
OFTAL	-0.0436	0.0221	-1.9769	-0.0156	0.0022	-7.1323
DR	-0.0138	0.0114	-1.2098	0.0103	0.0052	1.9931
DRA	0.0073	0.0127	0.5765	-0.0170	0.0057	-3.0022
OR	0.0265	0.0185	1.4316	-0.0132	0.0063	-2.0927
ORA	-0.0214	0.0139	-1.5359	0.0071	0.0067	1.0670

to confirm that regressions that do not eliminate component 12 give results that are very similar to the ordinary least squares regression and are of little interest. We therefore present here the results of the incomplete principal component regression eliminating components 11 and 12.

Table 5.29 gives the estimates, standard errors, and t statistics for both ordinary least squares and the incomplete principal component regression using 10 components. It is immediately obvious that the standard errors of the estimated coefficients are much smaller: Instead of only two coefficients having t statistics greater than 2.5 (an admittedly arbitrary value), six now enjoy that distinction, with one additional one having a t value of 2.495. All coefficients have the expected signs; that is, positive for accomplishments of the team and negative for accomplishments of the opponents. Thus, it appears that the incomplete principal component regression has provided useful information on the importance of the individual independent variables.

5.5 Summary

This chapter considers a data condition called **multicollinearity**, which is defined as the existence of strong correlations among the independent variables of a regression model. Although multicollinearity is not strictly a violation of assumptions, its presence does decrease the precision of the estimated regression coefficients, thereby making interpretations more difficult.

The existence and nature of multicollinearity are studied by using variance inflation factors and variance proportions. Additional information can be obtained through the use of principal components, as well as other multivariate analyses that are not covered in this book.

Because multicollinearity is often caused by having too many independent variables in the model, it is tempting to "solve" the multicollinearity problem by using a statistically based variable selection procedure. Although this method will usually provide models with less multicollinearity, it may not provide insight into the structure of the regression relationships. And because variable selection is actually more useful when multicollinearities are not present, its presentation is deferred to Chapter 6.

In this chapter we presented several remedial methods that may be used to provide additional information about the structure of the regression relationships:

1. The use of information about known relationships to redefine variables in the model.
2. The use of the results of principal components to create a new set of independent variables. Similar procedures are available using other multivariate methods.
3. The use of biased estimation methods. The method of incomplete principal component regression is presented in some detail. Another method, known as ridge regression, is presented without examples.

The reader may be frustrated by the lack of definitive results of the use of these methods on the "real" data examples in this chapter. Unfortunately, this will be an almost universal occurrence, because we are dealing with a situation where the data are simply inadequate to obtain the desired results.

CHAPTER EXERCISES

In each of these exercises there is some degree of multicollinearity. For each exercise:

1. Determine the extent and nature of the multicollinearity.
2. Discuss possible practical reasons why the multicollinearity exists.
3. Implement at least one form of remedial action and determine if these methods provided more useful results.

1. In Exercise 2 of Chapter 3 we used data from a set of tournaments featuring professional putters. In that exercise we related players' "Player of the Year" point scores to various variables characterizing the players' performances. At the end of the tour there is an "auction" that places a "price" on the players. In this exercise we will estimate the price using a regression with the following independent variables:

TNMT: The number of tournaments participated in by the player
WINS: The number of tournaments won

AVGMON: Average money won per tournament

PNTS: The number of "Player of the Year" points accumulated by the player

ASA: The player's "adjusted" point average; the adjustment reflects the difficulties of the courses played

The data set is in File REG05P01.

2. In Exercise 1 of Chapter 3 we related gasoline mileage (MPG) to several characteristics of cars:

WT: Weight in pounds

ESIZE: Engine piston displacement in cubic inches

HP: Engine horsepower rating

BARR: Number of barrels in carburetor

There is already a modest amount of multicollinearity among these variables. For this exercise we add another variable:

TIME: Time taken to a quarter mile from a dead stop (in seconds)

This additional variable creates somewhat greater multicollinearity. The data set is in File REG05P02.

3. Pork bellies are a surprisingly valuable portion of a pig, as indicated by the fact that prices of pork bellies are frequently quoted in the media. This means that their price is a factor of the price of the pig carcass. Unfortunately, the weight of this item is not readily discerned from an examination of the whole carcass as it is in the process of being sold, and there is interest in being able to estimate pork-belly weight based on some more readily available carcass dimensions. We will therefore investigate a regression to estimate belly weight (BELWT) from the following measurements:

AVBF: Average of three measures of back fat thickness

MUS: A muscling score for the carcass; higher numbers mean relatively more muscle

LEA: The loin area

DEP: Average of three measures of fat opposite the tenth rib

LWT: Live weight of the carcass

CWT: Weight of the slaughtered carcass

WTWAT: A measure of specific gravity of the carcass

DPSL: Average of three determinations of the depth of the belly

LESL: Average of three measures of leanness of belly cross-sections

Obviously, many of these variables are correlated. The data on 48 carcasses are available in File REG05P03.

4. Water evaporation is a major concern in planning irrigation. Data are collected daily from June 6 through July 21 in a central Texas location on the following variables that may affect the amount of evaporation:

MAXAT: Maximum daily air temperature

MINAT: Minimum daily air temperature

AVAT: The integrated area under the daily air temperature curve, a measure of average air temperature

MAXST: Maximum daily soil temperature

MINST: Minimum daily soil temperature

AVST: The integrated area under the daily soil temperature curve, a measure of average soil temperature

MAXH: Maximum daily humidity

MINH: Minimum daily humidity

AVH: The integrated area under the daily humidity curve, a measure of average humidity

WIND: Total wind, measured in miles per day

The response is:

EVAP: Daily total evaporation from soil

The data are in File REG05P04. The natural grouping of variables (MAX, MIN, AV) of three factors would appear to provide for redefining variables, either by using common sense or principal components.

5. Data on the following variables are obtained from the U.S. Bureau of the Census *State and Metropolitan Area Data Book, 1986 (A Statistical Abstract Supplement)* and given in File REG05P05. A set of 53 primary metropolitan areas (PMSAs) are selected for this data set. The variables are as follows:

PMSA: Identification of PMSA

AREA: Land area in square miles

POP: Total population, 1980, in thousands

YOUNG: 1980 population ages 18–24 in thousands

DIV: Total number of divorces in 1982 in thousands

OLD: 1982 total number of Social Security benefit recipients in thousands

EDUC: Number of adults, 25 years or older, having completed 12 or more years of school

POV: Total number of persons below poverty level in 1979

UNEMP: Total number unemployed, 1980

CRIME: Total number of serious crimes in 1980

The purpose of the analysis is to determine factors affecting crime. The independent variables as presented have a high degree of multicollinearity. There is, however, a good reason for this and determining the reason before performing any analyses should be helpful in developing remedial action.

6. Exercise 4 of Chapter 4 dealt with factors affecting state expenditures on criminal activities (courts, police, etc.):

STATE: The standard two-letter abbreviation (DC is included)

EXPEND: State expenditures on criminal activities ($1000)

BAD: The number of persons under criminal supervision

CRIME: Crime rate per 100,000

LAWYERS: The number of lawyers in the state

EMPLOY: The number of persons employed in the state
POP: The population of the state (1000)

The data are available in File REG04P04. Examine for and, if necessary, implement remedial procedures for multicollinearity both with and without possible outliers and/or influential observations found in that data set. Comment on results.

7. Freund and Wilson (1997) give an example where the size of squid eaten by sharks and tuna is to be estimated based on dimensions of the beaks (mouth). The beaks are indigestible and can be recovered and used to estimate the weight of the squid. To do this, a sample of 22 specimens was taken and the following measurements taken:

WT: Weight (response variable)
RL: Rostral length
WL: Wing length
RNL: Rostral to notch length
NWL: Notch to wing length
W: Width

Because of the nature of the response variable, it was decided that natural logarithm of the original data would be appropriate for the analysis. The data are given in Table 5.30 and File REG05P07.

TABLE 5.30

DATA FOR EXERCISE 7

OBS	RL	WL	RNL	NWL	W	WT
1	0.27003	0.06766	−0.82098	−0.28768	−1.04982	0.66783
2	0.43825	0.39878	−0.63488	−0.10536	−0.75502	1.06471
3	−0.01005	−0.17435	−1.07881	−0.56212	−1.13943	−0.32850
4	−0.01005	−0.18633	−1.07881	−0.61619	−1.30933	−0.21072
5	0.04879	−0.10536	−1.02165	−0.44629	−1.20397	0.08618
6	0.08618	−0.07257	−0.86750	−0.49430	−1.17118	0.19885
7	0.07696	−0.10536	−0.91629	−0.67334	−1.17118	0.01980
8	0.23902	0.07696	−0.82098	−0.26136	−1.07881	0.65752
9	−0.01005	−0.16252	−1.02165	−0.57982	−1.23787	−0.44629
10	0.29267	0.12222	−0.79851	−0.26136	−0.99425	0.73237
11	0.26236	0.09531	−0.79851	−0.27444	−0.96758	0.68310
12	0.28518	0.09531	−0.73397	−0.26136	−0.96758	0.64185
13	0.62058	0.38526	−0.51083	0.00995	−0.43078	2.14710
14	0.45742	0.29267	−0.65393	−0.05129	−0.69315	1.50185
15	0.67803	0.46373	−0.40048	0.18232	−0.52763	2.13889
16	0.58779	0.44469	−0.41552	0.01980	−0.52763	1.81970
17	0.55962	0.45742	−0.46204	0.08618	−0.52763	2.02022
18	0.54232	0.35767	−0.44629	0.01980	−0.46204	1.85003
19	0.51879	0.45108	−0.32850	−0.04082	−0.38566	2.03209
20	0.55962	0.46373	−0.38566	0.07696	−0.47804	2.05156
21	0.78390	0.62058	−0.28768	0.21511	−0.32850	2.31747
22	0.54812	0.51282	−0.44629	0.13103	−0.59784	1.92862

(a) Use the data as is and perform the regression.
(b) Retain the variable W and express the rest as ratios of this value. Perform the regression. Compare it to the results in (a).
(c) Perform a principal component analysis and evaluate it.
(d) Perform the regression on the principal components found in (c).
(c) Can any of the principal components be omitted from the analysis? If so, perform the regression without them. Explain the results.

Problems with the Model

6.1 Introduction

We have repeatedly noted that the analysis of a regression model assumes that the model is specified before the analysis is performed. However, we have also seen that many statistical analyses are of an exploratory nature, which means that the results of a statistical analysis may be used to find a suitable model. Of course, any regression analysis must begin with some proposed model, and that model may or may not be suitable. The proposed model may not contain all necessary independent variables, or it may be of the wrong form; that is, it may specify a straight-line relationship when the true relationship is curvilinear. If these conditions have occurred, the model is said to be inadequately specified and the results of using that model are subject to **specification error**. Effects of specification error and some diagnostic tools are presented in Section 6.2. Another diagnostic tool, called the lack of fit test, is presented in Section 6.3.

As we will see, inadequately specified models can produce badly biased estimates. Because this result is well known, many investigators will add as many variables as possible to the model, even if the effect on the response variable is not expected to be significant. A model that has extraneous variables is considered **overspecified**. As we have already seen in Chapter 5, overspecified models may suffer from multicollinearity. Additional discussion of the effects of over-specified models is presented in Sections 6.4 through 6.8, where a set of procedures for **variable selection** is introduced for determining which variables should be included in a model. Because the results of variable selection are often misunderstood and subsequently misused, we devote three sections to assist in more appropriate uses of variable selection. In Section 6.6 we show the potential lack of reliable results of such procedures, in Section 6.7 we suggest situations for which variable selection may or may not be useful, and in Section 6.8 we show how influential outliers may affect this procedure.

6.2 Specification Error

Specification error occurs when the model used for a regression analysis does not contain sufficient parameters to adequately describe the behavior of the data. Two primary causes of specification error are as follows:

1. Omitting independent variables that should be in the model
2. Failing to account for relationships that are not strictly linear; that is, where the relationships need to be described by curves, which often requires the use of additional parameters

These two causes are not mutually exclusive.

There are two main effects of specification error:

1. Because the error mean square contains the effects of omitted parameters, it is inflated. That is, the estimate of the random error is biased upward.
2. The resulting estimated coefficients are biased estimates of the population parameters. We have already noted this result in Sections 3.2 and 3.4.

Detection of the omission of important independent variables may be done by a residual plot, but, as we will see, is not always straightforward. The existence of curved responses is more easily detected by residual or partial residual plots, and if adequate data are available, by a formal lack of fit test. We will illustrate with two simple artificially generated examples.

EXAMPLE 6.1   Ignoring a Variable   We return to Example 3.2 to illustrate the effect of ignoring a variable. The data are reproduced in Table 6.1.

The results of performing regressions with the underspecified models using only X1 and X2, as well as the correctly specified model using both variables, are summarized in Table 6.2.

Obviously, the coefficients of the two underspecified models are biased and their residual mean squares are larger. Of course, we do not normally know that we have not specified the correct model, and if we look at either of the two inadequate models, there is little hint of trouble, since both models produce statistically significant regressions, and unless we have some prior knowledge of the magnitude of the variance, the fit of each of the models appears adequate.

TABLE 6.1

SPECIFICATION ERRORS

OBS	X1	X2	Y
1	0	2	2
2	2	6	3
3	2	7	2
4	2	5	7
5	4	9	6
6	4	8	8
7	4	7	10
8	6	10	7
9	6	11	8
10	6	9	12
11	8	15	11
12	8	13	14

TABLE 6.2

REGRESSIONS WITH DIFFERENT MODELS

CONTRADICTORY PARTIAL AND TOTAL REGRESSIONS

MODEL	RMSE	INTERCEP	X1	X2
Using X1 only	2.12665	1.85849	1.30189	.
Using X2 only	2.85361	0.86131	.	0.78102
Using X1 and X2	1.68190	5.37539	3.01183	-1.28549

We have already discussed the use of residual plots to investigate data and model problems. The residual plot for the incorrectly specified model using only X2, using values of X1 as the plotting symbol, is shown in Figure 6.1. Note that the values of X1 increase from lower left to upper right. If we did not know about X1, the plot would not appear suspicious. ◆

These results suggest that when we do not have knowledge of the specification error, a study of other factors for observations corresponding to the "corners" or other extreme points of the residual plot may help to discover other variables that may lead to a more correctly specified model. However, this is a rather subjective process and, in general, we can see that without other knowledge it is not easy to detect the omission of independent variables from a regression model.

E X A M P L E 6 . 2   Ignoring a Curved Response   We again use an artificial data set containing 16 observations generated using the model

$$y = 10 + 2x_1 - 0.3x_1^2 + \epsilon,$$

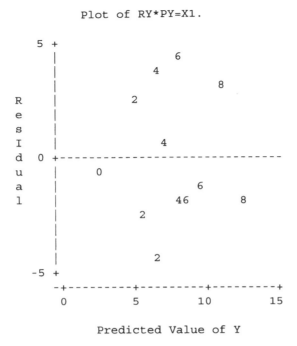

FIGURE 6.1 Residual Plot for Incorrect Model

where ϵ is normally distributed with mean 0 and a standard deviation of 3. This model describes a curved line that has a positive slope for the small values of x and becomes negative for the larger values of x. This is called a polynomial model, and its use is presented in detail in Chapter 7. The data are shown in Table 6.3 and the plot of the data and the population regression in Figure 6.2.

Assume for now that we do not know that the response is curved and use the simple linear model

$$y = \beta_0 + \beta_1 x_1 + \epsilon.$$

The results of the analysis produced by PROC REG of the SAS System are shown in Table 6.4.

Since we know the correct model, it is easy to see that the residual standard deviation of 3.8 is moderately larger than the true standard deviation of 3.0, and furthermore, the regression line has a negative slope, whereas the true line starts with a positive slope. Thus, we can see that the misspecified model can lead to incorrect conclusions.

We again turn to the residual plot; that is, the plot of residuals against the predicted values. The residual plot for this model is shown in Figure 6.3. The pattern of the residuals is typical of this type of misspecification: Residuals of one sign (positive in this example) are located largely in the middle while the residuals with the opposite sign are at the ends. Such a pattern suggests that the model be modified to include some sort of curved response.   ◆

TABLE 6.3

DATA FOR
EXAMPLE 6.2

X1	Y
0	8.3
0	7.1
1	12.4
9	3.7
9	2.7
3	20.9
3	10.3
2	13.2
9	4.1
4	10.6
7	8.1
0	8.8
7	9.5
0	12.1
9	4.8
2	12.9

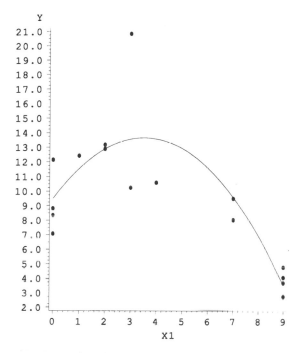

FIGURE 6.2 Data and Plot of Regression Line

TABLE 6.4

RESULTS WITH INCORRECT MODEL

Analysis of Variance

Source	DF	Sum of Squares	Mean Square	F Value	Prob>F
Model	1	105.85580	105.85580	7.168	0.0180
Error	14	206.76358	14.76883		
C Total	15	312.61938			

Root MSE	3.84302	R-square	0.3386	
Dep. Mean	9.34375	Adj R-sq	0.2914	
C.V.	41.12934			

Parameter Estimates

Variable	DF	Parameter Estimate	Standard Error	T for H0: Parameter=0	Prob > |T|
INTERCEP	1	12.292379	1.46153422	8.411	0.0001
X	1	-0.725816	0.27110811	-2.677	0.0180

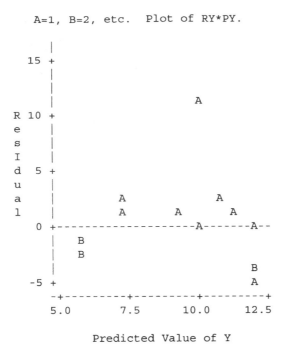

FIGURE 6.3 Residual Plot

Residual plots are also useful for models with more than one independent variable, although the patterns may not be as obvious. Identification of variables having a curved relationship may also be provided by the partial residual plots (Section 4.2).

6.3 Lack of Fit Test

We have already seen that one of the effects of an inadequately specified model is that the error mean square becomes large, but "large" is only relevant to some knowledge about the magnitude of true or "normal" variation. If we have knowledge of the magnitude of the true error variance, it is possible to construct a test to see if the error mean square is indeed an inflated estimate of that variance. In some applications, knowledge of the true error variance may indeed be available from previous experiments or other data, and a test, such as the χ^2 test, may be useful to detect a biased estimate of this variance.

In some cases, the data may provide information on the true error variance. Remember that the random error measures unassigned variability, which is equivalent to the variability, among units treated alike, often referred to as "pure" error. In a regression situation, this is the variability among observations having the same value of the independent variable(s). Thus, for data containing multiple observations for one or more values of the independent variables, a variance computed from such observations will provide an estimate of this variance. This quantity is obtained from the within (often called error) mean square from an analysis of variance, using values of the independent variable(s) as factor levels.

EXAMPLE 6.2 REVISITED

An examination of Table 6.3 shows that there are indeed multiple observations for some of the values of x_1. The use of PROC ANOVA of the SAS System, using the individual values of x_1 as the factor levels, produces the results shown in Table 6.5.

The results do show that the error mean square (8.138) is smaller than that for the straight-line model (14.769) and quite similar to that for the true quadratic model (9). However, because all of these estimates are based on the same observations, they are not independent and cannot be directly used for an F test for the equality of variances.

The formal test for comparing these estimates is obtained by the comparison of unrestricted and restricted models presented in Chapter 1. The six degrees of freedom "model" sum of squares from the analysis of variance indicates that the variation among the seven means identified by different values of x_1 can be described by some unspecified six-parameter model. This sum of squares is the unrestricted model sum of squares, as it accounts for all of the variability among the seven means. The one degree of freedom sum of squares for the regression model (Table 6.4) is the sum of squares for the restricted model, which specifies that the means of the response must be linearly related to x_1. The difference between the sums of squares of the unrestricted and restricted models is due to other parameters that may be needed to describe the relationship among means. As shown in Chapter 1, the resulting mean square is used to test the hypothesis that no other parameters are needed.

The quantities required for the test are available from Tables 6.4 and 6.5.

Unrestricted model	Table 6.5	df = 6	SS = 239.38
Restricted model	Table 6.4	df = 1	SS = 105.86
Difference		df = 5	SS = 133.52

TABLE 6.5

ANALYSIS OF VARIANCE USING X AS FACTOR LEVELS

Dependent Variable: Y

Source	DF	Sum of Squares	Mean Square	F Value	Pr > F
Model	6	239.379375	39.896563	4.90	0.0171
Error	9	73.240000	8.137778		
Corrected Total	15	312.619375			

This difference is referred to as the sum of squares for the **lack of fit**, since it measures how well or poorly the regression fits the data. The lack of fit mean square is obtained by dividing by degrees of freedom,

$$MS(\text{lack of fit}) = 133.52/5 = 26.70,$$

which is divided by the unrestricted model error mean square (Table 6.5) to provide the F ratio:

$$F = 26.70/8.14 = 3.28.$$

The p-value for this test is 0.061. Although this is greater than the commonly used value of 0.05, it may be considered sufficiently small to justify the search for a better, in this case quadratic, model.

Using the data from Table 6.3 and a quadratic model, we get

$$\hat{\mu}_x = 9.482 + 2.389x_1 - 0.339x_1^2,$$

with a residual standard deviation of 2.790. This estimated regression equation, as well as the estimated standard deviation, is consistent with the model used to generate the data.

The same procedure can now be used to determine if the quadratic model is adequate. Notice, however, that the error mean square for the quadratic model is $(2.790)^2 = 7.78$, which is almost identical to the unrestricted model error mean square of 8.14. Therefore, the test is not needed to determine that the quadratic model fits well. ◆

Comments

The objective lack of fit test is, of course, more powerful than the subjective examination of residuals. Unfortunately, it can only be used if there exist multiple observations for some values of the independent variables.[1] Finding such observations obviously becomes more difficult as the number of independent variables increases. Of course, multiple observations may be created as the result of an experiment, where it may often be more efficient to replicate only at selected combinations of the independent variables.

Unfortunately, the lack of fit test is not typically used for determining the need for additional factor or independent variables, since observations with identical values for the included variables may not have identical values for the not-included variables.

EXAMPLE 6.3   **Example 4.5 Revisited**   This example concerned the loads on the individual lines in a block and tackle. The picture of the assembly and resulting data are shown in Figure 4.4 and Table 4.18. This example was used in Section 4.3 to show how multiple observations for individual values of line numbers can be

[1] It has been suggested that the "pure" error can be estimated using "near neighbors" as repeated observations (Montgomery and Peck, 1982, Section 4.9.4).

used to implement a weighted regression that is being done as a remedy for unequal variances. In this section we will ignore the unequal variances and show how the correct model is obtained using residual plots and a lack of fit test.

If friction is present, the load on the lines should increase from line 1 to line 6, and as a first approximation, should increase uniformly, suggesting a linear regression of LOAD on LINE. The results of the regression are given in Table 6.6.

The regression is certainly significant, indicating an increase of almost 43 units of force with each line. However, the residual plot, shown in Figure 6.4, clearly indicates that the relationship is not well described by a straight line, but that some sort of curved line may be needed.

Since in this example we do have repeated observations, we can perform a lack of fit test. The unrestricted model sums of squares are obtained by the analysis of variance, with results shown in Table 6.7.

Since the regression was significant, the overall results are not surprising. The primary feature of interest is that the unrestricted model error mean square, the "pure" error, is 69.99, as compared with 219.3 for the restricted regression model, indicating that the regression model may be inadequate. We compute the lack of fit test using the model sums of squares as follows:

Unrestricted model	df = 5	SS = 330,542.5
Restricted model	df = 1	SS = 321,600.0
Lack of fit	df = 4	SS = 8,942.5

TABLE 6.6

REGRESSION OF LOAD ON LINE

Analysis of Variance

Source	DF	Sum of Squares	Mean Square	F Value	Prob>F
Model	1	321600.02286	321600.02286	1466.218	0.0001
Error	58	12721.71048	219.33984		
C Total	59	334321.73333			

Root MSE	14.81013	R-square	0.9619	
Dep Mean	412.26667	Adj R-sq	0.9613	
C.V.	3.59237			

Parameter Estimates

Variable	DF	Parameter Estimate	Standard Error	T for H0: Parameter=0	Prob > |T|
INTERCEP	1	262.226667	4.35998308	60.144	0.0001
LINE	1	42.868571	1.11954030	38.291	0.0001

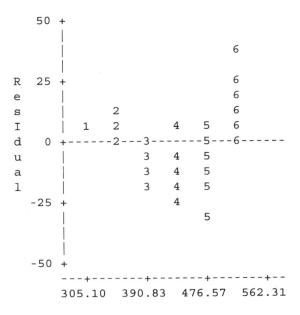

FIGURE 6.4 Residual Plot for Linear Regression

TABLE 6.7

ANALYSIS OF VARIANCE FOR THE FULL MODEL

Analysis of Variance Procedure
Dependent Variable: LOAD

Source	DF	Sum of Squares	Mean Square	F Value	Pr > F
Model	5	330542.5333	66108.5067	944.61	0.0001
Error	54	3779.2000	69.9852		
Corrected Total	59	334321.7333			

R-Square	C.V.	Root MSE	LOAD Mean
0.988696	2.029200	8.365715	412.266667

The lack of fit mean square is 8942.5/4 = 2235.6, which is divided by the unrestricted model error mean square to provide an F ratio of 31.94 with (4, 54) degrees of freedom, giving overwhelming evidence that the linear regression is not adequate.

The simplest curved line regression is obtained by adding a quadratic term[2] to provide for a curved line. This model provides the estimated curve as

$$\hat{\mu}_{y|x} = 296.66 + 17.04x + 3.689x^2,$$

which describes an upward-sloping convex curve. The added quadratic term is indeed significant, and the regression has a residual mean square of 134.04, which is certainly smaller than that for the linear regression but is still considerably larger than the pure error mean square, indicating that this model may also not be satisfactory. In fact the lack of fit test for this model (as the reader may verify) produces an F ratio of 18.39 with $(3, 54)$ degrees of freedom, confirming that suspicion.

Now the quadratic regression is often chosen for convenience, as it was here, but usually does not have any basis in theory. Actually, as we noted, a linear regression is logical here, assuming that the pulleys are equivalent. However, there is some reason to believe that line 6, which is the one being pulled by the drum, may behave differently. We can check this idea by using a linear regression and adding an indicator variable[3] that allows line 6 to deviate from the linear regression. This variable, which is labeled C1, has the value of zero for lines 1 through 5 and one for line 6. The addition of this variable produces the following models:

$$\widehat{LOAD} = \hat{\beta}_0 + \hat{\beta}_1 \, LINE$$

for lines 1 through 5, and

$$\widehat{LOAD} = \hat{\beta}_0 + \hat{\beta}_1 \, LINE + \hat{\beta}_2$$
$$= (\hat{\beta}_0 + \hat{\beta}_2) + \hat{\beta}_1 \, LINE$$

for line 6, where $\hat{\beta}_2$ is the coefficient for C1. The results of fitting this model are shown in Table 6.8.

This model certainly fits better than the quadratic, and the error mean square is now only slightly larger than the pure error mean square. The reader may verify that the lack of fit test produces an F ratio of 2.74, which has a p-value of 0.054. This result indicates some evidence of lack of fit, but considering the small increase in the error mean square for the restricted model, it is probably not worth investigating. This analysis is, in fact, the one used to determine the model in Example 4.5, Table 4.19.

The regression coefficient for LINE estimates an increase of 36.88 units of load for each line number, with an additional increase of 41.92 units for line 6. The plots of these regression equations are shown in Figure 6.5, where the points represent the data, the solid line shows the linear regression, and the dashed line shows the final model with the indicator variable.

In Section 4.3 we fitted this model using weighted regression because of the unequal variances, which are quite evident in Figure 6.5. The plots and lack of fit tests we have presented here can be performed with weighted regression and

---

[2] A complete coverage of this type of model is provided in Chapter 7.

[3] A more complete discussion of the use of such variables as slope and intercept shifters is presented in Chapter 9.

TABLE 6.8

REGRESSION WITH DUMMY VARIABLE

Analysis of Variance

Source	DF	Sum of Squares	Mean Square	F Value	Prob>F
Model	2	329968.05333	164984.02667	2160.032	0.0001
Error	57	4353.68000	76.38035		
C Total	59	334321.73333			

Root MSE	8.73959	R-square	0.9870
Dep Mean	412.26667	Adj R-sq	0.9865
C.V.	2.11989		

Parameter Estimates

Variable	DF	Parameter Estimate	Standard Error	T for H0: Parameter=0	Prob > |T|
INTERCEP	1	276.200000	2.89859252	95.288	0.0001
LINE	1	36.880000	0.87395853	42.199	0.0001
C1	1	41.920000	4.00498111	10.467	0.0001

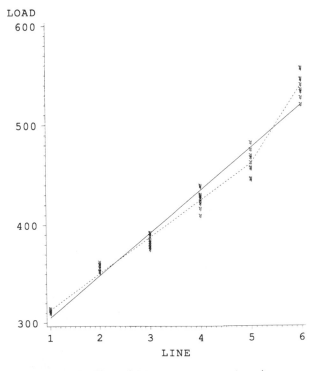

FIGURE 6.5 Plot of Models for Line Loads

analysis of variance, and the results would be equivalent. However, as we have noted, the estimates of the coefficients change very little with the weighted regression and therefore using the more easily implemented unweighted analyses will produce the form of the correct final model, which can then be re-estimated using weighted regression. ◆

6.4 Overspecification: Too Many Variables

We have seen that underspecification, that is, leaving out variables that should be in the model, results in inflation of the error variance. Since this is a rather well-known result, one common practice for avoiding such results is to put into an initial model all conceivably relevant variables, with the number of variables often restricted only by the availability of data.

When this procedure is followed, it usually happens that the initial model contains too many variables; that is, some of these variables are not needed in the sense that they do not contribute to the fit of the model. Such a model is said to be **overspecified**. Now it is indeed true that an overspecified model does not lead to biased estimates of either the variance or the regression parameters; however, there are two adverse effects of such overspecification:

1. Because overspecified models tend to exhibit increased multicollinearity, the variances of the coefficients tend to become large.
2. Estimated mean responses or predicted values tend to have large variances. This result occurs even when there is little or no multicollinearity.

Remedial methods to counteract the effects of multicollinearity were presented in Chapter 5, where we noted that these methods were not always successful. One reason for this lack of success is that all remedial methods use initial models, which are in many cases subject to overspecification, and analyses based on overspecified models are more likely to provide unsatisfactory analyses and interpretation.

It then seems natural to ask: Why not reduce the overspecification by deleting unnecessary variables from the model? Now if we knew which variables were not needed, such deletions would be easy to make, but then we would not put these variables into the model to begin with. Therefore, if we do not know which variables are not needed, any selection of variables must be based on the data. In other words, variables are chosen or deleted based on statistics such as p-values of coefficients estimated by the data being analyzed. Because these methods use statistics from the analysis of the data, we will call these **data-driven** procedures.

Several procedures for this type of data-driven **variable selection** are described in Section 6.5. These methods have become increasingly sophisticated and are universally available with today's computing hardware and software, which has led to widespread misuse and abuse of these procedures. Therefore, it is important to discuss some practical aspects associated with data-driven variable selection procedures.

There are three primary cautions that must be noted when using data-driven variable selection procedures:

1. *p*-values for statistical tests on coefficients of the finally chosen model do not have their intended interpretation. This is because the *p*-values are strictly valid only for a single analysis of a set of data. When the data are then used for further analyses, the *p*-values may still be printed on the computer output, but are no longer valid. That does not mean that they are useless, but they must be used with caution.
2. Because sample data are used to drive these procedures, any finally chosen model is, strictly speaking, valid only for that sample. Unfortunately, there are no statistics that indicate how good the final model is for inferences on the population.
3. If data-driven procedures are driven to the extreme, it may be possible to produce any desired result. It is for this reason that data-driven variable selection is often called "data dredging," which has led to the famous quote[4]

> If you torture the data enough, it will confess.

One criterion for deciding on the appropriateness of variable selection is related to the purpose of the regression analysis. For this discussion, we restate the two distinct purposes for performing a regression analysis:

1. To provide a vehicle for estimating the behavior of the response; that is, to estimate values of the conditional mean or predict future values of the response variable.
2. To provide information on the structure of the regression relationship. That is, we want to specify the effect of the individual independent variables on the behavior of the response variable.

These two purposes are certainly not mutually exclusive and in many analyses there is some interest in both. However, one or the other of these is often of primary importance. For example, when we want to estimate the potential lumber harvest of a forest (Example 2.1), the primary purpose is to use the model equation to predict tree volumes from some easily measured tree characteristics. On the other hand, in the NBA data (Example 5.4), a coach may want to use the model to decide if it is better to concentrate on attempting more field goals or on increasing the success ratio of such attempts.

Now if estimation or prediction is the primary purpose, variable selection is certainly justified: When there are fewer variables, the data collection process is less burdensome and, furthermore, the variances of the estimated means do indeed tend to be smaller (assuming little or no increase in the residual mean square). Of course, we have already noted that the model selected from a sam-

---

[4] We have not been able to find a reliable source for the origin of this quote.

ple really applies only to that sample, but some help is available in this area (Section 6.6).

On the other hand, if we are primarily interested in the structure of the model, variable selection can indeed become counterproductive: Because of sampling fluctuations and multicollinearity, we may delete variables that could provide the very information we seek.

In conclusion, then, data-driven model selection should be used as a last resort; that is, if no other information can be used to select a model. And even then, it must be used and its results interpreted with great care and caution.

6.5 Variable Selection Procedures

In this section we will assume that each coefficient corresponds to a specific variable. This condition does not occur, for example, in polynomial models, where one coefficient corresponds to, say, x_1, another to x_1^2, and so forth. As we will see in Chapter 7, selection of variables and the appropriate powers of variables to use in a polynomial regression is also important, but the selection procedures are done in a different manner. These procedures are also presented in Chapter 7.

We start with an initial model with m variables and want to select a model with a subset of $p \leq m$ variables. The selection process consists of two related questions:

1. What is the magnitude of p; that is, how many variables should we select?
2. Given a value of p, which combination of variables should be selected?

Of these two, finding the combination of variables is the more difficult. What we want is an optimum subset, which is defined as follows

---

DEFINITION

An **optimum subset** is that variable combination which, for a given value of p, gives the **maximum R-square**, which is, of course, the same as the minimum error mean square.

Although the criterion for an optimum combination is obvious, the mechanics of obtaining that combination are not easy. In fact, for all practical purposes, it is necessary to examine all possible combinations to guarantee finding that optimum combination.[5] This is not a trivial task; for example, with an initial model containing only 10 variables, more than a thousand regressions will need to be

---

[5] Recent research has developed algorithms that allow some shortcuts. However, even with these methods, a large number of combinations must be examined.

computed for optimum combinations for all 10 subset sizes. Fortunately, vast increases in computing power and the development of efficient algorithms have made this method readily available for moderate-sized models with that many or more variables. However, for very large models, such as ones used for weather forecasting, this procedure is still not directly available. We will later present some computationally less demanding methods that will often perform quite well. However, not only do these methods not guarantee finding optimum subsets, but they also do not reveal if the resulting combination is optimal.

One benefit from examining all possible variable combinations is that this procedure also provides the second-best combination, and so forth. Being able to examine near-optimum variable combinations provides additional insight into the model as follows:

- A near-optimum model may contain a more attractive variable combination. For example, a near-optimal set may contain variables for which data are easier to collect, or the variable combination may contain variables whose existence in the model more nearly conforms to prior knowledge, or may exhibit less multicollinearity. Remember that the optimum combination applies only to the particular data set being analyzed and there is no guarantee that it is optimum for the population. Therefore, there may be little or no loss in using a near-optimum set.
- The existence of a number of near-optimum subsets with almost identical R-square values is a clear indication of multicollinearity and usually indicates that additional variables may be deleted without serious loss in precision.

Size of Subset

There is no objective criterion for the choice of the optimum subset size. Intuitively, one would choose the optimum subset size to be the minimum size that does not cause a meaningful decrease in R-square, residual mean square, or some other statistic describing the predictive quality of the model. However, because this is an exploratory analysis, there are no formal statistical tests for significant or meaningful changes. For these reasons, a common practice for choosing the subset size is as follows:

1. Obtain the optimum and several near-optimum subsets for each subset size and compute a goodness of fit statistic.
2. Then graph the statistic against the number of variables, and choose that subset size to be (a) the minimum size before the statistic shows a tendency for a poorer fit and (b) one where the second-best model is demonstrably poorer.

Various goodness of fit statistics will be discussed later.

EXAMPLE 6.4   **Baseball Data**   Table 6.9 gives data on some team performance variables for major league baseball teams for the 1975 through 1980 seasons. A regression analysis is performed to ascertain the effect of these performance variables on the percentage of games a team wins (PER).

TABLE 6.9

BASEBALL DATA

OBS	Year	LEAGUE	PER	RUNS	DOUBLE	TRIPLE	HR	BA	ERROR	DP	WLK	SO
1	1976	EAST	62.3	770	259	45	110	272	115	148	397	918
2	1976	EAST	56.8	708	249	56	110	267	163	142	460	762
3	1976	EAST	53.1	615	198	34	102	246	131	116	419	1025
4	1976	EAST	46.3	611	216	24	105	251	140	145	490	850
5	1976	EAST	44.4	629	243	57	63	260	174	163	581	731
6	1976	EAST	34.0	531	243	32	94	235	155	179	659	783
7	1976	WEST	63.0	857	271	63	141	280	102	157	491	790
8	1976	WEST	56.8	608	200	34	91	251	128	154	479	747
9	1976	WEST	49.4	625	195	50	66	256	140	155	662	780
10	1976	WEST	45.7	595	211	37	85	246	186	153	518	746
11	1976	WEST	45.1	570	216	37	64	247	141	148	543	652
12	1976	WEST	43.2	620	170	30	82	245	167	151	564	818
13	1977	EAST	62.3	847	266	56	186	279	120	168	482	856
14	1977	EAST	59.3	734	278	57	133	274	145	137	485	890
15	1977	EAST	51.2	737	252	56	96	270	139	174	532	768
16	1977	EAST	50.0	692	271	37	111	266	153	147	489	942
17	1977	EAST	46.3	665	294	50	138	260	129	128	579	856
18	1977	EAST	39.5	587	227	30	88	244	134	132	490	911
19	1977	WEST	60.5	769	223	28	191	266	124	160	438	930
20	1977	WEST	54.3	802	269	42	181	274	95	154	544	868
21	1977	WEST	50.0	680	263	60	114	254	142	136	545	871
22	1977	WEST	46.3	673	227	41	134	253	179	136	529	854
23	1977	WEST	42.6	692	245	49	120	249	189	142	673	827
24	1977	WEST	37.7	678	218	20	139	254	175	127	701	915
25	1978	EAST	55.6	708	248	32	133	258	104	155	393	813
26	1978	EAST	54.7	684	239	54	115	257	167	133	499	880
27	1978	EAST	48.8	664	224	48	72	264	144	154	539	768
28	1978	EAST	46.9	633	269	31	121	254	234	150	572	740
29	1978	EAST	42.6	600	263	44	79	249	136	155	600	859
30	1978	EAST	40.7	607	227	47	86	245	132	159	531	775
31	1978	WEST	58.6	727	251	27	149	264	140	138	440	800
32	1978	WEST	57.1	710	270	32	136	256	134	120	567	908
33	1978	WEST	54.9	613	240	41	117	248	146	118	453	840
34	1978	WEST	51.9	591	208	42	75	252	160	171	483	744
35	1978	WEST	45.7	605	231	45	70	258	133	109	578	930
36	1978	WEST	42.6	600	191	39	123	244	153	126	624	848
37	1979	EAST	60.5	775	264	52	148	272	134	163	504	904
38	1979	EAST	59.4	701	273	42	143	264	131	123	450	813
39	1979	EAST	53.1	731	279	63	100	278	132	166	501	788
40	1979	EAST	51.9	683	250	53	119	266	106	148	477	787
41	1979	EAST	49.4	706	250	43	135	269	159	163	521	933
42	1979	EAST	38.9	593	255	41	74	250	140	168	607	819
43	1979	WEST	55.9	731	266	31	132	264	124	152	485	773
44	1979	WEST	54.9	583	224	52	49	256	138	146	504	854
45	1979	WEST	48.8	739	220	24	183	263	118	123	555	811
46	1979	WEST	43.8	672	192	36	125	246	163	138	577	880
47	1979	WEST	42.2	603	193	53	93	242	141	154	513	779
48	1979	WEST	41.3	669	220	28	126	256	183	139	494	779

(Continued)

TABLE 6.9 (*Continued*)

BASEBALL DATA

OBS	Year	LEAGUE	PER	RUNS	DOUBLE	TRIPLE	HR	BA	ERROR	DP	WLK	SO
49	1980	EAST	56.2	728	272	54	117	270	136	136	530	889
50	1980	EAST	55.6	694	250	61	114	257	144	126	460	823
51	1980	EAST	51.2	666	249	38	116	266	137	154	451	832
52	1980	EAST	45.7	738	300	49	101	275	122	174	495	664
53	1980	EAST	41.4	611	218	41	61	257	154	132	510	886
54	1980	EAST	39.5	614	251	35	107	251	174	149	589	923
55	1980	WEST	57.1	637	231	67	75	261	140	145	466	929
56	1980	WEST	56.4	663	209	24	148	263	123	149	480	835
57	1980	WEST	54.9	707	256	45	113	262	106	144	506	833
58	1980	WEST	50.3	630	226	22	144	250	162	156	454	696
59	1980	WEST	46.6	573	199	44	80	244	159	124	492	811
60	1980	WEST	45.1	591	195	43	67	255	132	157	536	728

The variables[6] are as follows:

LEAGUE: East or West (not used as a variable at this point)
RUNS: Number of runs scored
DOUBLE: Number of doubles
TRIPLE: Number of triples
HR: Number of home runs
BA: Batting average
ERROR: Number of fielding errors
DP: Number of double plays (defensive)
WLK: Number of walks given to opposing team
SO: Number of strikeouts for opposing team

The results of the linear regression using these variables are given in Table 6.10.

The regression is certainly significant ($P < 0.0001$) and the coefficient of determination shows a reasonable fit, although the residual standard deviation of 3.9 indicates that about one-third of the estimated winning percentages will be off by as much as four percentage points. Only the coefficient for walks is significant and has the expected negative sign. However, multicollinearity is not serious, with only moderate VIF values for runs, home runs, and batting average, a result that appears to be reasonable.

Somehow, one would expect somewhat better indications on the effects of the individual performance variables. Because multicollinearity is not very serious, remedial measures suggested for that problem would not be very helpful.[7] Therefore, it seems reasonable that if we had fewer variables, we could have a more useful model. We therefore perform an all-possible variable combinations procedure (R-SQUARE selection using PROC REG of the SAS System), requesting that the four best models be printed for each subset size. The results are shown in Table 6.11.

---

[6] In order to provide a manageable presentation, these variables were subjectively chosen from a larger set.

[7] However, investigating multicollinearity for this example would be a good exercise.

TABLE 6.10

REGRESSION FOR BASEBALL DATA

Analysis of Variance

Source	DF	Sum of Squares	Mean Square	F Value	Prob>F
Model	9	2177.73127	241.97014	15.651	0.0001
Error	50	773.03723	15.46074		
C Total	59	2950.76850			

Root MSE	3.93202	R-square	0.7380
Dep Mean	50.00500	Adj R-sq	0.6909
C.V.	7.86324		

Parameter Estimates

Variable	DF	Parameter Estimate	Standard Error	T for H0: Parameter=0	Prob > |T|	Variance Inflation
INTERCEP	1	24.676979	24.00041151	1.028	0.3088	0.00000000
RUNS	1	0.021562	0.02287285	0.943	0.3504	9.70575638
DOUBLE	1	-0.027054	0.02408689	-1.123	0.2667	1.84097750
TRIPLE	1	0.119568	0.06248849	1.913	0.0614	1.99295746
HR	1	0.031931	0.03331433	0.958	0.3424	4.53083322
BA	1	0.171720	0.12084133	1.421	0.1615	6.07177806
ERROR	1	-0.020062	0.02450871	-0.819	0.4169	1.36209724
DP	1	-0.047247	0.03860665	-1.224	0.2268	1.43214007
WLK	1	-0.049407	0.00893003	-5.533	0.0001	1.29375661
SO	1	-0.000140	0.00812946	-0.017	0.9864	1.36068328

The output gives the four best (highest R-square) models in order of decreasing fit, for all subset sizes (Number in Model). Given for each model is the R-square value; C_p, an assessment statistic we discuss later; and a listing of the variables in the model. Thus, for example, the best one-variable model is a function of batting average and produces an R-square value of 0.458; the best two-variable model uses runs and walks and has an R-square value of 0.690; and so forth. Looking at the R-square values, we can see that this statistic increases rather rapidly going from one to three variables, increases somewhat when a fourth variable is added, and then changes very little as further variables are added. Therefore, we may argue that a three- or four-variable model may be adequate. ◆

A number of statistics, such as R-square, the error mean square, or residual standard deviation, may be used to help in deciding how many variables to retain. For example, the documentation for PROC REG from the SAS System lists 13 such statistics. Most of these are related to the residual mean square, and although arguments favoring each can be made, we will present only one of these here—the Mallows C_p statistic, which is given under the heading C(p) in Table 6.11.

TABLE 6.11

VARIABLE SELECTION FOR BASEBALL DATA

```
N = 60      Regression Models for Dependent Variable: PER

Number in
  Model     R-square        C(p)   Variables in Model

     1     0.45826324    47.39345  BA
     1     0.45451468    48.10889  RUNS
     1     0.44046161    50.79099  WLK
     1     0.22525896    91.86361  ERROR
    ------------------------------------------
     2     0.69030270     5.10744  RUNS WLK
     2     0.66330564    10.25998  BA WLK
     2     0.54107373    33.58861  HR WLK
     2     0.51488356    38.58715  DOUBLE WLK
    ------------------------------------------------
     3     0.70982875     3.38079  RUNS TRIPLE WLK
     3     0.69977858     5.29892  RUNS BA WLK
     3     0.69674469     5.87795  RUNS DP WLK
     3     0.69608770     6.00334  RUNS ERROR WLK
    ------------------------------------------------
     4     0.71942898     3.54853  RUNS TRIPLE DP WLK
     4     0.71393040     4.59796  RUNS TRIPLE ERROR WLK
     4     0.71270881     4.83111  RUNS TRIPLE WLK SO
     4     0.71264668     4.84297  RUNS DOUBLE TRIPLE WLK
    ------------------------------------------------------
     5     0.72472666     4.53744  RUNS TRIPLE BA DP WLK
     5     0.72406640     4.66345  RUNS TRIPLE ERROR DP WLK
     5     0.72169243     5.11654  RUNS DOUBLE TRIPLE DP WLK
     5     0.72100271     5.24817  TRIPLE HR BA DP WLK
    ------------------------------------------------------
     6     0.72990371     5.54937  RUNS DOUBLE TRIPLE BA DP WLK
     6     0.72914230     5.69469  DOUBLE TRIPLE HR BA DP WLK
     6     0.72839207     5.83787  RUNS TRIPLE BA ERROR DP WLK
     6     0.72750286     6.00758  RUNS TRIPLE HR BA DP WLK
    ------------------------------------------------------------
     7     0.73449478     6.67314  RUNS DOUBLE TRIPLE HR BA DP WLK
     7     0.73336343     6.88906  DOUBLE TRIPLE HR BA ERROR DP WLK
     7     0.73319937     6.92037  RUNS DOUBLE TRIPLE BA ERROR DP WLK
     7     0.73141090     7.26171  RUNS TRIPLE HR BA ERROR DP WLK
    ------------------------------------------------------------
     8     0.73802018     8.00030  RUNS DOUBLE TRIPLE HR BA ERROR DP WLK
     8     0.73451106     8.67003  RUNS DOUBLE TRIPLE HR BA DP WLK SO
     8     0.73336536     8.88869  DOUBLE TRIPLE HR BA ERROR DP WLK SO
     8     0.73320840     8.91865  RUNS DOUBLE TRIPLE BA ERROR DP WLK SO
    ------------------------------------------------------------------
     9     0.73802173    10.00000  RUNS DOUBLE TRIPLE HR BA ERROR DP WLK SO
    ------------------------------------------------------------------------
```

One slightly different statistic sometimes used for assessing variable selection is the PRESS statistic (Section 4.2), which is a function of both the residual mean square and the existence of influential observations. This statistic may be useful for finding variable combinations that are less affected by influential observations. We will, however, not use this statistic here.

The C_p Statistic

The C_p statistic, proposed by Mallows (1973), is a measure of total squared error for a subset model containing p independent variables. As we noted in Chapter 5, the effectiveness of a biased estimator is measured by a total squared error that is a measure of the error variance plus the bias introduced by not including important variables in a model. Such a measure may therefore indicate when variable selection is deleting too many variables. The C_p statistic is computed as follows:

$$C_p = \frac{\text{SSE}(p)}{\text{MSE}} - (n - 2p) + 2,$$

where

> MSE is the error mean square for the full model (or some other estimate of pure error)
> SSE(p) is the error sum of squares for the subset model containing p independent variables[8]
> n is the sample size

For any given number of selected variables, larger C_p values indicate equations with larger error mean squares. By definition, $C_p = (m + 1)$ for the full model. When for any *subset* model $C_p > (p + 1)$, there is evidence of bias due to an incompletely specified model. On the other hand, if there are values of $C_p < (p + 1)$, that model is said to be overspecified; that is, it contains too many variables.

Mallows recommends that C_p be plotted against p, and further recommends selecting that subset size where the minimum C_p first approaches $(p + 1)$ starting from the full model. Also of interest are differences in the C_p statistic between the optimum and near-optimum models for each subset size. The C_p plot for the best four models for each subset size of the baseball data is shown in Figure 6.6.[9]

In this plot we can clearly see that models near the full model are indeed overspecified because several C_p values are clearly less than $(p + 1)$, while the two-variable model is clearly underspecified ($C_p > 3$). According to Mallow's criterion, the optimum subset model should contain three variables. Furthermore, for three variables there is a relatively large difference in the C_p

---

[8] In the original presentation of the C_p statistic (Mallows, 1973), the intercept coefficient is also considered as a candidate for selection; hence, in that presentation the number of variables in the model is one more than what is defined here and results in the "+2" in the equation shown here and in the plot.

[9] In this plot, the horizonal axis shows the number of variables in the model, *not* including the intercept.

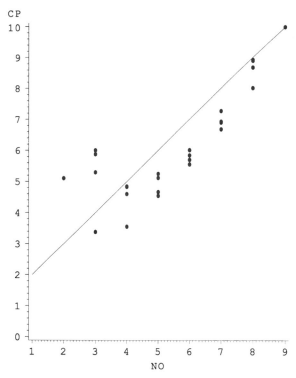

FIGURE 6.6 C_p Plot for Baseball Data

values between the best and second-best model, indicating that the model using RUNS, TRIPLE, and WLK should be most useful. The results for this model are shown in Table 6.12.

The model statistics show that the R-square value has decreased only slightly from that of the full model. However, the error mean square has been *reduced*, which is confirmed by the virtual lack of change in the adjusted R-square value. Looking at the coefficients, we see that in spite of the fact that p-values should not be taken literally, the coefficients for RUNS and WLK certainly appear to be important and do have the expected sign. The coefficient for TRIPLE, with a listed p-value of 0.0573, may not be very useful, but with the lack of multicollinearity, little is lost by retaining that variable in the model.

Other Selection Methods

Recent developments in computer algorithms and increased speed of computers have made it possible to use the all-possible-models procedure for most applications. However, in some applications an initial model may contain 100 or even 200 variables. Such large models arise, for example, in weather forecasting models, where it is desired to predict a weather variable at one weather station using

TABLE 6.12

OPTIMUM SUBSET MODEL FOR BASEBALL DATA

```
                        Analysis of Variance

                        Sum of         Mean
        Source      DF   Squares       Square      F Value     Prob>F

        Model        3   2094.54032   698.18011    45.663      0.0001
        Error       56    856.22818    15.28979
        C Total     59   2950.76850

            Root MSE      3.91022    R-square     0.7098
            Dep Mean     50.00500    Adj R-sq     0.6943
            C.V.          7.81965

                        Parameter Estimates

                  Parameter    Standard     T for H0:                    Variance
  Variable   DF    Estimate      Error     Parameter=0   Prob > |T|     Inflation

  INTERCEP    1    42.300404    7.58098301     5.541       0.0001       0.00000000
  RUNS        1     0.049160    0.00790803     6.217       0.0001       1.17315271
  TRIPLE      1     0.088503    0.04559148     1.941       0.0573       1.07273946
  WLK         1    -0.055596    0.00817854    -6.798       0.0001       1.09730531
```

as predictors various weather variables at a previous time period at a large number of other stations. In such situations, the all-possible-models selection requires too much computer time.

When the all-possible-subset selection is not feasible, other less computationally demanding methods may be useful. The two most frequently used methods are as follows:

Backward elimination: This method starts by considering the full model, which includes all candidate variables. The single variable contributing least to this model, as evidenced by the t or F statistics for testing each partial coefficient, is deleted. The coefficients for the resulting $(m - 1)$ variable model are then examined and the variable contributing least is eliminated. The process is then repeated and deletion is terminated when all remaining variables are "significant" at some specified p-value. Useful information may be obtained by continuing the process until only one variable remains and using the C_p plot to evaluate subset sizes. Highly efficient matrix algorithms make this procedure very fast for models with many variables. The important feature of this model is that once a variable has been dropped, it can never re-enter the model.

Forward selection: In this procedure we start with the "best" one-variable model, which is that variable having the highest simple correlation with the response variable. The second variable is picked that gives the maximum improvement in fit, which is revealed by the maximum of the partial correlations of all independent variables with the response variable, holding constant the already-chosen variable. Variables are added in the same manner, one at a time, until no additions provide adequate reduction in the error mean square as evidenced by a stated p-value, or the process can be continued until all variables are included and a C_p plot used to evaluate subset sizes. This method is also computationally efficient, and especially so if selection is discontinued when no additions are deemed "significant." As in the backward elimination procedure, once a variable is chosen, it can never be deleted from the model.

Other selection methods are indeed available. For example, the **stepwise** procedure starts like forward selection but allows deletion of a single variable at any stage before another variable is added. The success of this method is highly dependent on the "p-values" for deciding on addition and elimination. Although this method has intuitive appeal, it usually does not provide appreciably more nearly optimum models than are obtained by the backward elimination and forward selection methods.

To save space, we will not present full results of these methods, but instead compare the selection processes with the results of the all-possible-regressions procedure for the baseball data.

- The **backward elimination** procedure does indeed select the optimum model for models with eight down to two variables, but for the one-variable model it selects the second-best model (Table 6.11) using RUNS, which has a C_p value of 48.1, compared with the optimum model with BA, which has a C_p value of 47.4. However, since a three-variable model appears to be most suitable, this procedure will indeed select the "best" model.

- The **forward selection** procedure (obviously) selects the optimum one-variable model. At this point the selection proceeds as follows:

  Two variables: BA, WLK, the second-best model with $C_p = 10.26$, compared with the optimum subset model $C_p = 5.11$.
  Three variables: RUNS, TRIPLE, WLK, the second-best model with $C_p = 5.30$, compared with the optimum subset $C_p = 3.38$.
  Four variables: RUNS, TRIPLE, BA, WLK, the fifth-best model with $C_p = 4.94$, compared with the optimum subset $C_p = 3.55$.
  For five or more variables, the procedure selects the optimum subsets.

Looking at the C_p plot, the forward selection procedure would suggest using its four-variable choice, which is actually the fifth best. The resulting model is shown in Table 6.13 as produced as a portion of the forward selection option of the PROC REG procedure in the SAS System. We can readily see that this model contains two variables that may not be considered useful.

TABLE 6.13

MODEL CHOSEN BY FORWARD SELECTION

Step 4   Variable TRIPLE Entered   R-square = 0.71212847   C(p) = 4.94187165

	DF	Sum of Squares	Mean Square	F	Prob>F
Regression	4	2101.32625402	525.33156350	34.01	0.0001
Error	55	849.44224598	15.44440447		
Total	59	2950.76850000			

Variable	Parameter Estimate	Standard Error	Type II Sum of Squares	F	Prob>F
INTERCEP	29.67132746	20.53076432	32.25781038	2.09	0.1541
RUNS	0.04125215	0.01433551	127.89037074	8.28	0.0057
TRIPLE	0.07606146	0.04951664	36.44166098	2.36	0.1303
BA	0.06887111	0.10390058	6.78593028	0.44	0.5102
WLK	-0.05430107	0.00844870	637.98128620	41.31	0.0001

6.6 Reliability of Variable Selection

We have already noted that the results of data-based variable selection procedures are optimal only for the data set being used, and there are no statistics that tell us about the reliability of the variable selection processes. We therefore use simulations to illustrate the nature of the variability of results of these procedures. The simulations are based on Examples 5.1 and 5.2, which were used to illustrate multicollinearity.

The model has been modified to be

$$y = 3.0x_1 + 3.0x_2 + 3.0x_3 + 2.0x_4 + 2.0x_5 + 2.0x_6 + 2.0x_7 + \epsilon,$$

where ϵ is the random error. Because the independent variables are all generated to have equal dispersion, the choice of coefficients provides a convenient visual guide to the relative importance of the variables, with x_1 to x_3 being more important, and the remainder somewhat less so. Because we will investigate how well the selection procedure chooses models with three variables, this choice of coefficients dictates that choosing x_1 to x_3 is optimum, although not overwhelmingly so.

Because selection effectiveness is influenced by multicollinearity and the fit of the model, we will perform variable selections under two multicollinearity scenarios:

1. No multicollinearity, as in Example 5.1
2. The correlation pattern of Example 5.2, with correlations among adjacent variables being 0.9

For each multicollinearity scenario we will generate samples with full-model R-square values of approximately 0.95, 0.75, and 0.50, respectively.

The optimum three-variable model is selected from each of the samples from the six multicollinearity patterns and R-square combinations. Table 6.14 shows how many times (or percent, since there are 100 samples) each variable occurs among the three picked for the selected model; the last line shows how many times the selected model contains the three most important variables, x_1, x_2, and x_3.

> *No multicollinearity:* We would expect that the procedure will pick the three more important variables with greater frequency. This is indeed true, but the number of times one or more "wrong" variables are selected is certainly not trivial and increases with decreasing R^2. Also, the correct selection occurs less than 50%, even when the model R^2 is 0.95.
>
> *With multicollinearity:* As expected, the results are not very encouraging. There is almost no tendency to pick the more important variables, but the results do not appear to get worse as R^2 gets smaller. Note also that in only 11 cases in 300 does the procedure pick the three most important variables.

These results are certainly not encouraging. Admittedly, the first three variables are not much more important than the last four, and if we had included one or more very large coefficients, the results would have been "better."

Cross Validation

Even though there are no reliable statistics for evaluating the validity of a variable selection, there do exist some tools that may be used to provide some information on the usefulness of the selected model. Generally, these methods involve **data splitting**, where the data are divided into two parts. One set is used

TABLE 6.14

SIMULATION OF VARIABLE SELECTION

Variable Selected	No multicollinearity			With multicollinearity		
	$R^2=0.95$	$R^2=0.75$	$R^2=0.5$	$R^2=0.95$	$R^2=0.75$	$R^2=0.5$
x_1	77	68	58	39	43	40
x_2	80	72	62	48	48	54
x_3	77	66	57	51	48	46
x_4	22	28	38	41	38	40
x_5	15	20	28	36	32	35
x_6	19	27	31	42	44	42
x_7	10	19	26	43	47	43
x_1, x_2, x_3	43	23	10	1	5	5

for the variable selection process and the results of that selection are evaluated on the second set. In other words, this type of analysis shows how well a variable selection based on one set of data performs when applied to another equivalent set of data. Most frequently the data are divided in some random fashion, but other divisions may be appropriate. Of course this procedure depends on having sufficient observations so that each portion is large enough to provide reasonably definitive results. We illustrate this method with the baseball data.

EXAMPLE 6.4
REVISITED

Data Splitting   The baseball data naturally divide into two portions: the East and West leagues. One may argue that although the two leagues are different, the rules are the same; hence, the division should not be of major concern. In this example, we will actually perform the selection for each league and see how well it works on the data for the other.

Table 6.15 shows the optimum selection for each subset size for the data from each league. It can be seen that for the East league the clear choice is the three-variable model using RUNS, DP, WLK, whereas for the West league the choice is the three-variable model using DOUBLE, ERROR, WLK. Both are three-variable models, and both use WLK, but there the similarity ends. Not only are the other two variable choices different, but we also see that for other subset sizes the choices are quite different. Also, the R-square values for the East league are considerably higher than those for the West league.

One way of comparing selections is to use the coefficients of the selected models to predict the responses and compute the resulting error sums of squares. For observations in the East league, the chosen model gives coefficients and p-values (given below the coefficients):

$$\hat{PER} = 21.318 + 0.0868*RUNS - 0.1164*DP - 0.0249*WLK,$$
$$(0.0001) \qquad (0.0053) \qquad (0.0461)$$

with a residual standard deviation of 2.816. The model chosen for the West league data gives coefficients and p-values:

$$\hat{PER} = 73.992 + 0.0713*DOUBLE - 0.1131*ERROR - 0.0445*WLK,$$
$$(0.0123) \qquad (0.0010) \qquad (0.0002)$$

with a residual standard deviation of 3.556.

We now use each of these equations to predict values, that is, $\hat{\mu}_{y|x}$, for both leagues, and compute the error sum of squares, $\sum(y - \hat{\mu}_{y|x})^2$, for the data for each league. The results are as follows:

		Using data from	
		East	West
Using model estimated from	East	206.11	1272.01
	West	1132.48	328.75

TABLE 6.15

VARIABLE SELECTIONS FOR EAST AND WEST LEAGUES

East League

N = 30   Regression Models for Dependent Variable: PER

Number in Model	R-square	C(p)	Variables in Model
1	0.73714983	27.23227	RUNS
2	0.85715486	4.92892	RUNS DP
3	0.87777124	2.75370	RUNS DP WLK
4	0.88253161	3.78963	RUNS HR DP WLK
5	0.88940145	4.39836	RUNS TRIPLE HR DP WLK
6	0.89622276	5.01691	RUNS TRIPLE HR ERROR DP WLK
7	0.90067754	6.11473	RUNS DOUBLE TRIPLE HR ERROR DP WLK
8	0.90101293	8.04680	RUNS DOUBLE TRIPLE HR BA ERROR DP WLK
9	0.90124404	10.00000	RUNS DOUBLE TRIPLE HR BA ERROR DP WLK SO

West League

Number in Model	R-square	C(p)	Variables in Model
1	0.46493698	25.91413	ERROR
2	0.69526415	5.56679	BA WLK
3	0.73992927	3.23319	DOUBLE ERROR WLK
4	0.75755861	3.52271	DOUBLE BA ERROR WLK
5	0.78443458	2.91509	DOUBLE ERROR DP WLK SO
6	0.78624470	4.73946	DOUBLE HR ERROR DP WLK SO
7	0.78989459	6.38534	DOUBLE HR BA ERROR DP WLK SO
8	0.79328633	8.05625	RUNS DOUBLE TRIPLE HR ERROR DP WLK SO
9	0.79386613	10.00000	RUNS DOUBLE TRIPLE HR BA ERROR DP WLK SO

Obviously, the models selected from data of one league do not provide a good model for estimating winning percentages for the other.

Another way of comparing the selections is to see how the selected variables from one set work with the data from the other set. Using the selected model *variables* based on the West league data on the data from the East league, we have

$$\hat{PER}_{East} = 64.530 + 0.1240*DOUBLE + 0.032*ERROR - 0.0990*WLK,$$

$$(0.0038) \qquad (0.4455) \qquad (0.0001)$$

with a residual standard deviation of 4.751. Using the East league selection on the West league data,

$$\hat{PER}_{West} = 51.029 + 0.041*RUNS + 0.019*DP - 0.058*WLK,$$

$$(0.0001) \qquad (0.750) \qquad (0.0001)$$

with residual standard deviations of 4.394. The equations are obviously quite different and the fit much poorer. It is of interest to note that the East league choice provides the third-best and the West league choice the 13th-best model for the *entire* data set. ◆

One might argue that better results would be obtained by a splitting at random rather than by league, but this is not necessarily the case, as we will see.

Resampling

Current research in statistics is focusing on the use of computers to develop new procedures to aid in obtaining sampling distributions of results of statistical analyses. One such method employs repeated sampling of the data, obtaining various estimates or other results, and using these results to generate empirical sampling distributions. We will employ a simple resampling procedure here to investigate the behavior of variable selection for the baseball data.

To select a random sample from the data set, we generate for each observation a random number from the uniform distribution with range from 0 to 1. If the random number is less than some specified value, say, r, the observation is sampled; otherwise, it is not. If $r = 0.5$, the total sample will contain, on the average, half of the observations. There are conflicting consequences in choosing r: If it is small, the sample sizes may be inadequate, whereas if r is too large, the samples will be too similar.

We generate 100 samples using $r = 0.7$, and subject each sample to the all-possible-combinations procedure for subset sizes from 1 to 5. For each subset size, we count how many times each variable is selected, and since there are 100 samples, this is also the percentage of models containing each variable. The last line in the table is the mean of the C_p values for the selections. The results are shown in Table 6.16.

Looking first at the average C_p values, it appears that the optimum subset size is indeed 3. In most samples the number of walks, triples, and runs are chosen,

TABLE 6.16

VARIABLE SELECTIONS FOR REPEATED SAMPLES

Variable	One	Two	Three	Four	Five
			Variables in Model		
		Number of Times Variable Selected			
RUNS	36	77	77	74	61
DOUBLE	0	0	3	17	46
TRIPLE	0	0	51	69	80
HR	0	0	6	22	42
BA	36	23	27	44	69
ERROR	0	2	9	27	37
DP	0	2	20	41	52
WLK	28	100	100	100	100
SO	0	0	7	6	13
Mean C_p	30.6	5.02	2.34	3.34	3.69

but batting averages and double plays are used quite often. The picture becomes muddier as more variables are chosen; the only consistent choices are to include walks and exclude strikeouts. It appears that multicollinearity is doing more harm than we thought.

The results do suggest that we should choose models with RUNS, WLK, and either BA or TRIPLE, which are indeed the two best three-variable models selected, as seen in Table 6.11. Abbreviated outputs for these models are shown in Table 6.17.

In terms of overall fit, there is not much difference, but the model with TRIPLE has less multicollinearity; hence, the p-values for coefficients are smaller, although, as we have noted, these statistics should not be taken too seriously.

6.7 Usefulness of Variable Selection

We have noted that the usefulness of variable selection depends on the purpose of the statistical analysis. That is, if the primary purpose of the analysis is to provide for the estimation of values of the response variable, then variable selection may be appropriate, whereas if the purpose is to determine the structure of the relationship of the response variable and the independent variables, variable selection may not be useful as the primary analytic tool. We will illustrate this principle with the NBA data originally presented as Example 5.4 (data in Table 5.8).

TABLE 6.17

BEST THREE-VARIABLE MODELS

Model with RUNS, BA, WLK
Analysis of Variance

Source	DF	Sum of Squares	Mean Square	F Value	Prob>F
Model	3	2064.88459	688.29486	43.510	0.0001
Error	56	885.88391	15.81936		
C Total	59	2950.76850			

Parameter Estimates

Variable	DF	Parameter Estimate	Standard Error	T for H0: Parameter=0	Prob > |T|	Variance Inflation
INTERCEP	1	19.235824	19.60795925	0.981	0.3308	0.00000000
RUNS	1	0.037195	0.01426005	2.608	0.0116	3.68699888
BA	1	0.129368	0.09730702	1.329	0.1891	3.84782069
WLK	1	-0.052834	0.00849585	-6.219	0.0001	1.14446375

Model with RUNS, TRIPLE, WLK
Analysis of Variance

Source	DF	Sum of Squares	Mean Square	F Value	Prob>F
Model	3	2094.54032	698.18011	45.663	0.0001
Error	56	856.22818	15.28979		
C Total	59	2950.76850			

Parameter Estimates

Variable	DF	Parameter Estimate	Standard Error	T for H0: Parameter=0	Prob > |T|	Variance Inflation
INTERCEP	1	42.308424	7.58098201	5.581	0.0001	0.00000000
RUNS	1	0.049160	0.00790803	6.217	0.0001	1.17315271
TRIPLE	1	0.088503	0.04559148	1.941	0.0573	1.07273946
WLK	1	-0.055596	0.00817854	-6.798	0.0001	1.09730531

EXAMPLE 6.5   **Example 5.4, Revisited** Variable selection using the originally observed variables is performed by the all-possible-subsets procedure with results shown in the top portion of Table 6.18. The results are quite obvious: The four-variable model using FGM, FTM, FGAL, FTAL is clearly optimum. The final equation is shown in the bottom portion of Table 6.18.

TABLE 6.18

VARIABLE SELECTION AND OPTIMUM SUBSET MODEL FOR NBA DATA

NBA STATISTICS

N = 66   Regression Models for Dependent Variable: WINS

Number in Model	R-square	C(p)	Variables in Model
1	0.16754535	248.21833	FGM
1	0.14862208	255.27018	DRA
2	0.29474832	202.81552	FGM OFGAL
2	0.28878493	205.03781	DR DRA
3	0.52468768	119.12748	FGM OFGAL OFTAL
3	0.50302502	127.20018	FGM OFGAL OFTAT
4	0.83142420	6.82060	FGM FTM OFGAL OFTAL
4	0.82511898	9.17027	FGM FTM OFGAL OFTAT
5	0.83874450	6.09266	FGM FTAT FTM OFGAL OFTAL
5	0.83752115	6.54854	FGM FTM OFGAL OFTAL OR
6	0.84430335	6.02113	FGM FTM OFGAT OFGAL OFTAL DRA
6	0.84179912	6.95434	FGM FTAT FTM OFGAL OFTAL DRA
7	0.84992103	5.92767	FGM FTAT FTM OFGAT OFGAL OFTAL DRA
7	0.84816601	6.58169	FGAT FGM FTM OFGAT OFGAL OFTAL OR
8	0.85203476	7.13998	FGAT FGM FTM OFGAT OFGAL OFTAL OR ORA
8	0.85123320	7.43868	FGM FTAT FTM OFGAT OFGAL OFTAL DRA ORA
9	0.85470683	8.14422	FGAT FGM FTM OFGAT OFGAL OFTAL DR OR ORA
9	0.85290669	8.81505	FGAT FGM FTM OFGAT OFGAL OFTAT OFTAL OR ORA
10	0.85686915	9.33842	FGAT FGM FTM OFGAT OFGAL OFTAT OFTAL DR OR ORA
10	0.85538290	9.89228	FGAT FGM FTM OFGAT OFGAL OFTAL DR DRA OR ORA
11	0.85746389	11.11678	FGAT FGM FTM OFGAT OFGAL OFTAT OFTAL DR DRA OR ORA
11	0.85688545	11.33234	FGAT FGM FTAT FTM OFGAT OFGAL OFTAT OFTAL DR OR ORA
12	0.85777728	13.00000	FGAT FGM FTAT FTM OFGAT OFGAL OFTAT OFTAL DR DRA OR ORA

Selected Model
Analysis of Variance

Source	DF	Sum of Squares	Mean Square	F Value	Prob>F
Model	4	3846.16835	961.54209	75.214	0.0001
Error	61	779.83165	12.78413		
C Total	65	4626.00000			

(Continued)

TABLE 6.18 (*Continued*)

VARIABLE SELECTION AND OPTIMUM SUBSET MODEL FOR NBA DATA

```
          Root MSE        3.57549     R-square       0.8314
          Dep Mean       41.00000     Adj R-sq       0.8204
          C.V.            8.72071


                         Parameter Estimates


                       Parameter      Standard     T for H0:
          Variable  DF   Estimate        Error    Parameter=0     Prob > |T|

          INTERCEP   1   45.881271   13.94537817       3.290        0.0017
          FGM        1    0.058313    0.00379236      15.377        0.0001
          FTM        1    0.031047    0.00294690      10.535        0.0001
          OFGAL      1   -0.059212    0.00423221     -13.991        0.0001
          OFTAL      1   -0.031998    0.00279179     -11.461        0.0001
```

The results indicate that wins are a function of field goals and free throws made by the team and allowed by the other team. The coefficients show almost identical effects for goals made and allowed, with higher coefficients for field goals and with expected opposite signs for those made and those allowed.

Now this model is indeed quite "good," but it is extremely uninteresting. After all, scores are an exact function of goals made, and it should not take a sophisticated statistical analysis to discover this well-known fact. In other words, the model can be effectively used to estimate the number of wins for a team, but does not tell us anything about what various team performance statistics may be contributing to the number of wins.

In Section 5.4 we showed how we could reduce the multicollinearity in this example by redefining variables into attempts and percentage of attempts made. In that model we saw, for example, that for field goals, the success percentages were important, whereas for free throws the number of attempts was important. Because there are a number of nonsignificant variables in that model, a variable selection may now be used to obtain a clearer picture of the effects of these variables. The results of this selection are shown in the top portion of Table 6.19.

The results are quite different: Instead of a model with only four variables, the optimum model appears to require nine. The results of using that model are shown in the bottom portion of Table 6.19. We can see that this model contains all goal variables, and defensive rebounds allowed. The very small *p*-values for all but opponents' free-throw percentage and defensive rebounds and their "correct" signs suggest that these are indeed important. Also, as in the full model, the percentages, especially for field goals, are more important than the number of attempts. In other words, teams may profit by focusing on accuracy of field goal attempts—which may very well be obvious, but it is nice to know that a statistical analysis agrees. ◆

TABLE 6.19

VARIABLE SELECTION AND SELECTED MODEL USING DERIVED VARIABLES

Number in Model	R-square	C(p)	Variables in Model
1	0.23430809	215.65761	FGPC
1	0.14862208	246.72934	DRA
2	0.55019583	103.10940	FGPC OFGPC
2	0.40911445	154.26877	OFGPC DRA
3	0.65561600	66.88161	FGPC OFGPC OR
3	0.59963741	87.18074	FGPC FTAT OFGPC
4	0.70670198	50.35665	FGPC OFGPC OR ORA
4	0.67444183	62.05493	FGPC OFGAT OFGPC OR
5	0.73167424	43.30113	FGPC FTAT OFGPC OR ORA
5	0.73123141	43.46171	FGPC OFGPC OFTPC OR ORA
6	0.82167231	12.66575	FGAT FGPC FTAT OFGAT OFGPC OFTAT
6	0.74676576	39.82860	FGPC FTAT OFGPC OFTPC OR ORA
7	0.83328370	10.45519	FGAT FGPC FTAT FTPC OFGAT OFGPC OFTAT
7	0.82834987	12.24431	FGAT FGPC FTAT OFGAT OFGPC OFTAT OFTPC
8	0.83882391	10.44618	FGAT FGPC FTAT FTPC OFGAT OFGPC OFTAT DRA
8	0.83881947	10.44779	FGAT FGPC FTAT FTPC OFGAT OFGPC OFTAT OFTPC
9	0.84817827	9.05407	FGAT FGPC FTAT FTPC OFGAT OFGPC OFTAT OFTPC DRA
9	0.84600978	9.84042	FGAT FGPC FTAT FTPC OFGAT OFGPC OFTAT OFTPC OR
10	0.85064271	10.16041	FGAT FGPC FTAT FTPC OFGAT OFGPC OFTAT OFTPC OR ORA
10	0.84961420	10.53337	FGAT FGPC FTAT FTPC OFGAT OFGPC OFTAT OFTPC DRA ORA
11	0.85344503	11.14422	FGAT FGPC FTAT FTPC OFGAT OFGPC OFTAT OFTPC DR OR ORA
11	0.85114245	11.97919	FGAT FGPC FTAT FTPC OFGAT OFGPC OFTAT OFTPC DRA OR ORA
12	0.85384275	13.00000	FGAT FGPC FTAT FTPC OFGAT OFGPC OFTAT OFTPC DR DRA OR ORA

Selected Model
Analysis of Variance

Source	DF	Sum of Squares	Mean Square	F Value	Prob>F
Model	9	3923.67267	435.96363	34.762	0.0001
Error	56	702.32733	12.54156		
C Total	65	4626.00000			

Root MSE	3.54141	R-square	0.8482	
Dep Mean	41.00000	Adj R-sq	0.8238	
C.V.	8.63758			

(Continued)

TABLE 6.19 (*Continued*)

VARIABLE SELECTION AND SELECTED MODEL USING DERIVED VARIABLES

Parameter Estimates

Variable	DF	Parameter Estimate	Standard Error	T for H0: Parameter=0	Prob > |T|
INTERCEP	1	124.737567	57.12689848	2.184	0.0332
FGAT	1	0.026205	0.00336783	7.781	0.0001
FGPC	1	3.954196	0.42716042	9.257	0.0001
FTAT	1	0.020938	0.00277240	7.552	0.0001
FTPC	1	0.378898	0.19783733	1.915	0.0606
OFGAT	1	-0.024017	0.00299845	-8.010	0.0001
OFGPC	1	-4.696650	0.37789280	-12.429	0.0001
OFTAT	1	-0.020164	0.00289705	-6.960	0.0001
OFTPC	1	-0.923071	0.49693658	-1.858	0.0685
DRA	1	-0.010064	0.00541687	-1.858	0.0684

6.8 Variable Selection and Influential Observations

We noted in Chapter 4 that outliers and influential observations can affect the degree of multicollinearity as well as the estimates and standard errors of coefficients. Thus, we can expect that the existence of such observations could affect the results of a variable selection. We illustrate with the data on average lifespan used in Example 4.3.

EXAMPLE 6.8   **Example 4.3 Revisited**   The original model contained six variables, of which only one had a *p*-value exceeding 0.05. The variable selection, shown in the first portion of Table 6.20, confirms this by pointing to the five-variable model as being optimum. The estimates for this five-variable model are shown in the bottom of Table 6.20. The estimated coefficients and their standard error have not changed much in the selected model.

In Chapter 4 we noted that the District of Columbia was an influential outlier, largely because this "state" is atypical in many respects and especially so because of the large number of hospital beds that exist in federal institutions. It may therefore be argued that an analysis omitting this observation may be useful. To illustrate the effect of omitting this influential outlier, we perform a variable selection on the data, omitting the District of Columbia. The results of the variable selection are shown in the top portion of Table 6.21. It now appears that we only need four variables in the model and that the model does not include BEDS, whose coefficient had the second smallest *p*-value in the model using all observations. In fact, BEDS is the first variable to be deleted. The estimates for that model are shown in the bottom portion of Table 6.21.

The coefficients retain their original signs, but now education and birth rates are the dominant coefficients while the percentage of males and the divorce rates appear to contribute very little. This model does appear to be more useful.   ◆

TABLE 6.20

VARIABLE SELECTION FOR LIFE EXPECTANCY DATA

VARIABLE SELECTION

Number in Model	R-square	C(p)	Variables in Model
1	0.09657506	27.78862	BIRTH
1	0.06944192	30.03479	BEDS
2	0.25326742	16.81708	MALE BIRTH
2	0.23287541	18.50520	BIRTH BEDS
3	0.31881552	13.39079	MALE BIRTH BEDS
3	0.30324604	14.67968	BIRTH BEDS EDUC
4	0.40110986	8.57818	MALE BIRTH DIVO BEDS
4	0.38238339	10.12842	BIRTH DIVO BEDS EDUC
5	0.46103802	5.61712	MALE BIRTH DIVO BEDS EDUC
5	0.41352849	9.55012	MALE BIRTH DIVO BEDS INCO
6	0.46849267	7.00000	MALE BIRTH DIVO BEDS EDUC INCO

SELECTED MODEL

Analysis of Variance

Source	DF	Sum of Squares	Mean Square	F Value	Prob>F
Model	5	52.74146	10.54829	7.699	0.0001
Error	45	61.65574	1.37013		
C Total	50	114.39720			

Root MSE	1.17052	R-square	0.4610	
Dep Mean	70.78804	Adj R-sq	0.4012	
C.V.	1.65356			

Parameter Estimates

Variable	DF	Parameter Estimate	Standard Error	T for H0: Parameter=0	Prob > |T|
INTERCEP	1	70.158910	4.24143214	16.541	0.0001
MALE	1	0.115416	0.04503775	2.563	0.0138
BIRTH	1	-0.468656	0.10013612	-4.680	0.0001
DIVO	1	-0.207270	0.07237049	-2.864	0.0063
BEDS	1	-0.003465	0.00096213	-3.602	0.0008
EDUC	1	0.175312	0.07837350	2.237	0.0303

TABLE 6.21

VARIABLE SELECTION OMITTING THE DISTRICT OF COLUMBIA

VARIABLE SELECTION

Number in Model	R-square	C(p)	Variables in Model
1	0.11477202	14.21836	EDUC
1	0.09263914	15.72397	BIRTH
2	0.26745938	5.83168	BIRTH EDUC
2	0.18095644	11.71611	DIVO EDUC
3	0.29862181	5.71183	BIRTH DIVO EDUC
3	0.28962796	6.32364	MALE BIRTH EDUC
4	0.33608169	5.16359	MALE BIRTH DIVO EDUC
4	0.32129451	6.16950	MALE BIRTH EDUC INCO
5	0.35039370	5.64580	MALE BIRTH DIVO EDUC INCO
5	0.35175187	6.09761	MALE BIRTH DIVO BEDS EDUC
6	0.36788711	7.00000	MALE BIRTH DIVO BEDS EDUC INCO

SELECTED MODEL

Analysis of Variance

Source	DF	Sum of Squares	Mean Square	F Value	Prob>F
Model	4	29.60711	7.40178	5.695	0.0009
Error	45	58.48788	1.29973		
C Total	49	88.09499			

Root MSE	1.14006	R square	0.3361
Dep Mean	70.88960	Adj R-sq	0.2771
C.V.	1.60822		

Parameter Estimates

Variable	DF	Parameter Estimate	Standard Error	T for H0: Parameter=0	Prob > |T|
INTERCEP	1	67.574126	3.77578704	17.897	0.0001
MALE	1	0.075467	0.04736153	1.593	0.1181
BIRTH	1	-0.319672	0.09864082	-3.241	0.0022
DIVO	1	-0.119141	0.06714336	-1.774	0.0828
EDUC	1	0.236254	0.08199487	2.881	0.0060

Comments

The interplay between outliers or influential observations and multicollinearity and variable selection poses an obvious question: Which problem do we tackle first? As is true with virtually any exploratory analysis, there is no intrinsically correct procedure. It can be argued that one should investigate outliers and influential observations first, and then analyze for multicollinearity and also possibly perform variable selection. But it may not be wise to do this with outliers and/or influential observations deleted, unless there are good reasons to do so, as in the previous example

As we have noted, one tool that has been found useful in this connection is the PRESS sum of squares; that is, $\sum(y_i - \hat{\mu}_{y|x-i})^2$. Remember that the difference between that statistic and the error sum of squares is an indicator of the effect of influential outliers. Now if there are several competing models from a variable selection, one can see which are more affected by such observations.

In Example 6.8 (life expectancy data), the selected five-variable model that was obtained with the use of DC, the PRESS statistic, is 126.02 and the error mean square is 61.66; whereas for the optimum four-variable model obtained when DC is deleted, the PRESS statistic is 104.36 and the error mean square is 58.48. Thus, in both models the PRESS statistic is about twice the error sum of squares, and it seems that deleting DC has not materially reduced the effects of influential outliers. Actually, a PRESS statistic that is only twice the error mean square usually does not indicate a very serious outlier problem; hence, there may not be much outlier effect to cure, although deleting DC did produce a different model. Also, it should be added that the R-squares for all of these models are quite low; hence, we do not really have a very good model in any case.

6.9 Summary

Not many years ago, before computers expanded the horizons of statistical data analysis, the emphasis in discussions of variable selection was on the mechanics; that is, how close to an optimum model can we get with available computer resources? Now that this problem has become essentially irrelevant for all but very large models, we can concentrate our efforts on more meaningful topics. Of particular interest are the following questions:

1. Under what circumstance is variable selection a useful procedure?
2. How reliable are the results of variable selection; that is, how good is the selected model for the population?
3. How is variable selection affected by outliers and influential observations?
4. Finally, we know that the p-values obtained by variable selection are not valid. How can we evaluate the effectiveness of the coefficients of a selected model?

For these reasons we have only briefly covered the mechanics of variable selection, especially since they are easily implemented by most statistical software packages. Instead, we have spent more effort in addressing the foregoing concerns. Unfortunately, as is the case for most statistical methods employed in exploratory analyses, there are no unique and correct procedures and answers. We can only hope that this chapter will cause those who employ these methods to do so carefully, bearing in mind that any results must be carefully evaluated.

CHAPTER EXERCISES

Several of these exercises use data from exercises in Chapter 5, where variable selection was not an option. For this chapter, perform variable selection and compare results with those obtained in Chapter 5. In some cases, variable selection using derived variables may be useful. We should note that because principal components are uncorrelated, variable selection for principal component regression will indeed proceed according to the t statistics of the coefficients. Outlier detection and other diagnostics may also be useful for some of these. As a final exercise, suggest the most appropriate model and detail its usefulness.

1. The gas mileage data for a selection of cars (Chapter 5, Exercise 2) provide a compact data set with reasonably few variables. The data are in File REG05P02. The obvious multicollinearity is that bigger cars have more of everything.

2. The pork belly data (Chapter 5, Exercise 3, File REG05P03) is an interesting example even if the interpretation of variables is somewhat obscure. Because the focus is on estimation, variable selection may be more useful here.

3. The purpose of collecting the evaporation data (Chapter 5, Exercise 4, File REG05P04) was to determine what factors influence evaporation of water from soil. Therefore, variable selection is not normally recommended. However, because of the extreme multicollinearity, it may be useful to eliminate some redundant variables. After eliminating some variables, a study of structure may still be useful.

4. The purpose of Exercise 5 of Chapter 5 (File REG05P05) was to determine factors associated with crime; hence, variable selection may not be useful. Furthermore, remedies for the multicollinearity may have produced a better model. Nevertheless, variable selection, either on the original or the modified model, may help to clarify the relationships.

5. In Exercise 5 of Chapter 3 the overall CPI was related to some of its components using data for the years 1960 through 1994. Using this data and a variable selection procedure, determine the best model to predict the ALL variable. What condition might affect the results of this analysis?

6. These data from Draper and Smith (1988, p. 374) provide an excellent exercise for comparing several approaches for the analysis of a very small data set. A parcel packing crew consists of five workers identified by numbers 1 through 5, plus a foreman who works at all times. For each day we record:

$X_j = 1$ if worker j is on duty, and 0 otherwise
$Y = $ number of parcels dispatched

The data are shown in Table 6.21 and are available in File REG06P06.

(a) State some purposes for collecting and analyzing this data set.
(b) Perform regression of Y on X_1–X_5. The computer output may have some unexpected results. Explain the reasons for this.
(c) Perform a regression of Y on X_1, X_2, X_3, and X_4. Why do we not use X_5? Interpret results. Is the intercept of any value? Be careful in any interpretations.
(d) Create a new variable NUMBER $= X_1 + X_2 + X_3 + X_4 + X_5$, and perform the regression of Y on NUMBER. Make a formal (hypothesis test) comparison of the results with those of part (c). What do the results suggest?
(e) Would variable selection be useful here? Explain your answer.

7. Myers (1990) uses an example (Example 5.2) that attempts to estimate the manpower needs for operating Bachelor Officers Quarters (BOQ) for the U.S. Navy. Data for 25 such installations are given in Table 6.22 and are available in File REG06P07.

TABLE 6.21

PACKING DATA

X_1	X_2	X_3	X_4	X_5	Y
1	1	1	0	1	246
1	0	1	0	1	252
1	1	1	0	1	253
0	1	1	1	0	164
1	1	0	0	1	203
0	1	1	1	0	173
1	1	0	0	1	210
1	0	1	0	1	247
0	1	0	1	0	120
0	1	1	1	0	171
0	1	1	1	0	167
0	0	1	1	0	172
1	1	1	0	1	247
1	1	1	0	1	252
1	0	1	0	1	248
0	1	1	1	0	169
0	1	0	0	0	104
0	1	1	1	0	166
0	1	1	1	0	168
0	1	1	0	0	148

TABLE 6.22

BOQ DATA

OCCUP	CHECKIN	HOURS	COMMON	WINGS	CAP	ROOMS	MANH
2.00	4.00	4.0	1.26	1	6	6	180.23
3.00	1.58	40.0	1.25	1	5	5	182.61
16.60	23.78	40.0	1.00	1	13	13	164.38
7.00	2.37	168.0	1.00	1	7	8	284.55
5.30	1.67	42.5	7.79	3	25	25	199.92
16.50	8.25	168.0	1.12	2	19	19	267.38
25.89	3.00	40.0	0.00	3	36	36	999.09
44.42	159.75	168.0	0.60	18	48	48	1103.24
39.63	50.86	40.0	27.37	10	77	77	944.21
31.92	40.08	168.0	5.52	6	47	47	931.84
97.33	255.08	168.0	19.00	6	165	130	2268.06
56.63	373.42	168.0	6.03	4	36	37	1489.50
96.67	206.67	168.0	17.86	14	120	120	1891.70
54.58	207.08	168.0	7.77	6	66	66	1387.82
113.88	981.00	168.0	24.48	6	166	179	3559.92
149.58	233.83	168.0	31.07	14	185	202	3115.29
134.32	145.82	168.0	25.99	12	192	192	2227.76
188.74	937.00	168.0	45.44	26	237	237	4804.24
110.24	410.00	168.0	20.05	12	115	115	2628.32
96.83	677.33	168.0	20.31	10	302	210	1880.84
102.33	288.83	168.0	21.01	14	131	131	3036.63
274.92	695.25	168.0	46.63	58	363	363	5539.98
811.08	714.33	168.0	22.76	17	242	242	3534.49
384.50	1473.66	168.0	7.36	24	540	453	8266.77
95.00	368.00	168.0	30.26	9	292	196	1845.89

The variables are

OCCUP: Average daily occupancy
CHECKIN: Monthly average number of check-ins
HOURS: Weekly hours of service desk operation
COMMON: Square feet of common-use area
WINGS: Number of building wings
CAP: Operational berthing capacity
ROOMS: Number of rooms

and the response variable,

MANH: Monthly man-hours required to operate

(a) Perform the regression to estimate MANH.
(b) Perform variable selection. Is the resulting model useful?
(c) The existence of multicollinearity is obvious. Determine its nature and try to remedy it. You may want to use more than one method. If suitable, perform variable selection on an alternative set of variables.
(d) After all of this, there is an outlier. Identify it and suggest additional analyses.

PART III

ADDITIONAL USES OF REGRESSION

The final section of this book presents additional applications of the regression model. The primary emphasis is on models that do not describe a straight line relationship of the response to the independent or factor variables. In this part we must distinguish between linear models that are linear in the parameters but may describe relationships that are not described by straight lines and nonlinear models that are not linear in the parameters. Also included is the use of models where the response variable is not a continuous variable.

Chapter 7 introduces the fitting of curved response functions by the use of polynomial models. Because polynomial functions rarely have foundations in theory, they are primarily used as approximations to more complex, usually nonlinear, functions. The chapter presents the use of polynomials in one variable, segmented polynomials in one variable, and polynomials with several variables. We will see that model selection and visual presentation of results are important aspects in the use of polynomial models.

Chapter 8 deals with intrinsically linear models, which are nonlinear models that can be linearized through transformations, as well as with intrinsically nonlinear models for which different methodology must be used. Examples of nonlinear models include exponential decay and growth models and the logistic growth model.

Chapter 9 presents the general linear model that is used to show the equivalence of regression, which normally relates the response to continuous factor variables, and the analysis of variance, which relates the response to categorical factors. Being able to pose an analysis of variance as a regression is essential when dealing with unbalanced data in a factorial data structure. This methodology is also required when a model must contain both quantitative and qualitative factors.

Chapter 10 introduces the special case when the response variable is categorical. Standard linear models methodology may be adapted for the situation where the response is binary, but special methodology is required when the response has more than two categories.

Polynomial Models

7.1 Introduction

Up to this point all the models we have used have either described straight-line (or planar) relationships or have been slight modifications of such models. For example, in Example 4.5 we saw that we could estimate a linear equation with a break, that is, a sudden change in the slope, and in Chapter 6 we presented an example in which the addition of the square of an independent variable produced a curved response line. Obviously, not all physical phenomena can be modeled with a straight line, at least for extended values of the independent variables. Therefore, it is important to be able to use models that are graphically described with curved lines.

As we have already seen, a simple nonlinear model occurs when we include the squared value of an independent variable. The resulting model is a simple example of a large class of mathematical functions called polynomials. The graph of a model with a squared independent variable is a parabola and can be used to fit a number of physical processes. For example, if we were modeling the sales of ice cream using the month of the year as an independent variable, the following model would be a good choice:

$$y = \beta_0 + \beta_1 x + \beta_2 x^2 + \epsilon$$

where y is sales of ice cream and x is the month of the year, labeled 1 through 12.

We would expect the sales of ice cream to follow a parabolic trend, with the maximum occurring in July or August, the minimum in December or January. Since this is a relatively simple mathematical expression, an elementary exercise in calculus can easily be used to solve for the month at which maximum sales occur. This is typical of sales data for a number of commodities, and many economic models use the square term to reflect the cyclic nature of this type of data.

Unfortunately, theory often dictates a model that is more complex and cannot be estimated using the straightforward techniques discussed thus far. For example, theories of physics specify that the radiation of a radioactive substance over time can be modeled by the function

$$y = \beta e^{-\gamma t} + \epsilon$$

where

y = observed radiation
t = time
β = initial count at time zero
γ = rate of decay, which is related to the half-life of the substance
ϵ = random error

Notice that the model contains a random error term, which means that a statistical model is appropriate and we should be able to use a sample to obtain estimates for the unknown coefficients to predict and/or explain the behavior of the radiation over time. Unfortunately, the mathematical form of the deterministic portion of the model is not linear in the parameters and cannot be estimated with the methodology presented so far. Therefore, we are presented with two options:

1. We can use more sophisticated methodology and obtain estimates of the coefficients directly from the model. We will discuss this methodology in detail in Chapter 8, where we will obtain the estimates for the coefficients in this decay model.
2. We can find an equation that is linear in the parameters, that closely approximates the nonlinear model, and whose graph is a smooth curve (or plane in several dimensions). This equation will lend itself to straightforward regression methodology to obtain the needed estimates. This type of regression analysis is usually called **curve fitting** or **smoothing**. In this approach we are usually not able to obtain estimates of the coefficients of the original model, but we can obtain reasonable estimates of the response variable and construct a graph of the response curve.

The most popular functional form for curve fitting is the polynomial model; that is, a model that is a linear function of powers of one or more independent variables. Implementing polynomial models is the primary topic of discussion in this chapter, and we will present techniques for using polynomial models for curve fitting. We will also present polynomial models in which some model parameters are different for ranges of the independent variables.

7.2 Polynomial Models with One Independent Variable

A polynomial model with one independent variable, x, is written

$$y = \beta_0 + \beta_1 x + \beta_2 x^2 + \beta_3 x^3 + \ldots + \beta_m x^m + \epsilon,$$

where

y is the response variable
x is the independent variable
β_i, $i = 0, 1, 2, \ldots, m$, are the coefficients for the ith power of x
ϵ is the random error, defined as usual

The model as written is called an mth-order polynomial,[1] and m may take any value, although values less than 3 or 4 are most commonly used in practice.
Reasons for the popularity of the polynomial model are as follows:

- It is easy to implement because it can be performed as a linear regression
- According to various series expansions, any continuous function can be approximated over a limited range by a polynomial function

Another reason why the polynomial model is popular is that it allows for a sequential fitting of increasingly complex curves. Figure 7.1 shows the basic shapes of the linear, quadratic, cubic, and fourth-order (often called quartic) polynomial curves. As we can see, the linear function is a straight line, the quadratic is a parabola with one "hump," the cubic has two humps and one inflection point, and the quartic has three humps and two inflection points. Adding additional terms simply increases the number of humps and inflection points.

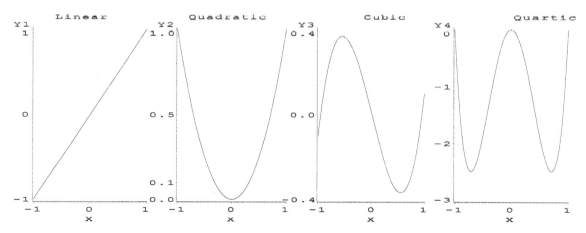

FIGURE 7.1 Polynomial Shapes

[1] Polynomial models as used in this chapter are restricted to having integer-valued exponents.

The actual shape of a polynomial response curve depends on the relative magnitudes of the coefficients of the various polynomial terms. For example, a hump may look more like a shoulder; however, there will still be the correct number of inflection points, although in some cases, some humps and inflection points are outside the range of the data.

The polynomial model is easy to implement. Simply define w_i as x^i and specify the model

$$y = \beta_0 + \beta_1 w_1 + \beta_2 w_2 + \beta_3 w_3 + \ldots + \beta_m w_m + \epsilon,$$

and proceed as with any linear regression model. Although a few computer programs are specifically designed for fitting polynomial models, most programs require that the powers of the independent variable, that is, the w_i, be computed before executing the regression.

As we have noted, an important issue in fitting polynomial models is the specification of m, the degree of polynomial. Obviously, the greater the value of m, the better the fit. In fact, if $m = n - 1$, the regression will fit perfectly! However, as with most regressions, models with fewer parameters are preferred. The obvious answer is variable selection. That is, fit a model with arbitrarily large m and use a variable selection procedure to determine the appropriate number of parameters. Unfortunately, such a procedure is not recommended because in virtually all polynomial models it is customary to include all lower powers[2] of x less than m.

This means that polynomial models are built sequentially. An initial model with an arbitrary number of terms is computed as usual. However, instead of computing partial sums of squares for the coefficients, we compute *sequential* sums of squares, which show the sum of squares obtained by adding each term to a model containing all lower-order terms. Thus, the sequential sum of squares for the linear term shows the reduction in the error sum of squares as the linear term is added to the intercept, the sequential sum of squares for the quadratic is that obtained by adding that term to the model containing the intercept and linear term, and so forth. Terms are normally included in the model as they are significant (at some predetermined level) until two successive terms are deemed nonsignificant. Because the Type I error of including a term when it is not needed is not of great consequence, a larger level of significance may be used.

Before continuing with some examples, a few words of caution:

- Because polynomial models are simply a curve-fitting process and normally do not correspond to some physical model, the usual warning about extrapolation is extremely relevant.
- Values of powers of a variable tend to be highly correlated. These correlations tend to become very high (i) as the range of the variable is small compared to the mean and (ii) as higher powers are used. In other words, there may be extreme multicollinearity among the variables in a polynomial regression. Because with polynomial regression we are usually inter-

---

[2] Exceptions may occur if one is fitting a function for which the series expansion specifies that either even or odd powers be included.

ested in estimation of the response rather than interpreting coefficients, this multicollinearity poses no real problem. However, extreme multi-collinearity does tend to cause roundoff error in computing the inverse of $X'X$. This effect can be ameliorated by linear transformations of the independent variables. For example, if the independent variable is calendar years, roundoff error problems will be reduced by transforming that variable to start with the year 1.

EXAMPLE 7.1   Fitting a Normal Curve   We will illustrate the method of curve fitting by using a polynomial equation to approximate the standard normal distribution. The model we wish to fit is

$$y = \frac{1}{\sqrt{2\pi}} e^{-(x^2/2)} + \epsilon.$$

The deterministic portion of the model is the equation for the normal distribution with mean 0 and unit variance. For data, we generate values of the normal function for 31 x values from -3 to $+3$ in increments of 0.2 and add a normally distributed random error with mean 0 and standard deviation of 0.05. The data points are shown in Figure 7.2 along with the actual normal curve.[3] Note that the actual normal curve does extend beyond the generated data.

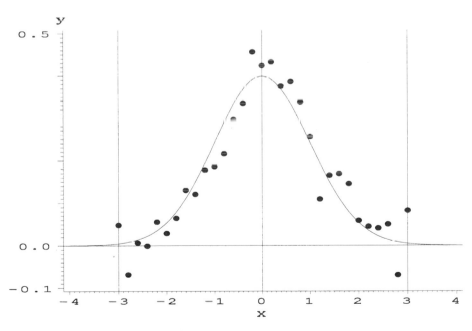

FIGURE 7.2 Data for Fitting Normal Curve

[3] The actual data are available on the data diskette in File REG07X01.

We begin by fitting a sixth-degree polynomial, that is,

$$y = \beta_0 + \beta_1 x + \beta_2 x^2 + \beta_3 x^3 + \beta_4 x^4 + \beta_5 x^5 + \beta_6 x^6 + \epsilon.$$

We use PROC GLM of the SAS System, which directly allows the use of powers of variables and also provides the sequential (labeled Type I) sums of squares and the corresponding F values. Because the coefficients and other results are of no interest at this point, only the sequential sums of squares are shown in Table 7.1.

The model is certainly significant and the error mean square is very close to the true error variance. The sequential (Type I) sums of squares are used to determine the minimum degree of polynomial model required. We can immediately see that the fifth- and sixth-degree terms are not needed; hence, we will use the fourth-degree polynomial. The results are shown in Table 7.2.

Again, the model is significant and the residual mean square close to the true value of 0.0025. Although the coefficients in a polynomial model are usually of little interest, an interesting result here is that coefficients of the odd powers are not significant. This is because the normal curve is symmetric about 0 and the odd powers reflect lack of symmetry about 0. This, then, is one example where one may be justified in using only even powers in the model; however, it gains us very little. Actually, in most applications such symmetry does not exist.

The predicted curve (solid line) is shown in Figure 7.3 along with the normal curve (solid line with dots). The curves have been extrapolated beyond the

TABLE 7.1

SIXTH-DEGREE POLYNOMIAL

Source	DF	Sum of Squares	Mean Square	F Value	Pr > F
Model	6	0.62713235	0.10452206	41.25	0.0001
Error	24	0.06081401	0.00253392		
Corrected Total	30	0.68794636			

	R-Square	C.V.	Root MSE	Y Mean
	0.911601	30.87710	0.050338	0.16302708

Source	DF	Type I SS	Mean Square	F Value	Pr > F
X	1	0.00595140	0.00595140	2.35	0.1385
X*X	1	0.48130767	0.48130767	189.95	0.0001
X*X*X	1	0.00263020	0.00263020	1.04	0.3184
X*X*X*X	1	0.12736637	0.12736637	50.26	0.0001
X*X*X*X*X	1	0.00155382	0.00155382	0.61	0.4412
X*X*X*X*X*X	1	0.00832288	0.00832288	3.28	0.0825

$\rightarrow$

TABLE 7.2

FOURTH-DEGREE POLYNOMIAL

Dependent Variable: Y

Source	DF	Sum of Squares	Mean Square	F Value	Pr > F
Model	4	0.61725565	0.15431391	56.76	0.0001
Error	26	0.07069071	0.00271887		
Corrected Total	30	0.68794636			

R-Square	C.V.	Root MSE	Y Mean
0.897244	31.98415	0.052143	0.16302708

Parameter	Estimate	T for H0: Parameter=0	Pr > |T|	Std Error of Estimate
INTERCEPT	0.3750468415	21.31	0.0001	0.01760274
X	0.0195953922	1.49	0.1478	0.01313621
X*X	.1194840059	10.34	0.0001	0.01156088
X*X*X	-.0020601205	-0.98	0.3344	0.00209456
X*X*X*X	0.0092537953	6.84	0.0001	0.00135203

range of the data (-3 to $+3$) to show why extrapolation is not justified for polynomial models. ◆

In the previous example we used sequential tests using the residual mean square to determine the degree of polynomial model required. If multiple

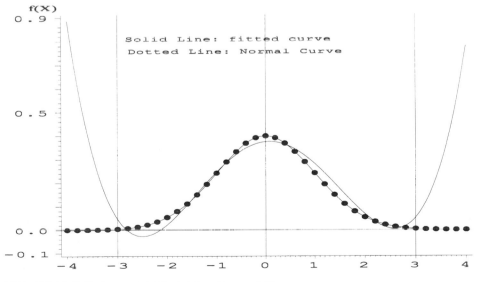

FIGURE 7.3 Polynomial Model for Normal Curve

observations exist for individual values of the independent variable, a lack of fit test (Section 6.3) may be used.

The decay model presented in Section 7.1 can be approximated by a polynomial in a very similar manner. Example 8.5 in Section 8.3 presents data from such a model (Table 8.12). Using the nonlinear methods presented in Chapter 8, the estimated model is

$$\hat{\mu}_{y|t} = 517.3e^{-0.04t}.$$

The procedure discussed previously to fit a polynomial model resulted in a cubic polynomial being appropriate to approximate the decay model. The results of the cubic polynomial regression are given in Table 8.15 and resulted in the polynomial equation

$$\hat{\mu}_{y|t} = 566.81 - 31.56t + 0.78t^2 - 0.007t^3.$$

This polynomial fits the data well; however, it does not give us any idea as to the values of the rate of decay or the intitial count at time zero (parameters in the original model). A comparison of the polynomial with other models is given in Section 8.3.

EXAMPLE 7.2   **Cooling Degree Days**   Suppliers of energy use climatic data to estimate how much energy may be needed for heating or cooling. Cooling degree days, defined as the sum of the excess of mean daily temperatures above 75°F for a period, such as a month, is used as a measure to ascertain air conditioning requirements. Table 7.4 shows monthly cooling degree days for a Texas city for the five years 1983 through 1987. We will use these data to fit a curve to describe the pattern of cooling degree days over the months.

The first step of the analysis is to fit the unrestricted model. In this case the analysis of variance, using month and years as the sources of variation, is the un-

TABLE 7.3

COOLING DEGREE DAYS

Month	No.	1983	1984	1985	1986	1987
January	1	1	2	1	4	4
February	2	1	13	10	47	1
March	3	19	70	93	36	14
April	4	78	154	160	197	146
May	5	230	368	305	274	360
June	6	387	462	498	514	471
July	7	484	579	578	634	581
August	8	508	594	673	571	664
September	9	395	411	464	518	424
October	10	215	255	233	149	162
November	11	65	47	92	79	62
December	12	8	47	5	3	15→

restricted model,[4] which provides 44 degrees of freedom for the pure error. The results are shown in Table 7.4.

Although the model is certainly significant, the results are not very interesting. Obviously we would expect MONTH to be significant; the fact that YEAR is also significant suggests that the use of years as blocks is justified. The important number here is the error variance: 1371.8, which is the pure error for the lack of fit test. We will compare this value with the residual mean squares from a polynomial fit.

As before, we start with a sixth-degree polynomial. This time we will use SAS PROC REG, where the variables M, M2, and so forth are the computer names for MONTH, MONTH^2, and so forth. We have requested the sequential (Type I) sums of squares and also the coefficients of each step in the polynomial building process (Sequential Parameter Estimates). The results are shown in Table 7.5.

The lack of fit test is performed using the ANOVA sum of squares for MONTH as the unrestricted model and the regression as the restricted model:

Unrestricted	SS = 2,859,790,	df = 11
Restricted	SS = 2,846,194,	df = 6
Lack of fit	SS = 13,596,	df = 5

$MS = 2719.2$, $F = 2719.2/1371.8 = 1.982$, df = (5, 44); not significant. The sixth-degree model is adequate.

TABLE 7.4

ANALYSIS OF VARIANCE FOR COOLING DEGREE DAYS

Dependent Variable: CDD

Source	DF	Sum of Squares	Mean Square	F Value	Pr > F
Model	15	2887246.91666667	192483.12777778	140.31	0.0001
Error	44	60361.66666667	1371.85606061		
Corrected Total	59	2947608.58333333			

R-Square	C.V.	Root MSE	CDD Mean
0.979522	15.39532	37.03857530	240.58333333

Source	DF	Anova SS	Mean Square	F Value	Pr > F
YEAR	4	27456.33333333	6864.08333333	5.00	0.0021
MONTH	11	2859790.58333333	259980.96212121	189.51	0.0001

[4] We treat years as blocks, although the data could be considered a completely randomized design with years as independent samples.

TABLE 7.5

POLYNOMIAL REGRESSION FOR COOLING DEGREE DAYS

Analysis of Variance

Source	DF	Sum of Squares	Mean Square	F Value	Prob>F
Model	6	2846193.5288	474365.58813	247.906	0.0001
Error	53	101415.05455	1913.49160		
C Total	59	2947608.5833			

Root MSE	43.74347	R-square	0.9656
Dep Mean	240.58333	Adj R-sq	0.9617
C.V.	18.18225		

Parameter Estimates

Variable	DF	Parameter Estimate	Standard Error	T for H0: Parameter=0	Prob > |T|	Type I SS
INTERCEP	1	-265.300001	154.80643613	-1.714	0.0924	3472820
M	1	520.601746	258.76374769	2.012	0.0493	153976
M2	1	-343.096323	148.58309995	-2.309	0.0249	2003899
M3	1	100.023998	39.49811828	2.532	0.0143	271801
M4	1	-12.706640	5.30267844	-2.396	0.0201	313902
M5	1	0.704359	0.34875857	2.020	0.0485	98045
M6	1	-0.013791	0.00892261	-1.546	0.1282	4571.140137

Sequential Parameter Estimates

INTERCEP	M	M2	M3	M4	M5	M6
240.58333333	0	0	0	0	0	0
145.1969697	14.674825175	0	0	0	0	0
-380.4409091	239.9482018	-17.32872128	0	0	0	0
-84.73333333	11.397657898	24.915218115	-2.166355866	0	0	0
424.85757576	-567.9586733	207.49529429	-23.39931041	0.8166520979	0	0
-65.80909095	161.02915988	-126.2301959	40.61677808	-4.595112608	0.1665158371	0
-265.300001	520.60174625	-343.0963229	100.02399761	-12.70663968	0.7043589768	-0.01379085

The next step is to see if the sixth-degree term is needed. From the sequential sums of squares we obtain

$$F = 4571.1/1371.8 = 3.33, \text{ df} = (1, 44), \text{ not significant at the 0.05 level.}$$

The fifth-degree term has a sequential sum of squares of 98,045, which will obviously lead to rejection; hence, the fifth-degree polynomial model is required. The equation is obtained from the fifth line of the sequential parameter estimates:

$$\hat{\mu}_{y|x} = -65.809 - 161.03M - 126.23M^2 + 40.62M^3 - 4.595M^4 + 0.1655M^5.$$

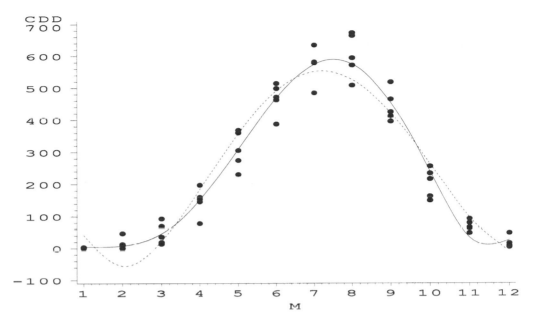

FIGURE 7.4 Polynomial Fit for Cooling Degree Days

The fifth-degree curve and the original data are plotted in Figure 7.4, which also includes the fourth-degree curve (dotted line), which does indeed show that adding the fifth-degree term improves the fit, especially for February. ◆

7.3 Segmented Polynomials with Known Knots

In Example 4.5 we used an indicator variable to allow a break in an otherwise straight-line regression. This was a simple example of segmented polynomial regression, sometimes called spline regression, where different polynomial models are used for different ranges of a single independent variable. A spline must join at points where the model specification changes. These points where the different models join are known as *knots*.[5] Before we proceed to implement such a model, some comments are in order:

- It is sometimes important to know if the response curve is continuous at a knot (Smith, 1979). We will not concern ourselves with this problem.
- If the location of knots is known, the model can be fitted using linear regression; if not, it is a nonlinear regression problem. We will only consider the case for known knots in this chapter.

---

[5] Segmented polynomials are hypothetically possible with more than one independent variable but are very difficult to implement.

Segmented Straight Lines

We want to fit the following model:

$$y = \beta_{01} + \beta_1 x_1 + \epsilon, \text{ for } x_1 \leq c$$

$$y = \beta_{02} + \beta_2 x_1 + \epsilon, \text{ for } x_1 \geq c.$$

The single knot occurs at $x_1 = c$, where $\hat{\mu}_{y|x}$ has the same value for both functions. Note that β_2 may take any value. If it is equal to β_1 we have a straight-line regression over the entire range of x_1. This model is readily fitted by defining a new variable:

$$x_2 = 0, \text{ for } x_1 \leq c$$

$$x_2 = (x_1 - c), \text{ for } x_1 > c,$$

and using the model $y = \gamma_0 + \gamma_1 x_1 + \gamma_2 x_2 + \epsilon$. This results in fitting the models

$$y = \gamma_0 + \gamma_1 x_1 + \epsilon, \text{ for } x_1 \leq c$$

$$y = (\gamma_0 - \gamma_2 c) + (\gamma_1 + \gamma_2) x_1 + \epsilon, \text{ for } x_1 > c.$$

In other words:

$$\beta_{01} = \gamma_0$$

$$\beta_1 = \gamma_1$$

$$\beta_{02} = \gamma_0 - \gamma_2 c$$

$$\beta_2 = \gamma_1 + \gamma_2.$$

Note that the test for $\gamma_2 = 0$ is the test for a straight-line regression.

Segmented Polynomials

The foregoing procedure is readily extended to polynomial models. For spline regression applications, quadratic polynomials are most frequently used. The quadratic spline regression with a single knot at $x_1 = c$ has the model

$$y = \beta_{01} + \beta_1 x_1 + \beta_2 x_1^2 + \epsilon \text{ for } x_1 \leq c$$

$$y = \beta_{02} + \beta_3 x_1 + \beta_4 x_1^2 + \epsilon \text{ for } x_1 > c.$$

Defining x_2 as before,

$$x_2 = 0, \text{ for } x_1 \leq c$$

$$x_2 = (x_1 - c), \text{ for } x_1 > c,$$

we fit the model

$$y = \gamma_0 + \gamma_1 x_1 + \gamma_2 x_1^2 + \gamma_3 x_2 + \gamma_4 x_2^2 + \epsilon,$$

which results in fitting the models

$$y = \gamma_0 + \gamma_1 x + \gamma_2 x^2 + \epsilon, \text{ for } x \leq c$$

$$y = (\gamma_0 - \gamma_3 c + \gamma_4 c^2) + (\gamma_1 + \gamma_3 - 2c\gamma_4)x + (\gamma_2 + \gamma_4)x^2 + \epsilon, \text{ for } x > c.$$

In other words:

$$\beta_{01} = \gamma_0$$

$$\beta_1 = \gamma_1$$

$$\beta_2 = \gamma_2$$

$$\beta_{02} = \gamma_0 - \gamma_3 c + \gamma_4 c^2$$

$$\beta_3 = \gamma_1 + \gamma_3 - 2c\gamma_1$$

$$\beta_4 = \gamma_2 + \gamma_4.$$

Furthermore, tests of the hypotheses

$$H_{01} : (\gamma_3 - 2c\gamma_1) = 0$$

and

$$H_{02} : \gamma_4 = 0$$

provide information on the differences between the linear and quadratic regression coefficients for the two segments. Many computer programs for multiple regression provide the preceding estimates, as well as standard errors and tests.

EXAMPLE 7.3   Simulated Data   Forty-one observations are generated for values of x from 0 to 10 in steps of 0.25, according to the model

$$y = x \quad 0.1x^2 + \epsilon \text{ for } x \leq 5$$

$$y = 2.5 + \epsilon \text{ for } x > 5.$$

Note that $\hat{\mu}_{y|x}$ has the value 2.5 at $x = 5$ for both functions. The variable ϵ is a normally distributed random variable with mean 0 and standard deviation of 0.2. This curve may actually be useful for describing the growth of animals that reach a mature size and then grow no more. The resulting data are shown in Table 7.6.
We define

$$x_1 = x,$$

$$x_2 = 0 \text{ for } x \leq 5$$

$$= (x - 5) \text{ for } x > 5,$$

and fit the model

$$y = \gamma_0 + \gamma_1 x_1 + \gamma_2 x_1^2 + \gamma_3 x_2 + \gamma_4 x_2^2 + \epsilon.$$

TABLE 7.6

DATA FOR SEGMENTED POLYNOMIAL

x	y	x	y	x	y	x	y
0.00	−.06	2.50	1.72	5.00	2.51	7.50	2.59
0.25	0.18	2.75	2.10	5.25	2.84	7.75	2.37
0.50	0.12	3.00	2.04	5.50	2.75	8.00	2.64
0.75	1.12	3.25	2.35	5.75	2.64	8.25	2.51
1.00	0.61	3.50	2.21	6.00	2.64	8.50	2.26
1.25	1.17	3.75	2.49	6.25	2.93	8.75	2.37
1.50	1.53	4.00	2.40	6.50	2.62	9.00	2.61
1.75	1.32	4.25	2.51	6.75	2.43	9.25	2.73
2.00	1.66	4.50	2.54	7.00	2.27	9.50	2.74
2.25	1.81	4.75	2.61	7.25	2.40	9.75	2.51

Remember that we want to make inferences on the coefficients of the segmented regression that are linear functions of the coefficients of the model we are actually fitting. Therefore, we use PROC GLM of the SAS System because it has provisions for providing estimates and standard errors for estimates of linear functions of parameters and also gives the F values for the sequential sums of squares. The output is shown in Table 7.7.

The model is obviously significant. The sequential (SAS Type I) sums of squares indicate that both quadratic coefficients are needed; normally this means all terms should be kept. The parameter estimates, which are denoted by γ_i in the text, are labeled by the variable names (X1 for x_1 and so forth). From our equivalences, we know that INTERCEPT, X1, and X1SQ are the estimated parameters for the function when $x \leq 5$; the estimates of -0.0418, 0.9998, and -0.0917 are quite close to the parameter values of 0, 1, and -0.1 for intercept, linear, and quadratic.

The estimates of the parameters for $x > 5$ are found in the last section, which provides the estimates and tests for specified linear functions of the model parameters. The parameter INT2 is the intercept for the second segment; the estimate of 3.2228 appears not to be too close to the actual parameter value of 2.5, but the standard error of 0.9351 makes the estimate within the 0.95 confidence interval. The reason for the large standard error is that this estimate is actually an extrapolation of the existing data for the second segment.

The parameters LIN2 and QUAD2 are those for the linear and quadratic of the second segment: The estimates -0.1472 and 0.0071 are both quite close to the parameter values of 0. The line labeled TESTLIN is the test for the difference of the linear coefficient between the two segments. In terms of the computer labels, this is the test that $(X2 - 10 \cdot QUAD2) = 0$; this hypothesis is re-

TABLE 7.7

RESULTS FOR SEGMENTED POLYNOMIAL

Source	DF	Sum of Squares	Mean Square	F Value	Pr > F
Model	4	22.36845012	5.59211253	162.16	0.0001
Error	36	1.24142793	0.03448411		
Corrected Total	40	23.60987805			

R-Square	C.V.	Root MSE	Y Mean
0.947419	8.882009	0.185699	2.090732

Source	DF	Type I SS	Mean Square	F Value	Pr > F
X1	1	13.37128251	13.37128251	387.75	0.0001
X1SQ	1	8.10256755	8.10256755	234.97	0.0001
X2	1	0.03573130	0.03573130	1.04	0.3155
X2SQ	1	0.85886877	0.85886877	24.91	0.0001

Parameter	Estimate	T for H0: Parameter=0	Pr > |T|	Std Error of Estimate
INTERCEPT	-.0418458407	-0.38	0.7039	0.10923909
X	0.9997612180	10.31	0.0001	0.09699696
X2	-.0917088463	-5.16	0.0001	0.01778169
XS	-.1588348849	-1.02	0.3155	0.15603822
XS2	0.0988185490	4.99	0.0001	0.01980088

Parameter	Estimate	T for H0: Parameter=0	Pr > |T|	Std Error of Estimate
int2	3.22279231	3.45	0.0015	0.93514650
lin2	-0.14725916	-0.56	0.5788	0.26284671
quad2	0.00710970	0.40	0.6916	0.01778169
testlin	-1.14702038	-4.55	0.0001	0.25210201

jected, and hence, we confirm the difference in the linear coefficient between the two segments. The test for the differences in the quadratics is obtained directly from the coefficient X2SQ, which is clearly rejected.

The plot of the estimated curve and the data points is shown in Figure 7.5. It is seen that the curve is reasonably close to what was specified. It is left to the reader to see if a simple quadratic or possibly cubic polynomial would fit as well.   ◆

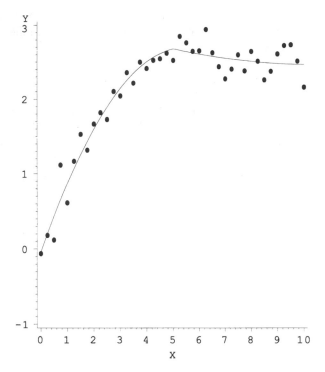

FIGURE 7.5 Segmented Polynomial Example

7.4 Polynomial Regression in Several Variables; Response Surfaces

When a polynomial model contains several independent variables, the model may include products of variables in addition to the powers of the individual variables. For example, the so-called quadratic polynomial response surface model with two variables x_1 and x_2 is

$$y = \beta_{00} + \beta_{10}x_1 + \beta_{20}x_1^2 + \beta_{01}x_2 + \beta_{02}x_2^2 + \beta_{11}x_1x_2 + \epsilon.$$

Note that the subscripts for the coefficients mirror the powers of the independent variables; this notation is especially useful if the model includes many variables and higher powers.

The interpretation of the terms that involve only one variable are the same as for the one-variable case. The interpretation of the cross-product term $\beta_{11}x_1x_2$ is aided by rewriting the model:

$$y = \beta_{00} + (\beta_{10} + \beta_{11}x_2)x_1 + \beta_{20}x_1^2 + \beta_{01}x_2 + \beta_{02}x_2^2 + \epsilon.$$

We can now see that the coefficient β_{11} shows that the response to x_1 is not constant: It increases linearly with x_2. For example, given the model

$$y_1 = x_1 + x_2 - 0.2x_1x_2,$$

(we have omitted the quadratic terms and the error for simplicity) the response to x_1 is $(1 - 0.2) = 0.8$ units when x_2 has the value 1, while the response is $(1 - 9 \times 0.2) = -0.8$ units when x_2 has the value 9. In the same fashion the response to x_2 changes linearly with changes in x_1. The response curve is illustrated in the left portion of Figure 7.6, where we can see that the slope in the x_1 direction changes from negative to positive as x_2 increases and vice versa for x_2. However, note that any cross-section of the surface in either direction is still strictly linear.

Interpretation of products involving quadratic terms is done in a similar manner. For example, in the function

$$y_2 = x_1 - 0.5x_1^2 + 0.15x_1^2x_2,$$

we recombine terms,

$$y_2 = x_1 + (-0.5 + 0.15x_2)x_1^2,$$

where we can see that the coefficient for the quadratic term in x_1 goes from -0.35 when x_2 is equal to unity to $+0.86$ when x_2 is equal to 9. This is illustrated in the right portion of Figure 7.6. In this case, however, the interpretation for the

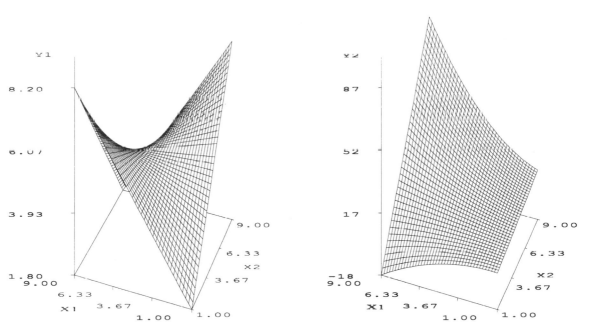

FIGURE 7.6 Interpretation of Cross-Product Terms

response to x_2 is not so straightforward: It indicates that the linear response to x_2 changes with the square of x_1. In fact, note that the response to x_2 is indeed linear for all values of x_1.

One problem with polynomial models with several variables is that of model building; that is, deciding how many terms should be included in a specific model. Because there are several variables, powers of variables as well as cross products, simple sequential model building is obviously not possible. Generally, one starts with a simple linear model, adds quadratics, then products of linear variables, and so forth. Some computer programs, such as PROC RSREG of the SAS System, provide information useful for such a process, but even these programs have some limitations, as we will see.

EXAMPLE 7.4   A Two-Factor Response Surface Model   (Freund and Wilson, 1997). A quality of steel called elasticity is affected by two operating conditions: quantity of a cleaning agent and the temperature used in the process. A 5×5 factorial experiment is conducted, for which all 25 combinations of 5 levels of the 2 operating conditions are each observed 3 times. The levels of the cleaning agent (CLEAN) are 0.0, 0.5, 1.0, 1.5, and 2.0 units, and levels of temperature (TEMPR) are 0.20, 0.93, 1.65, 2.38, and 3.10 (coded) units. The data are not reproduced here, but are available on the data diskette under the filename REG07X04.

We first perform the analysis of variance for the factorial experiment in order to obtain the estimate of the pure error for the lack of fit test and also to see how important the factors are. The results of that analysis are shown in Table 7.8.

All factors are highly significant and the estimated pure error variance is 17.45. We will first fit the standard quadratic response surface model,

$$y = \beta_{00} + \beta_{10}x_1 + \beta_{20}x_1^2 + \beta_{01}x_2 + \beta_{02}x_2^2 + \beta_{11}x_1x_2 + \epsilon,$$

TABLE 7.8

ANALYSIS OF VARIANCE

Source	DF	Sum of Squares	Mean Square	F Value	Pr > F
Model	24	132690.79040000	5528.78293333	316.77	0.0001
Error	50	872.69040000	17.45380800		
Corrected Total	74	133563.48080000			

R-Square	C.V.	Root MSE	ELAST Mean
0.993466	10.80087	4.17777548	38.68000000

Source	DF	Anova SS	Mean Square	F Value	Pr > F
TEMPR	4	22554.07993333	5638.51998333	323.05	0.0001
CLEAN	4	86093.77441333	21523.44360333	1233.173	0.0001
TEMPR*CLEAN	16	24042.93605333	1502.68350333	86.09	0.0001

where x_1 is TEMPR and x_2 is CLEAN. We use PROC REG of the SAS System, using mnemonic variable names TEMPR and T2 and so forth. The output is shown in Table 7.9.

The regression is obviously significant and it appears that all terms in the model are needed. However, we can immediately see that the error mean square of 192.95 is so much larger than the pure error of 17.45 that a formal lack of fit test is really not necessary, and we will need to fit a model with more terms. Often residual plots may be useful in determining what additional terms to add, but in this case they are of little help (the plots are not reproduced here; the reader may wish to verify). Lacking any other information, we will add terms for quadratic interactions; that is, $TEMPR^2 \times CLEAN$, $TEMPR \times CLEAN^2$, and $TEMPR^2 \times CLEAN^2$. The results are shown in Table 7.10.

We can see that the residual mean square is indeed much smaller, yet it is still sufficiently large that a formal lack of fit test is not needed. The sequential (SAS Type I) sums of squares are obviously significant for all higher-order terms, so all terms are needed. Again the residuals show very little, except possibly a cubic trend in TEMPR. We will leave it to the reader to try it.

Figure 7.7 shows the response surfaces for the quadratic response surface model on the left and that for the full quadratic polynomial model on the right.

TABLE 7.9

QUADRATIC RESPONSE SURFACE MODEL

Analysis of Variance

Source	DF	Sum of Squares	Mean Square	F Value	Prob>F
Model	5	120249.90737	24049.98147	124.643	0.0001
Error	69	13313.57343	192.95034		
C Total	74	133563.48080			

Root MSE	13.89066	R-square	0.9003	
Dep Mean	38.68000	Adj R-sq	0.8931	
C.V.	35.91173			

Parameter Estimates

Variable	DF	Parameter Estimate	Standard Error	T for H0: Parameter=0	Prob > |T|	Type I SS
INTERCEP	1	-8.035151	6.38250998	-1.259	0.2123	112211
TEMPR	1	36.275335	6.60146592	5.495	0.0001	12913
T2	1	-12.460423	1.82362047	-6.833	0.0001	9008.286136
CLEAN	1	-30.951947	8.79242724	-3.520	0.0008	78616
C2	1	23.787238	3.83418547	6.204	0.0001	7126.554138
TC	1	17.653917	2.21234522	7.980	0.0001	12286

TABLE 7.10

FULL QUADRATIC POLYNOMIAL MODEL

Analysis of Variance

Source	DF	Sum of Squares	Mean Square	F Value	Prob>F
Model	8	130740.50598	16342.56325	382.082	0.0001
Error	66	2822.97482	42.77235		
C Total	74	133563.48080			

Root MSE	6.54006	R-square	0.9789
Dep Mean	38.68000	Adj R-sq	0.9763
C.V.	16.90811		

Parameter Estimates

Variable	DF	Parameter Estimate	Standard Error	T for H0: Parameter=0	Prob > |T|	Type I SS
INTERCEP	1	8.121655	4.26834957	1.903	0.0614	112211
TEMPR	1	-3.299743	6.16256085	-0.535	0.5941	12913
T2	1	0.559825	1.80686225	0.310	0.7577	9008.286136
CLEAN	1	-7.204211	10.11238876	-0.712	0.4787	78616
C2	1	-2.815790	4.84851930	-0.581	0.5634	7426.554138
TC	1	1.570145	14.60007200	0.108	0.9147	12286
T2C	1	0.746663	4.28073972	0.174	0.8621	8995.530140
TC2	1	37.105900	7.00019872	5.301	0.0001	639.794700
T2C2	1	-9.177940	2.05245760	-4.472	0.0001	855.273768

From these we can see that the biggest difference between the two is that the full quadratic shows an almost flat response to TEMPR at low levels of CLEAN, whereas that response is quite dramatic for high levels of CLEAN. The standard quadratic response model simply does not have terms to describe this type of effect.

How are these results to be interpreted? Obviously, if we want to maximize the response, we would need to investigate higher levels of CLEAN. However, this may not be possible to do because of cost or other negative effects of too much CLEAN. Thus, the highest reasonable levels of CLEAN with TEMPR at about 2.0 would seem to provide maximum response. ◆

The number of factors need not be restricted to two; however, as the number of terms increases, model building and interpretation become more difficult. Fortunately, the problems may be alleviated by good computer programs, flexible graphics, and a little common sense. Normally we avoid using three variable

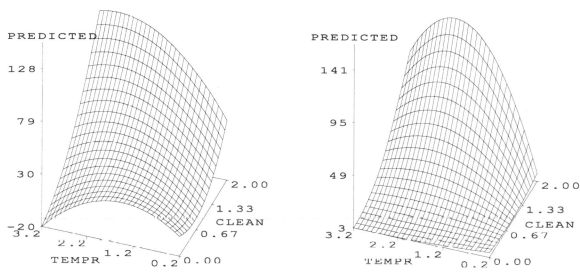

FIGURE 7.7 Comparison of Response Surfaces

products because three-factor interactions are difficult to interpret. We illustrate with a three-factor experiment.

EXAMPLE 7.5   A Three-Factor Experiment   The data for this example are from an experiment concerning a device for automatically shelling peanuts, reported in Dickens and Mason (1962). In the experiment, peanuts flow through stationary sheller bars and rest on a grid that has perforations just large enough to pass shelled kernels. The grid is reciprocated and the resulting forces on the peanuts between the moving grid and the stationary bars break open the hulls. The problem becomes one of determining the combination of bar grid spacing (SPACE), length of stroke (LENGTH), and frequency of stroke (FREQ) that would produce the most satisfactory performance. The performance criteria are (i) percent of kernel damage, (ii) shelling time, and (iii) the number of unshelled peanuts.

The paper just cited describes three separate experiments, one for each performance criterion. For this illustration we use the first experiment and kernel damage as the response variable. The experimental design is a three-factor composite design consisting of 15 factor-level combinations, with five additional observations at the center point (Myers, 1986). Figure 7.8 gives a three-dimensional representation of this design. We can see that the design consists of eight data points in a $2 \times 2 \times 2$ factorial (the box), one point at the center (SPACE = 0.86, LENGTH = 1.75, and FREQ = 175) that is replicated six times, and one point beyond the range of the factorial at each of six "spokes"

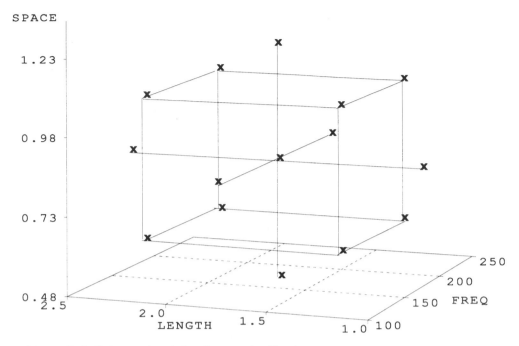

FIGURE 7.8 Schematic of the Composite Design

radiating in each direction from the center. This design has been specifically developed for efficient estimation of the quadratic response surface model. It requires only 15 individual data points as compared with 27 for the $3 \times 3 \times 3$ factorial experiment that is often used for this purpose. The six replications at the center are used to estimate the pure error for a lack of fit test. Of course the validity of this test rests on the assumption that the error variance is the same for all points of the experiment. Finally, this design does not allow for the usual factorial analysis of variance.

The data consist of responses resulting from the shelling of 1000 grams of peanuts. As previously noted, the factors of the experiment are:

LENGTH: Length of stroke (inches)
FREQ: Frequency of stroke (strokes/minute)
SPACE: Bar grid spacing (inches)

The response variable is:

DAMG: Percentage of damaged peanuts

The data are presented in Table 7.11 and are available on the data diskette as File REG07X05, which also contains the data for the other two response variables, labeled TIME and UNSHL.

TABLE 7.11

PEANUT SHELLER DATA

OBS	LENGTH	FREQ	SPACE	DAMG
1	1.00	175	0.86	3.55
2	1.25	130	0.63	8.23
3	1.25	130	1.09	3.15
4	1.25	220	0.63	5.26
5	1.25	220	1.09	4.23
6	1.75	100	0.86	3.54
7	1.75	175	0.48	8.16
8	1.75	175	0.86	3.27
9	1.75	175	0.86	4.38
10	1.75	175	0.86	3.26
11	1.75	175	0.86	3.57
12	1.75	175	0.86	4.65
13	1.75	175	0.86	4.02
14	1.75	175	1.23	3.80
15	1.75	250	0.86	4.05
16	2.25	130	0.63	9.02
17	2.25	130	1.09	3.00
18	2.25	220	0.63	7.41
19	2.25	220	1.09	3.78
20	2.50	175	0.86	3.72

We will use PROC RSREG of the SAS System for performing the regression for the response surface analysis. This program fits the quadratic response surface model, which includes linear, quadratic, and pairwise linear products. It also provides some results that are useful for determining the suitability of the model. The results are shown in Table 7.12.

The output addresses several aspects of the suitability of the model. The order of topics in this description is not the same as the order of relevant parts of the computer output, some of which have been manually labeled.

- *Is the model adequate?* This can only be answered if there are replications; in this case we have the six replicated observations at the center of the design providing a five degree-of-freedom estimate of the pure error. The lack of fit test is portion (3) and it shows that the p-value for the lack of fit is 0.1148. Thus, we can be reasonably satisfied that the model is adequate. It is of interest to note that if the model had not been adequate, the composite design allows only a very limited set of additional terms.
- *Do we need all three factors?* This question is answered by portion (5) of the output. Here are given the tests for the elimination of all terms involving each factor. In other words, the test for LENGTH is the test for the deletion of LENGTH, LENGTH², LENGTH × FREQ, and LENGTH × SPACE,

TABLE 7.12

RESPONSE SURFACE REGRESSION ANALYSIS

(1) Response Mean		4.702500	
Root MSE		0.846225	
R-Square		0.8946	
Coef. of Variation		17.9952	

(2) Regression	Degrees of Freedom	Type I Sum of Squares	R-Square	F-Ratio	Prob > F
Linear	3	40.257647	0.5926	18.739	0.0002
Quadratic	3	13.626260	0.2006	6.343	0.0111
Cross product	3	6.891100	0.1014	3.208	0.0704
Total Regress	9	60.775007	0.8946	9.430	0.0008

(3) Residual	Degrees of Freedom	Sum of Squares	Mean Square	F-Ratio	Prob > F
Lack of Fit	5	5.448685	1.089737	3.182	0.1148
Pure Error	5	1.712283	0.342457		
Total Error	10	7.160968	0.716097		

(4) Parameter	Degrees of Freedom	Parameter Estimate	Standard Error	T for H0: Parameter=0	Prob > |T|
INTERCEPT	1	36.895279	9.249104	3.989	0.0026
LENGTH	1	-0.172967	4.928866	-0.0351	0.9727
FREQ	1	-0.111699	0.051946	-2.150	0.0570
SPACE	1	-46.763375	10.216756	-4.577	0.0010
LENGTH*LENGTH	1	0.819069	1.055355	0.776	0.4556
FREQ*LENGTH	1	0.005889	0.013297	0.443	0.6673
FREQ*FREQ	1	0.000089827	0.000111	0.808	0.4377
SPACE*LENGTH	1	-3.847826	2.601615	-1.479	0.1699
SPACE*FREQ	1	0.077778	0.028907	2.691	0.0227
SPACE*SPACE	1	18.896464	4.405168	4.290	0.0016

(5) Factor	Degrees of Freedom	Sum of Squares	Mean Square	F-Ratio	Prob > F
LENGTH	4	2.676958	0.669240	0.935	0.4823
FREQ	4	6.050509	1.512627	2.112	0.1539
SPACE	4	59.414893	14.853723	20.743	0.0001

(6) Canonical Analysis of Response Surface

Factor	Critical Value
LENGTH	0.702776
FREQ	293.906908
SPACE	0.704050

Predicted value at stationary point   3.958022
Stationary point is a saddle point.

leaving a model with only the other two factors. In this output we can see that LENGTH may be omitted, and that FREQ is only marginally important.

● *Do we need quadratic and cross-product terms?* This question is answered by portion (2) of the output, which gives sequential (SAS Type I) sums of squares for using first only the three linear terms, then adding the three quadratics, and finally adding the three cross-product terms. Here we see that the linear terms are definitely needed and the quadratics are also needed, but the need for product terms is not definitely established. The final line shows that the overall model is definitely significant.

Additionally, portion (1) of the output gives some overall statistics, and portion (4) gives the coefficients and their statistics. Here we can see that one product term (SPACE*FREQ) is indeed significant at the 0.05 level; hence, a decision to omit all product terms would be a mistake.

Often response surface experiments are performed to find some optimum level of the response. In this application, for example, we would like to see what levels of the factors produce the minimum amount of damaged kernels. Portion (6) attempts to answer that question. The first statistics identify the "critical values," which give the factor levels and estimated response at a "stationary point," that is, a point at which the response surface has no slope. Now, by laws of geometry, there is for a quadratic response function only one stationary point, which can either be a maximum, minimum, or saddle point (where the surface increases along one axis and decreases at another; it looks like a saddle). For this experiment the output shows that the stationary point is a saddle point,[6] which is not useful for our purposes. Furthermore, it is outside the range of the experiment, and hence, it is of little use.

Although the experiment did not provide us with the desired minimum response, it is of interest to examine the nature of the response surface. This is, of course, somewhat difficult for a three-factor experiment. Basically, we need to examine all two-factor response curves for various levels of the third. However, in this example we have an easy way out: Since the factor LENGTH was seen to be of little importance, we can examine the response to the two other factors. However, just in case LENGTH has some effect, we will examine that response curve for two levels of LENGTH: 1.20 and 2.20. These plots are shown in Figure 7.9.

The graphs show that we were correct in not ignoring the effects of LENGTH, although obviously its effects are not very great. Both response surfaces have the same basic shape (a trough), but it is shifted somewhat for the higher value of LENGTH. Remember, we want to minimize the response; hence, it would seem that we should be in the neighborhood of 1.1 units of SPACE and FREQ of 100. Of course, the findings of the other responses would probably modify that recommendation. ◆

---

[6] This analysis is a relatively straightforward exercise in calculus. The stationary point occurs when all partial derivatives are 0. The nature of the stationary point is determined by the matrix of second partial derivatives; if it is positive definite, we have a minimum; if negative definite, a maximum; and if indefinite, a saddle point.

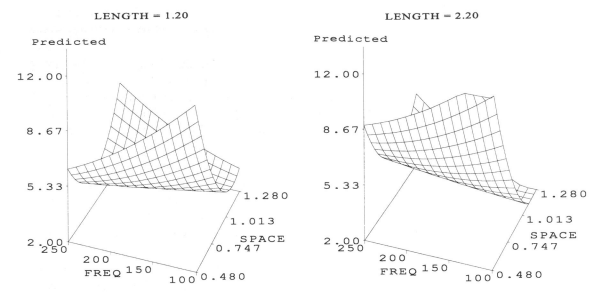

FIGURE 7.9 Response Surface Plots

7.5 Summary

In this chapter we have shown how to fit data to models that are functions of polynomials in one or more independent variables. All of the models presented in this chapter are analyzed by standard linear model methodology.

◁ CHAPTER EXERCISES ▷

1. Six beds containing 24 pine seedlings of drought-resistant pines were subjected to drought over a period of 12 days. The average weights of the seedlings in each bed were recorded each day. The data are available in File REG07P01. Fit a polynomial curve relating weight to day. Because there are six beds, a lack of fit may be performed, but it must be remembered that the data are *means* of 24 seedlings!

2. This exercise concerns an experiment on the effect of certain soil nutrients on the yield of ryegrass. The experimental unit is a pot with 20 ryegrass plants and the response variable is dry matter (YIELD) in grams. The nutrients in this study are calcium (CA), aluminum (AL), and phosphorus (P) in parts per million. The experimental design is a composite design, as shown in Figure 7.8, with eight replications at the center point. The data are shown in Table 7.13 and are available in File REG07P02. Fit a quadratic response surface and produce plots to interpret results.

3. The data for this exercise consist of mean weekly temperatures for 14 U.S. cities that lie roughly on a latitude/longitude grid coded 1–4 from south to north and 1–4 from east to west, for weeks 1 (early January), 13 (early April), and 25 (early June) for three successive years. The data are shown in Table 7.14, and a data file in a format more suitable for computer use is available in File REG07P03.

(a) Fit a response surface to show the trends of temperatures across latitude and longitude and week. The quadratic response surface is a good start, but that model may need to be modified. A final model should reveal several well-known climatological features.

(b) Check for outliers and determine what their cause may be. A good atlas will be useful.

4. These data result from an experiment to determine the effect of dietary supplements of calcium and phosphorus on the presence of these minerals and total ash content of the carapace (bony shell at head) of shrimp. The experiment consisted of two replications of a 4×4 factorial experiment with levels of calcium

TABLE 7.13

RYEGRASS DATA

CA	AL	P	YIELD
0	50	40	1.6273
120	30	24	1.4360
120	30	56	1.9227
120	70	24	0.3411
120	70	56	0.7790
200	0	40	2.5924
200	50	0	0.1502
200	50	40	0.9675
200	50	40	0.6115
200	50	40	0.3759
200	50	40	0.7094
200	50	40	0.6058
200	50	40	1.0180
200	50	40	0.8200
200	50	40	0.8077
200	50	80	1.3965
200	100	40	0.2221
280	30	24	0.6536
280	30	56	1.2839
280	70	24	0.2279
280	70	56	0.5592
400	50	40	0.4950

TABLE 7.14

TEMPERATURE DATA

CITY	LAT	LONG	1	2	3	1	2	3	1	2	3
	YEAR		WEEK 1			WEEK 2			WEEK 3		
Fargo, ND	1	1	10	9	−5	25	35	29	66	66	55
Marquette, MI	1	2	25	22	9	26	34	31	61	61	57
Burlington, VT	1	4	30	12	16	35	39	38	66	68	61
Lincoln, NE	2	1	34	18	18	34	53	39	69	73	70
Peoria, IL	2	2	35	23	16	41	47	39	69	73	70
Columbus, OH	2	3	42	29	22	41	42	44	69	77	71
Atlantic City, NJ	2	4	39	31	27	41	37	48	69	75	68
Oklahoma City, OK	3	1	51	31	30	51	60	44	75	79	77
Memphis, TN	3	2	52	37	32	53	59	46	76	81	81
Asheville, NC	3	3	42	34	34	45	51	50	66	70	70
Hatteras, NC	3	4	54	44	47	52	45	55	73	73	75
Austin, TX	4	1	60	45	43	60	69	55	81	84	84
New Orleans, LA	4	2	60	53	46	59	67	55	79	84	82
Talahassee, FL	4	3	60	53	49	60	63	56	78	81	84

supplement at 0, 1, 2, and 4% and phosphorus supplement at 0, 0.5, 1, and 2%. Chemical analyses for percent ash and calcium and phosphorus percentages were performed for four shrimp from each factor-level combination. The data, consisting of 128 observations, are in File REG07P04. Perform response surface analyses on at least one of the response variables, remembering that because we have data from an experiment there are estimates of pure error for lack of fit tests.

5. The *Annual Statistical Digest* of the Central Bank of Barbados gives data on various exports from that island nation. Table 7.15 gives annual total exports (EXPORT, in million BDS) for the years 1967 through 1993. The data are available in File REG07P05.

(a) Fit a polynomial curve over time. Plot residuals and determine if there are outliers.
(b) Since this is a time series, check for first-order autocorrelation and perform the indicated analysis as illustrated in Example 4.7.

6. An experiment was conducted to ascertain the effect of speed and acceleration on excessive carbon monoxide emissions from a truck. A truck was driven over a measured stretch with various combinations of initial speed (BSP), in 5-mph increments from 0 to 65, and amount of acceleration (ACCEL), which was computed as ending speed (ESP) minus beginning speed, in 5-mph increments from 5 to 65 mph with the restriction that the maximum ending

TABLE 7.15

BARBADOS EXPORTS

N	Year	EXPORT
1	1967	53.518
2	1968	59.649
3	1969	57.357
4	1970	62.106
5	1971	53.182
6	1972	63.103
7	1973	83.700
8	1974	125.555
9	1975	178.218
10	1976	137.638
11	1977	151.055
12	1978	186.450
13	1979	232.684
14	1980	337.291
15	1981	297.004
16	1982	372.627
17	1983	510.165
18	1984	583.668
19	1985	496.471
20	1986	420.614
21	1987	214.511
22	1988	248.029
23	1989	250.350
24	1990	244.820
25	1991	241.420
26	1992	271.384
27	1993	272.242

TABLE 7.16

REMOVING IRON

PHOS	IRON
0.05	0.33
0.10	0.19
0.15	0.10
0.20	0.25
0.25	0.17
0.30	0.12
0.35	0.12
0.40	0.12
0.50	0.12
0.60	0.12
0.80	0.12
0.90	0.07
1.00	0.18
1.50	0.14
2.00	0.17

speed is 70 mph. No combinations were replicated. A measure of excess carbon monoxide (TRCO) was recorded for each drive. The data are in File REG07P06. Perform a response surface analysis to describe the relationship of excess emissions to the speed factors. Interpret the results. Note that the model may use either BSP and ESP or BSP and ACCEL. Which is the more reasonable model? Use whichever appears more reasonable. Do the results make sense?

7. Removing iron from vegetable oil increases its shelf life. One method of reducing the amount of iron is to add some phosphoric acid in a water solution; iron will then precipitate out. Table 7.16 gives the amount of iron (IRON) remaining after adding various amounts of phosphoric acid (PHOS) in a fixed amount of water. Fit a polynomial curve to estimate the relationship between iron (the dependent variable) and phosphoric acid. Determine the amount of phosphoric acid that will give the maximum precipitation of the iron.

TABLE 7.17

EXPONENTIAL
GROWTH

t	y
0	5.1574
1	5.3057
2	6.7819
3	6.0239
4	7.2973
5	8.7392
6	9.4176
7	10.1920
8	11.5018
9	12.7548
10	13.4655
11	14.5894
12	16.8593
13	17.7358
14	19.1340
15	22.6170
16	24.4586
17	26.8496
18	30.0659
19	32.4246

8. To illustrate the use of a polynomial to approximate a nonlinear function, we generate 20 observations from the model

$$y = 5e^{0.1t} + \epsilon \ ,$$

where t takes on the values 0 through 19 and ϵ is normally distributed with mean 0 and standard deviation 0.5. Notice that the exponent is positive, making this a "growth" model rather than a decay model. The data are presented in Table 7.17.

(a) Use a polynomial to estimate the curve. Plot the curve.
(b) Compute the residuals from the true known function and compare the sum of squares of these residual to those of the polynomial. Does the polynomial appear to provide an adequate fit?
(c) Extrapolate the estimated curve to $t = 25$ and $t = 30$ and compare with the values from the known true function. Comment.

9. Example 7.5 illustrates a three-factor experiment using the percent of kernal damage (DAMG) as the response variable. Using the data given in REG07X05, do a similar analysis for each of the other response variables, TIME and UNSHL. Compare the results.

8

Intrinsically Linear and Nonlinear Models

8.1 Introduction

Up to now we have discussed models of the form

$$y = \beta_0 + \beta_1 Z_1 + \beta_2 Z_2 + \ldots + \beta_m Z_m + \epsilon ,$$

where the Z_i can represent any function of the basic predictor variables, x_i. A model that satisfies this relationship is said to be **linear in the parameters**. For example, in Chapter 7 we introduced the use of polynomials to "fit" certain nonlinear models, and saw that the ordinary least squares estimation procedure was appropriate for polynomial models. In fact, the procedures discussed in the first three chapters can be used on any model that is linear in the parameters. As we shall see in the first portion of this chapter, many nonlinear models can be "linearized" by an appropriate transformation on one or more of the variables. If the model can be put in the preceding form through suitable transformations, it is called **intrinsically linear**.

Obviously, not all models are intrinsically linear. If a model cannot be made linear by a suitable transformation, it is called **intrinsically nonlinear** or simply nonlinear. We saw in Chapter 7 that complicated nonlinear models can often be successfully approximated by using polynomial models. However, one advantage in being able to use the nonlinear model in its original form is that the coefficients in the more complicated model often represent meaningful physical properties. For example, the decay model presented in Section 7.1 is a nonlinear model that was estimated by a third-order polynomial in Section 7.2. The problem with the polynomial is that we cannot estimate the coefficients of the original model from the polynomial. This means that we have no estimate of the decay rate, which was probably the purpose of the exercise in the first place.

There are basically two strategies for dealing with nonlinear models, the first of which we have briefly discussed in Chapter 7. This strategy is to transform the variables (if the model is intrinsically linear), or to fit a polynomial model to the data. The second strategy is to try and fit the original model using more sophisticated methods. Obviously, we would like to follow the second strategy whenever possible; however, the methods necessary to estimate the parameters of the nonlinear model and do statistical inference on these estimates are often either very difficult or even impossible.

In this chapter we will discuss a few of the strategies employed to transform the more common intrinsically linear models to linear models and some of the more common transformations. We will then take a brief look at nonlinear models. In solving for estimates of parameters in nonlinear models, the theory behind the techniques usually requires a higher level of mathematics than is assumed for this text. A good discussion of many of these techniques, using calculus, is given in Draper and Smith (1988).

It is important to note that most physical processes cannot be modeled with a simple mathematical expression. In fact, many processes are so complex that the true relationships among the variables of the process may never be known. However, experience has shown that the nature of the responses can usually be reasonably well explained with readily implemented approximate models, such as the polynomial model. Therefore, it is usually a good idea to first try simple models to get ideas on the nature of the response. The information obtained by such preliminary analyses can then be used to evaluate results of fitting more complex models for which the validity is not always easily verified.

8.2 Intrinsically Linear Models

Our definition of an intrinsically linear model is one that can be made to look like the standard linear model of Chapter 3 by the use of one or more transformations. These transformations may involve one or more independent variables, the response variable, or all variables. For example, the multiplicative model,

$$y = \beta_0 x_1^{\beta_1} x_2^{\beta_2} \ldots x_m^{\beta_m} \epsilon,$$

has many uses in economics and the physical sciences. This model can be made linear by taking the logarithm[1] of both sides of the equality. This results in the model

$$\log(y) = \log(\beta_0) + \beta_1\log(x_1) + \beta_2\log(x_2) + \ldots + \beta_m\log(x_m) + \log(\epsilon),$$

---

[1] Usually the natural logarithm system is used; however, any base logarithm may be used.

which is indeed linear in the parameters and can be analyzed by the methods we have been using. Properties and uses of this model are presented later in this section.

Not all transforms involve all the variables. For example, the polynomial models covered in Chapter 7 used a transformation only on the independent variable. Other such transformations may involve square roots, reciprocals, logarithms, or other functions of the independent variables while leaving the response variable alone. Other transformations, such as the so-called *power transformation*, discussed later in this section, only involve the dependent variable. Other situations, such as the multiplicative model discussed earlier, require that both independent and response variables be transformed.

Sometimes simple plots of the dependent variable against the independent variables one at a time can be very informative. Certain patterns indicate a certain type of nonlinearity that can be corrected by the appropriate transformation. For example, Figure 8.1 shows several patterns that indicate a need for a transformation on the independent variable and the appropriate transformation. Notice that these patterns do not identify any violations of the assumption of equal variance, but do identify model specification problems.

Often, the nonlinearity of the relationship between x and y in a linear regression is accompanied by a violation of this equal variance assumption. In that case a transformation on the dependent variable may be indicated. Figure 8.2 shows several patterns that indicate a need for a simple transformation on the y variable and what that transformation should be. Notice that these patterns also identify violations of the equal variances assumption.

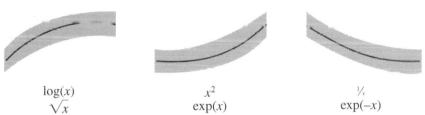

$\log(x)$
$\sqrt{x}$
x^2
$\exp(x)$
$\frac{1}{x}$
$\exp(-x)$

FIGURE 8.1 Patterns Suggesting Transformations on the x Variable

$\sqrt{y}$
$\log(y)$
$\frac{1}{y}$

FIGURE 8.2 Patterns Suggesting Transformations on the Response Variable

EXAMPLE 8.1   A large grocery chain conducted an experiment to determine the effect of the size of shelf display on the sale of canned spinach. The independent variable—shelf size (WIDTH) in feet—was varied in several of their stores, and the response—sales in cases of canned spinach (SALES)—was recorded for the month. The monthly sales was used as the response variable and the width of the shelf used as the independent variable. The data and the results of a linear regression analysis are shown in Table 8.1.

Figure 8.3 shows a plot of the response variable versus shelf width. The nature of this plot clearly shows that the effect of increasing shelf size diminishes with size. In this case, relating sales to the square root of shelf space may be justified by the fact that the visible portion of the display will be proportional to the square root of total space.

We now do the regression analysis using the square root of shelf space as the independent variable. The results of a simple linear regression using SALES as

TABLE 8.1

SALES OF CANNED SPINACH

DATA

Width	Sales	Width	Sales
0.5	42	1.5	100
0.5	50	2.0	105
1.0	68	2.0	112
1.0	80	2.5	112
1.5	89	2.5	128

ANALYSIS

Analysis of Variance

Source	DF	Sum of Squares	Mean Square	F Value	Prob>F
Model	1	6661.25000	6661.25000	97.753	0.0001
Error	8	545.15000	68.14375		
C Total	9	7206.40000			

Root MSE	8.25492	R-square	0.9244
Dep. Mean	88.60000	Adj R-sq	0.9149
C.V.	9.31707		

Parameter Estimates

Variable	DF	Parameter Estimate	Standard Error	T for H0: Parameter=0	Prob > |T|
INTERCEP	1	33.850000	6.12201458	5.529	0.0006
WIDTH	1	36.500000	3.69171369	9.887	0.0001

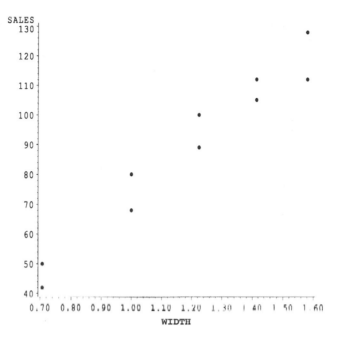

FIGURE 8.3 Sales of Canned Spinach

the dependent variable and the square root of shelf size (SQW) as the independent variable are shown in Table 8.2.

The linear regression obviously fits very well. Notice that the MSE using the transformed variable is smaller than that shown in Table 8.1, and the R-square is larger. And because the transformation is very straightforward, the interpretation of the results is also very straightforward. The estimated response is

$$\hat{SALES} = -12.25 + 85.07(SQW).$$

Therefore, for a shelf 2 feet wide, the expected sales become $-12.25 + 85.07(1.414)$ or 108.04 cases. ◆

It is often not a simple matter to determine what an appropriate transformation should be. Although trial-and-error methods work, they can be time-consuming. Examination of plots of the residuals can identify certain assumption violations, and an examination of the scatterplot can be enlightening. Other methods exist for identifying appropriate transformations. We will examine one of the most popular and flexible methods that can be used to identify which transformation of the response variable is most appropriate for correcting skewness of the distributions of the error terms, unequal error variances, and nonlinearity of the regression function.

Assume that the transformation to be used is a power of the response variable. Then we define a family of transformations as y^λ, where λ can take on any positive or negative value as the **power transformation**.

TABLE 8.2

REGRESSION USING SQUARE ROOT OF SHELF SPACE

Analysis of Variance

Source	DF	Sum of Squares	Mean Square	F Value	Prob>F
Model	1	6855.66658	6855.66658	156.373	0.0001
Error	8	350.73342	43.84168		
C Total	9	7206.40000			

Root MSE	6.62130	R-square	0.9513	
Dep. Mean	88.60000	Adj R-sq	0.9452	
C.V.	7.47326			

Parameter Estimates

Variable	DF	Parameter Estimate	Standard Error	T for H0: Parameter=0	Prob > |T|
INTERCEP	1	-12.246481	8.33192283	-1.470	0.1798
SQW	1	85.070870	6.80298650	12.505	0.0001

The problem now becomes one of identifying the appropriate value of λ. One way of doing this uses the fact that, most often, violations of the equal variance assumption result from the fact that there is a relationship between the variance and the mean value of the response. If the standard deviation is proportional to a power of the mean of y such that $\sigma \propto \mu^\alpha$, then the appropriate transformation will depend upon the value of α. Further, once we determine α, then $\lambda = 1 - \alpha$. Several of the common transformations are summarized below:

$$\alpha = -1, \qquad \lambda = 2, \qquad y_{new} = y^2$$

$$\alpha = 0.5, \qquad \lambda = 0.5, \qquad y_{new} = \sqrt{y}$$

$$\alpha = 0, \qquad \lambda = 1, \qquad \text{No transformation is needed}$$

$$\alpha = 1, \qquad \lambda = 0, \qquad y_{new} = \log_e(y), \text{ by definition}$$

$$\alpha = 1.5, \qquad \lambda = -0.5, \qquad y_{new} = 1/\sqrt{y}$$

$$\alpha = 2, \qquad \lambda = -1, \qquad y_{new} = 1/y$$

If multiple observations exist for individual values of the independent variable, we can empirically estimate α from the data. At the ith level of the independent variable $\sigma_{yi} \propto \mu_i^\alpha = \xi \mu_i^\alpha$, where ξ is a constant of proportionality. We can take the log of both sides and get:

$$\log \sigma_{yi} = \log \xi + \alpha \log \mu_i^\alpha.$$

Therefore, a plot of $\log \sigma_{yi}$ versus $\log \mu_i^\alpha$ would be a straight line with slope α. Since we do not know σ_{yi} or μ_i, we substitute reasonable estimates. An estimate of σ_{yi} can be obtained by calculating the sample standard deviation, s_i, of all responses at x_i. Similarly, an estimate of μ_i can be obtained by calculating the sample mean, $\bar{y}_i$, at each value of x_i. Once these estimates are obtained, then we can plot $\log s_i$ versus $\log \bar{y}_i$ and use the slope of the resulting straight-line fit as an estimate of α.

EXAMPLE 8.2   An experiment was done to examine the relationship between calcium and strength of fingernails. A sample of 8 college students were given calcium supplements in the amount of 10 mg, 20 mg, 30 mg, and 40 mg. At the end of the test period, fingernail strength was measured. The results, along with the means and standard deviations of each level of supplement, are presented in Table 8.3. The regression of strength on calcium is given in Table 8.4.

TABLE 8.3

CALCIUM DATA

Calcium	Fingernail Strength	Mean	Standard Deviation
10	14 46 24 14 65 59 30 31	35.375	19.42
20	116 74 27 135 99 82 57 31	77.625	38.57
30	44 70 109 133 55 115 85 66	84.625	31.50
40	77 311 79 107 89 174 106 72	126.875	81.22

TABLE 8.4

REGRESSION OF STRENGTH ON CALCIUM

Source	DF	Sum of Squares	Mean Square	F Value	Prob>F
Model	1	31696.90000	31696.90000	13.949	0.0008
Error	30	68168.60000	2272.28667		
C Total	31	99865.50000			

Root MSE	47.66851	R-square	0.3174	
Dep Mean	81.12500	Adj R-sq	0.2946	
C.V.	58.75933			

Parameter Estimates

Variable	DF	Parameter Estimate	Standard Error	T for H0: Parameter=0	Prob > |T|
INTERCEP	1	10.750000	20.64106950	0.521	0.6063
X	1	2.815000	0.75370529	3.735	0.0008

The regression is significant, but does not seem to be very strong (R-square of only 0.3174). An examination of the residuals indicates that there may be a problem with constant variance. A plot of the residuals against the predicted values is given in Figure 8.4.

These residuals indicate a problem with the equal variance assumption. Further, it seems that the amount of variation depends on the predicted values of y. To determine if a transformation will help, the logs of the standard deviations and the logs of the means are plotted in Figure 8.5.

Figure 8.5 indicates that a line with slope about 1.0 would fit best. Therefore, we will transform the response variable using natural logarithms. The resulting regression is given in Table 8.5 and the residuals in Figure 8.6. The residuals now indicate a constant variance. Further, the regression in Table 8.5 seems to indicate a much better fit.   ◆

If multiple observations are not available for any values of the independent variable, or if the variances are not proportional to the mean, the power transformation may still be useful. Identifying the value for λ is still a problem. Box and Cox (1964) have shown how the transformation parameter λ can be estimated simultaneously with the other model parameters. This procedure is called the Box–Cox method and uses maximum likelihood estimation. For example, consider the simple linear regression model written to represent an arbitrary member of the family of the power transformations:

$$y^\lambda = \beta_0 + \beta_1 x + \epsilon.$$

Notice that this model has an extra parameter, λ. The Box–Cox procedure assumes a normal error term and uses the method of maximum likelihood to estimate the value of λ along with the estimates of the regression parameters. Many statistical software packages offer this option. A discussion of the Box–Cox procedure can be found in Draper and Smith (1988).

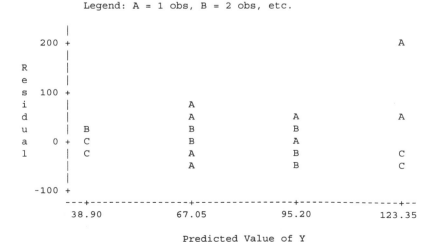

FIGURE 8.4 Residuals for Example 8.2

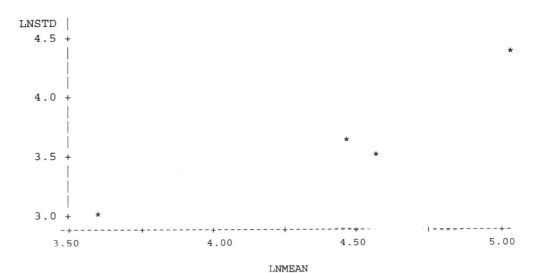

FIGURE 8.5 Plot of log (Standard Deviation) vs log (Mean) for Example 8.2

TABLE 8.5

REGRESSION OF LOG (STRENGTH) ON CALCIUM

Dependent Variable: LNY

Analysis of Variance

Source	DF	Sum of Squares	Mean Square	F Value	Prob>F
Model	1	6.51514	6.51514	23.361	0.0001
Error	30	8.36677	0.27889		
C Total	31	14.88191			

Root MSE	0.52810	R-square	0.4378	
Dep Mean	4.18194	Adj R-sq	0.4190	
C.V.	12.62816			

Parameter Estimates

Variable	DF	Parameter Estimate	Standard Error	T for H0: Parameter=0	Prob > |T|
INTERCEP	1	3.172988	0.22867511	13.876	0.0001
X	1	0.040358	0.00835003	4.833	0.0001

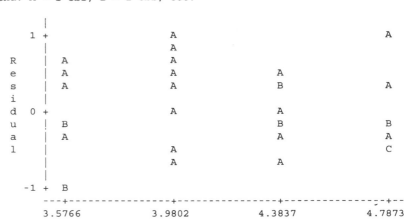

FIGURE 8.6 Residuals from Table 8.5

We now return to the multiplicative model introduced at the beginning of this chapter. The model is

$$y = \beta_0 \, x_1^{\beta_1} \, x_2^{\beta_2} \dots x_m^{\beta_m} \, \epsilon,$$

which when transformed becomes

$$\log(y) = \log(\beta_0) + \beta_1 \log(x_1) + \beta_2 \log(x_2) + \dots + \beta_m \log(x_m) + \log(\epsilon),$$

where the logarithms are base e, although base 10 may be used with equivalent results. The features of this model are as follows:

- The model is multiplicative,[2] that is, when an independent variable x_1, say, has the value x_1^*, then the model implies that the estimated response is multiplied by $(x_1^*)^{\beta_1}$, holding constant all other variables. For example, if the weight of an object is to be estimated by its dimensions (length, width, and height), the weight is logically a product of the dimensions. If the object is a cube, the exponents would all be unity, that is,

  Weight = Intercept × Length × Width × Height,

  where the intercept relates to the specific weight of the material of the cube. In other words, the β_i would all be unity. However, if the shape of the object is irregular, the coefficients could differ from that value.
- Another way to describe the model is to state that the coefficients represent proportional effects. That is, effects are proportional to the size or magnitude of the response variable. The coefficients represent the percent change in the

---

[2] The linear model using logarithms could logically be called a log-linear model; however, that nomenclature has been reserved for an analysis of a model for a categorical dependent variable (Chapter 10). We will therefore refer to it as the linear in logs model, as opposed to the multiplicative model in the original variables.

response associated with a 1% change in the independent variable, holding constant all other variables. Economists call these coefficients elasticities.

- The random error is also multiplicative. In other words, the magnitudes of the errors are proportional to values of the response variable.
- If the distribution of the error in the linear in logs model is normal, the distribution of the error in the multiplicative model is log-normal. This distribution is only defined for positive values and is highly skewed to the right. The equivalent of a zero error in the linear model is the value of unity in the multiplicative model. This means that the multiplicative deviations greater than unity will tend to be larger in absolute value than those less than unity. Again, this is a logical consequence of a multiplicative model.
- Although the estimates of the response in the linear in logs model are unbiased, the estimates of the conditional mean or predicted values in the multiplicative model are not. That is, if one performs the inverse log transformation on the estimates from the model, the residuals from the originally observed values will not sum to zero. This is a result of the skewness of the log-normal distribution.

Applications of the linear in logs model are quite numerous. They include those that estimate some function of size related to individual component sizes or dimensions, which is the property of Example 8.3. The model is also applied to many economic models where effects are proportional. The Cobb Douglas production function is a multiplicative model used to describe the relationship of product output to various inputs (labor, capital, etc.). In some engineering applications, this model is called a learning curve. Also frequently used are models involving variables that tend to vary by proportions or percentages, which is illustrated by Example 8.4.

EXAMPLE 8.3   Estimating the Price of Diamonds   We all know that prices of diamonds increase with weight, which is measured in carats. We will use data on the price and weight of a sample of diamonds sold at an auction. The data are shown in Table 8.6 and are available in File REG08X03 on the data disk.

We first use a linear model. The results of the analysis are shown in Table 8.7 and the residual plot in Figure 8.7. The results are not very satisfactory: The coefficient of determination is only 0.55, and the residual plot suggests that the straight-line fit is not satisfactory. In addition, the distribution of residuals strongly suggests that variance increases with price, and, as we have seen (Section 4.3), this condition results in incorrect confidence intervals for estimation and prediction. The results are, however, not surprising.

As we have seen in previous chapters, these deficiencies may be remedied by (i) using a polynomial or other transformation on the independent variable to describe the apparent curvilinear response or at least approximate it, or (ii) transforming the response variable. We will leave it to the reader to perform the polynomial regression analysis. Instead we will use the linear in logs model:

$$\log (\text{Price}) = \beta_0 + \beta_1 \log (\text{Carats}) + \epsilon.$$

With most computer programs such a model is implemented by creating new variables that are the logarithmic values of the variables and specifying these

TABLE 8.6

DIAMOND PRICES

Carats	Price	Carats	Price	Carats	Price
0.50	1918	0.75	5055	1.24	18095
0.52	2055	0.77	3951	1.25	19757
0.52	1976	0.79	4131	1.29	36161
0.53	1976	0.79	4184	1.35	15297
0.54	2134	0.91	4816	1.36	17432
0.60	2499	1.02	27264	1.41	19176
0.63	2324	1.02	12684	1.46	16596
0.63	2747	1.03	11372	1.66	16321
0.68	2324	1.06	13181	1.90	28473
0.73	3719	1.23	17958	1.92	100411

TABLE 8.7

LINEAR REGRESSION ANALYSIS

```
Dependent Variable: PRICE

                       Analysis of Variance

                          Sum of         Mean
     Source      DF       Squares        Square     F Value    Prob>F

     Model        1  5614124207.6  5614124207.6     33.654     0.0001
     Error       28  4670941237.8  166819329.92
     C Total     29   10285065445

          Root MSE    12915.85576     R-square      0.5459
          Dep Mean    13866.23333     Adj R-sq      0.5296
          C.V.           93.14610

                       Parameter Estimates

                    Parameter      Standard    T for H0:
     Variable   DF   Estimate        Error    Parameter=0    Prob > |T|

     INTERCEP    1      -19912   6282.0237300     -3.170        0.0037
     CARATS      1       33677   5805.2278581      5.801        0.0001
```

variables for the regression. The new variable names in this example are LPRICE and LCARATS. The results of this regression and the residual plot are shown in Table 8.8 and Figure 8.8, respectively.

The results indicate that the transformed model fits the data quite well. The coefficient of determination is much larger, although strict comparison is not advisable because of the different scales of measurement. The residual plot still

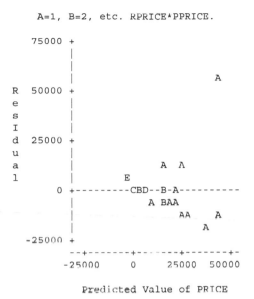

```
       A=1, B=2, etc. RPRICE*PPRICE.

       75000 +
             |
             |
             |                              A
   R   50000 +
   e         |
   s         |
   I         |
   d   25000 +
   u         |
   a         |                    A  A
   l         |           E
         0 +--------CBD--B-A----------
             |             A BAA
             |               AA    A
             |                A
      -25000 +
             --+-------+-------+-------+-
           -25000     0    25000  50000

            Predicted Value of PRICE
```

FIGURE 8.7 Residual Plot for Linear Regression

indicates somewhat larger variances of residuals with higher predicted values, but the evidence of nonlinearity is no longer there. The regression coefficient estimates a 2.5% increase (see later discussion of this model) in price associated with a 1% increase in weight (carats). This result reinforces the apparent

TABLE 8.8

ANALYSIS OF MODEL USING LOGARITHMS

Analysis of Variance

Source	DF	Sum of Squares	Mean Square	F Value	Prob>F
Model	1	30.86768	30.86768	219.753	0.0001
Error	28	3.93303	0.14047		
C Total	29	34.80072			

Root MSE	0.37479	R-square	0.8870
Dep Mean	8.94591	Adj R-sq	0.8829
C.V.	4.18948		

Parameter Estimates

Variable	DF	Parameter Estimate	Standard Error	T for H0: Parameter=0	Prob > |T|
INTERCEP	1	9.142210	0.06969600	131.173	0.0001
LCARATS	1	2.512166	0.16946543	14.824	0.0001

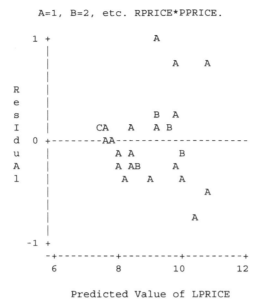

```
        A=1, B=2, etc. RPRICE*PPRICE.

     1 +                     A
       |
       |                 A       A
       |
   R   |
   e   |
   s   |                   B   A
   I   |        CA    A    A B
   d   0 +--------AA---------------------
   u   |           A A           B
   A   |           A AB      A
   l   |           A    A    A
       |                        A
       |
       |               A
       |
    -1 +
       -+---------+---------+---------+
        6         8        10        12
```

Predicted Value of LPRICE

FIGURE 8.8 Residuals for Logarithmic Model

nonlinearity of the relationship and is, of course, a well-known feature of diamond prices.

The problem with this analysis is that we are not really interested in estimating the logarithms of price; we want to see how this model describes the relationship of price to weight. We can do this by performing the reverse transformation on predicted values as well as various prediction intervals. The reverse
transformation is simply performed by exponentiating. That is,

$$\text{Predicted value} = e^{\text{Estimated value from logarithmic model}}.$$

Figure 8.9 shows the actual prices along with the predicted values and the 0.95
prediction intervals obtained in this manner.

The results appear to be quite reasonable. However, as was pointed out earlier, the estimated curve is not unbiased: The mean of the predicted values is
15,283 as compared to a mean of 13,866. Because this is a multiplicative model,
the bias is also multiplicative; that is, the bias is 15,283/13,866 = 1.10, or 10%.
The magnitude of the bias is known to be related to the coefficient of determination; it has been suggested that it is approximately $(1 - R^2)$, which in this example would be 11.3%. Dividing all predicted values by the bias (1.10 in this example) will, of course produce ad hoc unbiased estimates.

As suggested previously, a weighted least squares quadratic polynomial may
do as well and would be a good exercise for the reader. Such a model will, however, not provide useful regression coefficients. ◆

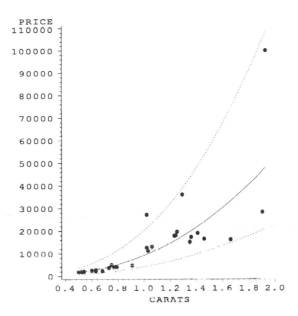

FIGURE 8.9 Statistics for the Retransformed Model

EXAMPLE 8.4   Airline Passenger Demand   The CAB (the airline-regulating body prior to deregulation) collected data relating to commercial airline passenger loads in 1966. For this example, data on number of passengers and number of airlines involved were arbitrarily selected for 74 pairs of cities from this database. Additional information that may relate to passenger loads was obtained from a standard atlas. The variables chosen for this study are:

> PASS: Number of passengers (in thousands) flying between cities in a sample week

This is the dependent variable. The following were used as independent variables:

> MILES: Air distance between the pair of cities
> INM: Median per capita income of the larger city
> INS: Median per capita income of the smaller city
> POPM: Population of the larger city (in thousands)
> POPS: Population of the smaller city (in thousands)
> AIRL: Number of airlines serving that route

The variables CITY1 and CITY2 (abbreviations for the pair of cities) are given for information only. They will not be used in the analysis. The object is to estimate the number of passengers. Table 8.9 gives selected observations from the data set; the entire data set is available as REG08X04 in the data file.

We first implement the linear model. Results are shown in Table 8.10 and the residual plot in Figure 8.10. The model is certainly significant and fits the data as

TABLE 8.9

AIRLINE PASSENGER DATA

CITY1	CITY2	PASS	MILES	INM	INS	POPM	POPS	AIRL
ATL	AGST	3.546	141	3.246	2.606	1270	279	3
ATL	TPA	7.463	413	3.246	2.586	1270	881	5
DC	NYC	150.970	205	3.962	2.524	11698	2637	12
LA	BOSTN	16.397	2591	3.759	3.423	7079	3516	4
LA	NYC	79.450	2446	3.962	3.759	11698	7079	5
MIA	DETR	18.537	1155	3.695	3.024	4063	1142	5
MIA	NYC	126.134	1094	3.962	3.024	11698	1142	7
MIA	PHIL	21.117	1021	3.243	3.024	4690	1142	7
MIA	TPA	18.674	205	3.024	2.586	1142	881	7
NYC	BOSTN	189.506	188	3.962	3.423	11698	3516	8
NYC	BUF	43.179	291	3.962	3.155	11698	1325	4
SANDG	CHIC	6.162	1731	3.982	3.149	6587	1173	3
SANDG	NYC	6.304	2429	3.962	3.149	11698	1173	4

TABLE 8.10

ANALYSIS OF LINEAR MODEL

```
                    Analysis of Variance

                         Sum of          Mean
        Source      DF    Squares        Square      F Value      Prob>F

        Model        6   72129.18435  12021.53073     25.187      0.0001
        Error       67   31978.77831    477.29520
        C Total     73  104107.96267

            Root MSE      21.84709    R-square     0.6928
            Dep Mean      27.36491    Adj R-sq     0.6653
            C.V.          79.83615


                     Parameter Estimates

                    Parameter      Standard     T for H0:
        Variable  DF  Estimate       Error     Parameter=0     Prob > |T|

        INTERCEP   1  -81.177928   41.81502872    -1.941         0.0564
        MILES      1   -0.016390    0.00427351    -3.835         0.0003
        INM        1   13.744915   12.49267870     1.100         0.2752
        INS        1    3.636293    8.21721889     0.443         0.6595
        POPM       1    0.002229    0.00110322     2.020         0.0474
        POPS       1    0.009687    0.00275194     3.520         0.0008
        AIRL       1    7.875925    1.80916995     4.353         0.0001
```

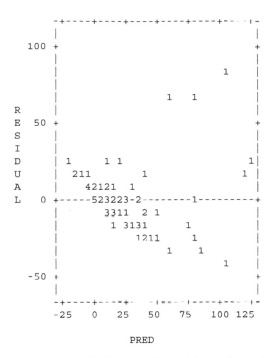

```
        -+----+----+----+----+----+-- -|-
         |                             |
 100 +                                 +
         |                             |
         |                     1       |
         |                             |
         |              1    1         |
  R      |                             |
  E   50 +                             +
  S      |                             |
  I      |                             |
  D      |  1       1 1              1 |
  U      |  211         1           1 |
  A      |    42121   1               |
  L    0 +-----523223-2--------1---------+
         |     3311   2 1               |
         |     1 3131       1           |
         |       1211     1             |
         |         1    1               |
         |              1               |
         |                             |
 -50 +                                 +
         |                             |
        -+----+----+-  -|----|----+---|-
        -25   0   25   50   75  100  125

                        PRED
```

FIGURE 8.10 Residuals for Linear Model

well as one might expect. The significance and signs of coefficients of distance and number of airlines are pretty well as expected. It is somewhat surprising that population of the smaller city is significant but that of the larger city is barely so ($\alpha = 0.05$). Also somewhat surprising is the nonsignificance of income.

Although these coefficients appear reasonable, they are really not very useful. For example, the coefficient for AIRL estimates that for an increase of one airline, the number of passengers should increase by 7.88 (thousand) for any pair of cities. The data in Table 8.9 show wide dispersions in the number of passengers. For a high-volume route such as Miami to New York (126,134 passengers), a change of 7880 would be negligible, whereas for a low-volume route like Atlanta to Augusta (3546 passengers), such a change would be unrealistic, and unmanageable if it occurred. A more likely scenario is that the *percentage* of passengers changes with the number of airlines. A similar argument applies to the other coefficients. In other words, we need a multiplicative model, which is provided by the linear in logs model.

The residual plot shows the typical pattern of increasing variances for larger values of the response variable, also suggesting that the logarithmic transformation may be advisable.

We next implement the model using the logarithms. We do this by creating new variables that are the logarithms, using the same variable names plus the prefix L. Thus, the logarithm of PASS is LPASS, etc. The results of the analysis are shown in Table 8.11 and the residuals in Figure 8.11.

TABLE 8.11

ANALYSIS OF THE LINEAR IN LOGS MODEL

Analysis of Variance

Source	DF	Sum of Squares	Mean Square	F Value	Prob>F
Model	6	67.65374	11.27562	36.542	0.0001
Error	67	20.67414	0.30857		
C Total	73	88.32789			

Root MSE	0.55549	R-square	0.7659
Dep Mean	2.67511	Adj R-sq	0.7450
C.V.	20.76512		

Parameter Estimates

Variable	DF	Parameter Estimate	Standard Error	T for H0: Parameter=0	Prob > |T|
INTERCEP	1	-6.473519	1.00488214	-6.442	0.0001
LMILES	1	-0.436167	0.10090749	-4.322	0.0001
LINM	1	2.914170	1.26729682	2.300	0.0246
LINS	1	0.778480	0.63181856	1.232	0.2222
LPOPM	1	0.429695	0.15249276	2.818	0.0063
LPOPS	1	0.392701	0.11964292	3.282	0.0016
LAIRL	1	0.712132	0.19455303	3.660	0.0005

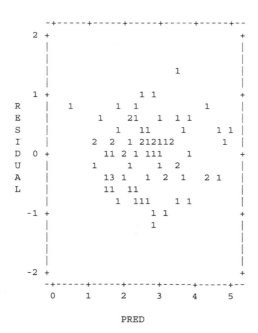

FIGURE 8.11 Residuals Using Linear in Logs Model

The overall model statistics are not too different from those of the linear model. Again, distance and number of airlines seem to be the most important factors. However, for this model the coefficients for populations of both cities are roughly equivalent. And in this model, income of the larger city is significant, although with a larger p-value than those of the other factors.

Furthermore, the coefficients now have a more reasonable interpretation. For example, the MILES coefficient now states that the number of passengers should decrease by 4.4% with a 10% increase in distance. ◆

8.3 Intrinsically Nonlinear Models

In Section 8.1 we noted that some models that are not linear in the parameters cannot be made so by the use of transformations. We called these *intrinsically nonlinear* models. Before we examine some methods that allow us to estimate coefficients in nonlinear models, we need to examine the general statistical model,

$$y = f(x_1, \ldots, x_m, \beta_1, \ldots, \beta_p) + \epsilon,$$

where f is some function of the m independent variables, $x_1, \ldots, x_m$, and p coefficients, $\beta_1, \ldots, \beta_p$. The value of m does not have to equal p. As usual, the error term ϵ is assumed to be normal with mean 0 and variance σ^2. Notice that if we define

$$f(x_1, \ldots, x_m, \beta_0, \ldots, \beta_m) = \beta_0 + \beta_1 x_1 + \ldots + \beta_m x_m,$$

we have described the linear regression model.

The procedure we used to estimate the unknown coefficients in the regression model was called the least squares principle and involved minimizing the SSE using calculus. This minimizing involved solving a set of linear equations. (See Appendix C.) We can use an analogous procedure to solve for unknown coefficients in almost any function f. That is, we can minimize the following:

$$\text{SSE} = \sum [y - f(x_1, \ldots, x_m; \beta_1, \ldots, \beta_p)]^2.$$

The problem arises from the fact that there may be more coefficients in the model than independent variables ($p > m$), or the minimizing procedure does not yield linear equations, thereby making closed forms or exact solutions impossible. Therefore, solutions are obtained by means of an *iterative search process*.

An iterative search process starts with some preliminary estimates of the parameters. These estimates are used to calculate a residual sum of squares and give an indication of what modifications of the parameter estimates may result in reducing this residual sum of squares. This process is repeated until no further changes in estimates result in a reduction, and we use these estimates.

Alternatively, we could use maximum likelihood methods to estimate the coefficients (see Appendix C for a brief discussion of maximum likelihood estimation). In fact, it can be shown that for many models, the two procedures give identical results as long as error terms are independent, normal, and have constant variances. (See, for example, Neter *et al.*, 1996.) Furthermore, maximum likelihood methods often require iterative numerical search methods to obtain solutions as well.

Researchers have developed a multitude of different iterative methods to solve the equations needed to fit nonlinear models. Considerations of efficiency, accuracy, and trying to avoid finding so-called local minima are important in choosing a specific method. Most computer software products offer several methods; for example PROC NLIN in SAS offers five different methods.

We will illustrate the procedure with several examples. The first will look at the exponential decay model given in Section 7.1; the second will look at a growth curve using the logistic growth model; and the third will use a segmented polynomial similar to that covered in Section 7.3, but with an unknown knot.

EXAMPLE 8.5   Radiation Decay   Table 8.12 lists the radiation count (y) taken at various times (t) immediately after exposure to a certain radioactive isotope. Theory indicates that these data should fit an exponential model known as the decay model. This model, shown in Section 7.1, is of the form

$$y = \beta_0 e^{-\beta_1 t} + \epsilon,$$

where β_0 is the initial count at $t = 0$, and β_1 is the exponential decay rate. The ϵ are assumed to be independent, normal errors. Notice that this model is intrinsically nonlinear because the error is not multiplicative.

In order to do the nonlinear regression, we need starting values for the iterative process that will estimate the coefficients. Fortunately, for this model we are able to obtain reasonable starting values. To get a starting value for β_0, we

TABLE 8.12

RADIATION DATA

Time (t)	Count (y)
0	540
5	501
5	496
10	255
10	242
15	221
20	205
25	210
30	165
35	156
40	137
45	119
50	109
55	100
60	53
65	41

observe that the first piece of data has a value of time $= 0$, thereby giving us a good estimate for β_0 of 540. To get a starting value for β_1 we choose the time $= 30$ where count $= 165$. This gives us the following equation:

$$165 = 540e^{-\beta_1(30)}.$$

Taking natural logarithms and solving gives us an estimate of $\beta_1 = 0.0395$ (the negative sign is part of the model; hence, the exponent is -0.0395). This gives us the necessary starting points of 540 and 0.0395. Table 8.13 gives a partial output using these starting values and the PROC NLIN nonlinear regression option in SAS.

We have omitted the summary of the iterative search procedure, which includes a statement that the search procedure did converge. The procedure used in PROC NLIN produces a sequence of parameter estimates that yield ever-decreasing residual sums of squares. The sequence converges when no further decrease appears possible. That means the procedure stops when it has found what it considers to be a minimum value for the residual sums of squares. The parameter values are given in the last two rows of Table 8.13. The estimates are $\hat{\beta}_1 = -0.0397$ and $\hat{\beta}_0 = 517.34$.

The first section of Table 8.13 gives the partitioning of the sum of squares that corresponds to the analysis of variance portion of the standard linear regression analysis. However, note that the partitioning starts with the uncorrected total sum of squares. This more closely resembled the analysis for the regression through the origin presented in Section 2.6. This is because in most nonlinear models there is no natural mean or intercept; hence, the corrected sum of squares may have no meaning. However, the corrected total sum of squares is provided in case the more "usual" analysis is desired.

TABLE 8.13

NONLINEAR REGRESSION

Non-Linear Least Squares Summary Statistics   Dependent Variable COUNT

Source	DF	Sum of Squares	Mean Square
Regression	2	1115336.3338	557668.1669
Residual	14	42697.6662	3049.8333
Uncorrected Total	16	1158034.0000	
(Corrected Total)	15	370377.7500	

Parameter	Estimate	Asymptotic Std. Error	Asymptotic 95 % Confidence Interval Lower	Upper
B0	517.3141078	36.425800343	439.18872007	595.43949555
B1	-0.0396592	0.005129728	-0.05066130	-0.02865700

There are no test statistics presented in the output for either the model or the parameters. This is because exact inference procedures are not available for nonlinear regression models with normal error terms due to the fact that least squares estimators for small sample sizes are not normally distributed, are not unbiased, and do not have minimum variance. Consequently, inferences about the regression parameters in nonlinear regression are usually based on large-sample or asymptotic theory. This theory tells us that the estimators are approximately normally distributed, almost unbiased, and at almost minimum variance when the sample size is large. As a result, inferences for nonlinear regression parameters are carried out in the same fashion as for linear regression when the sample size is reasonably large. These inference procedures, when applied to nonlinear regression, are only approximate. However, this approximation is often very good, and for some nonlinear regression models, the sample size can be quite small for the asymptotic inferences to be quite good. Unfortunately, for other nonlinear regression models, the sample size may need to be quite large. For a good discussion of when large-sample theory is applicable, see Neter *et al.* (1996).

The equivalent to the usual F test for the overall regression model can be calculated by using the corrected total sums of squares, subtracting the residual sums of squares, and taking the ratio of mean squares. This value will be approximately distributed as the F distribution for large samples. The standard errors of the estimated coefficients and the confidence intervals are also asymptotic (as indicated in Table 8.13). Therefore, we can be approximately 95% confident that the true value of β_0 is between 439.2 and 595.5, and the true value of β_1 is between -0.05 and -0.03. Notice that neither confidence interval contains 0, so we can conclude that both coefficients are significantly different from 0.

The estimated coefficients can be used to provide an estimated model:

$$\hat{\mu}_{y|t} = 517.3e^{-0.04t}.$$

The estimated initial count is 517.3 and the estimated exponential decay rate is 0.04. This means that the expected count at time t is $e^{-0.04} = 0.96$ times the count at time $(t - 1)$. In other words, the estimated rate of decay is $(1 - 0.96) = 0.04$, or approximately 4% per time period. We can obtain the estimated half-life, the time at which one-half of the radiation has occurred, by T2 = ln(2)/0.04 = 17.3 time periods.

The difference between this approach to solving the exponential model and that used in Example 8.2 is the assumption on the nature of the error term. If the error is multiplicative, we can treat the decay model as intrinsically linear and use the log transformation. That is, we define the model:

$$\log(\text{COUNT}) = \beta_0 + \beta_1(\text{TIME}) + \epsilon.$$

The result of using the transformed model in a linear regression on the data of Example 8.5 is given in Table 8.14.

Exponentiating both sides of the transformed model yields the following:

$$\text{COUNT} = e^{\beta_0}e^{\beta_1 t}e^{\epsilon}.$$

Using the results from Table 8.14 we get the following estimated model:

$$\widehat{\text{COUNT}} = e^{6.139}e^{-0.033t} = 463.6e^{-0.033t}.$$

TABLE 8.14

LOG MODEL FOR DECAY DATA

Dependent Variable: LOG

Analysis of Variance

Source	DF	Sum of Squares	Mean Square	F Value	Prob>F
Model	1	7.45418	7.45418	157.609	0.0001
Error	14	0.66214	0.04730		
C Total	15	8.11632			

Root MSE	0.21748	R-square	0.9184	
Dep Mean	5.16553	Adj R-sq	0.9126	
C.V.	4.21012			

Parameter Estimates

Variable	DF	Parameter Estimate	Standard Error	T for H0: Parameter=0	Prob > |T|
INTERCEP	1	6.138556	0.09467358	64.839	0.0001
TIME	1	-0.033124	0.00263849	-12.554	0.0001

Notice that the estimates do not differ that much from those found using the nonlinear model. Remember that the intrinsically linear model does make different assumptions about the random error.

Finally, we can use a polynomial model to "fit" the data. This model was considered in Section 7.2, where it was determined that a cubic polynomial best fits the data. The results of such a regression are given in Table 8.15.

All three of these approaches seem to do a creditable job of predicting the radioactive count over time, as is illustrated in Figure 8.12, where line 1 represents the nonlinear model, 2 the exponentiated log model, and 3 the polynomial model, and the dots are the data. However, the polynomial regression shows an increase in count toward the end, a result that is impossible, and the log model assumes a multiplicative error. Therefore, the only strictly correct analysis is the nonlinear model. By using this model we can account for the fact that the error term is probably additive, and we can estimate the needed parameters.

A more general form of the decay model is

$$y = \beta_0 + \beta_1 e^{-\beta_2 t} + \epsilon,$$

where

$(\beta_0 + \beta_1)$ is the initial count when $t = 0$
β_0 is the final count, when $t = \infty$
β_2 is the decay rate
ϵ is the normally distributed random error

TABLE 8.15

POLYNOMIAL REGRESSION FOR RADIATION DATA

Dependent Variable: COUNT

Source	DF	Type I SS	Mean Square	F Value	Pr > F
T	1	280082.937	280082.937	125.90	0.0001
T*T	1	39009.140	39009.140	17.53	0.0013
T*T*T	1	24589.031	24589.031	11.05	0.0061

General Linear Models Procedure

Dependent Variable: COUNT

Parameter	Estimate	T for H0: Parameter=0	Pr > |T|	Std Error of Estimate
INTERCEPT	566.8147957	15.69	0.0001	36.11788755
T	-31.5637828	-5.92	0.0001	5.33436752
TSQ	0.7825509	4.00	0.0018	0.19553224
TCUBIC	-0.0065666	-3.32	0.0061	0.00197517

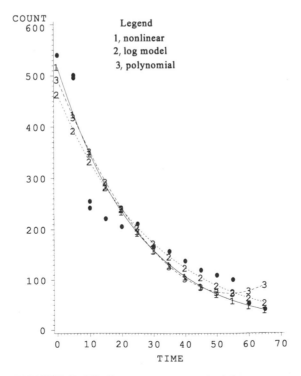

FIGURE 8.12 Comparison of Models

Note that the only difference is that this model does not require the curve to go to 0 as t gets large. Also, this model is intrinsically nonlinear even if we allow a multiplicative error.

Again, we need starting values for the iterative process that will estimate the coefficients. For this model, we are able to obtain reasonable starting values for the process using the same strategy as before. For example, we can use the value for $t = 0$ to estimate $\beta_0 + \beta_1$ by 540. Because the counts are decreasing up to the last time measured, it is reasonable to estimate β_0 by 0. This gives us a starting value for β_1 of 540.

We will use the observation for TIME = 30 and COUNT = 165 to solve for our estimate of β_2.

$$165 = 0 + 540e^{-30\beta_2}$$

$$\ln(165) - \ln(540) = 30\beta_2$$

$$5.10 = 6.29 - 30\beta_2,$$

which is solved to provide the estimate $\beta_2 = -0.04$.

Table 8.16 gives a partial output from PROC NLIN of SAS using these starting values.

The parameter estimates provide the model

$$\hat{\mu}_{y|x} = 83.28 + 486.74e^{-0.070t}.$$

Using these coefficients, the estimated initial count is $(83.28 + 486.74) = 570.02$, the estimated final count is 83.28, and the estimated exponential decay rate is 0.070. This means that the expected count at time t is $e^{-0.07} = 0.93$ times the count at time $(t - 1)$. In other words, the estimated rate of decay is $(1 - 0.93)$

TABLE 8.16

NONLINEAR REGRESSION

Non-Linear Least Squares Summary Statistics   Dependent Variable COUNT

Source	DF	Sum of Squares	Mean Square
Regression	3	1126004.0106	375334.6702
Residual	13	32029.9894	2463.8453
Uncorrected Total	16	1158034.0000	
(Corrected Total)	15	370377.7500	

Parameter	Estimate	Asymptotic Std. Error	Asymptotic 95 % Confidence Interval Lower	Upper
B0	83.2811078	26.205957197	26.66661319	139.89560248
B1	486.7398763	41.963485999	396.08333074	577.39642192
B2	-0.0701384	0.015400632	-0.10343320	-0.03684362

= 0.07, or approximately 7% per time period. We can obtain the estimated half-life, the time at which one-half of the radiation has occurred, by T2 = ln(2)/0.07 = 9.9 time periods.

The partitioning of sums of squares indicates that this model fits better than the two-parameter model. The standard error of the decay coefficient is three times as large as that for the two-parameter model. Using the formula for the variance of a sum of random variables provides the standard error of the starting value, $(\hat{\beta}_0 + \hat{\beta}_1)$ of 40.121, which compares favorably. The confidence interval for the asymptote is quite wide, but does not include 0.

Figure 8.13 shows a plot of the estimated model (line) and the observed values (dots). The model does appear to fit quite well.   ◆

Growth Models

Growth models are applied in many areas of research. In biology, botany, forestry, zoology, and ecology growth occurs in many different organisms, including humans. This growth is a function of the time the organism is alive. In chemistry, growth occurs as a result of chemical reactions. This growth is a function of the time the reaction has been taking place. In economics and political science, growth of organizations, supplies of commodities, manufactured goods, and even nations occurs as a function of time. Therefore, it is important to be able to model growth behavior over time. And because we need a model with parameters that can be interpreted, we will be using a nonlinear model.

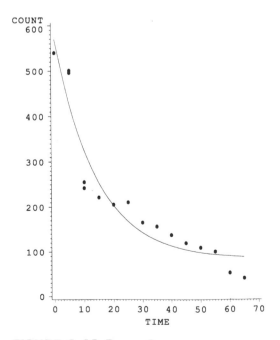

FIGURE 8.13 Decay Curve

The exponential model can be used for exponential growth as well as decay. The difference is in the sign of the coefficient β_1: If it is positive, the model describes a process that is growing; if negative, the model describes decay. The problem with using an exponential model as a growth model is that the expected value of y continues upward. Such a model usually does not fit data that are collected over a long period of time because most biological organisms stop growing when they reach maturity. Instead, most growth models allow for several different rates of growth at various time intervals and are usually S-shaped.

We will illustrate the application of a nonlinear growth model with an example. The model used will be one of the most commonly used growth models, the **logistic** or **autocatalytic** model,

$$ y = \frac{\beta_2}{1 + [(\beta_2 - \beta_0)/\beta_0]e^{\beta_1 t}} + \epsilon, $$

where

β_0 is the expected value of y at time $(t) = 0$ (initial value)
β_1 is a measure of the growth rate
β_2 is the expected value of y for very large values of time, often called the limiting value of y
ϵ is the random error, assumed to have mean 0 and variance σ^2

EXAMPLE 8.6   **Growth Model**   Table 8.17 lists data from an experiment attempting to raise Florida lobster in a controlled environment. The data show the overall length (LENGTH) of a certain species of lobster and the age (TIME) of the lobster.

We will use PROC NLIN in SAS to estimate the values of the parameters. Again, we need starting values for the estimation process. Although in this data set TIME does not start at 0, we will use the length at 14 months (59mm) to estimate the starting value, and the length at the last value of time (460mm) to estimate the limiting value. Because we really have no prior knowledge about the growth rate except that it is positive, we will arbitrarily choose a starting value for β_1 of 0.1.[3] A partial listing of the output is given in Table 8.18.

The initial portion of the output gives the usual partitioning of sums of squares. The model does appear to fit reasonably well.

The asymptotic 95% confidence intervals do not include zero; hence, for all coefficients we reject the hypothesis of no effect. Notice that the estimate of the limiting value of length is about 486.7. This implies that this species of lobster very rarely gets longer than 19 inches. The estimated initial value of 83.3 has little meaning. Figure 8.14 shows the predicted values (solid line) and the observed values (dots connected by broken line). Not only does the plot indicate quite a good fit to the data, but it illustrates the characteristic shape of the growth curve.   ◆

We now consider a segmented polynomial model, in which the location of the knot is unknown, for which the estimation of the parameters is more difficult.

---

[3] Most nonlinear regression programs allow the use of a grid search, which can provide better preliminary initial values.

TABLE 8.17

LOBSTER DATA

TIME (mo)	LENGTH (mm)
14	59
22	92
28	131
35	175
40	215
50	275
56	289
63	269
71	395
77	434
84	441
91	450
98	454
105	448
112	452
119	455
126	453
133	456
140	460
147	464
154	460

TABLE 8.18

NONLINEAR REGRESSION

```
Non-Linear Least Squares Summary Statistics    Dependent Variable COUNT

    Source              DF Sum of Squares    Mean Square

    Regression           3  1126004.0106     375334.6702
    Residual            13    32029.9894       2463.8453
    Uncorrected Total   16  1158034.0000

    (Corrected Total)   15   370377.7500


    Parameter    Estimate    Asymptotic              Asymptotic 95 %
                             Std. Error          Confidence Interval
                                                Lower          Upper
        B0    83.2811078   26.205957197   26.66661319   139.89560248
        B1    -0.0701384    0.015411632   -0.10343320    -0.03684362
        B2   486.7398763   41.963485999  396.08333074   577.39642192
```

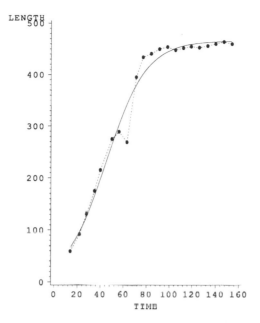

FIGURE 8.14 Plot of Growth Curve

EXAMPLE 8.7   Example 7.3 Revisited   This example consists of simulated data for a segmented polynomial with a knot at $x = 5$, presented in Table 7.6. The data were generated from the model

$$y = x - 0.1x^2 + \epsilon \text{ for } x \le 5$$

$$y = 2.5 + \epsilon \text{ for } x > 5.$$

The random error is normally distributed with mean 0 and standard deviation of 0.2.

If the location of the knot is known (which is $x = 5$ in this case), then the model parameters can be estimated with linear regression methodology. When the location of the knot is not known, the knot is a parameter of the model, and we must use nonlinear regression procedures to estimate its value.

We begin by using PROC NLIN and using the known population parameters as starting values. In terms of the computer names, and the name of the coefficient in PROC NLIN, these are as follows:

Parameter	Coefficient	Starting Value
INTERCEPT	B0	0.0
X1	B1	1.0
XSQ	B2	-0.1
X2	B3	0.0
X2SQ	B4	0.1
KNOT	KNOT	5.0

The results are shown in Table 8.19.

We begin to see that we may have a problem when we note that the error mean square (0.0412) is larger than that obtained with the linear regression (0.03455). The estimated coefficients are, however, quite close to the known population values; in fact, they are closer than those from the linear regression used when the location of the knot was known (Example 7.3). Finally, we see that the standard errors of two coefficients, B3 and KNOT, are so large as to make those estimates useless. Apparently the minimum error mean square here is what is known as a "local" minimum, as opposed to the "global" minimum that defines the true least squares estimate.

As we have noted, this model is linear in all parameters except for the knot. In order to get an idea of what the problem may be, we can estimate the linear portion for several values of the knot, and by inspection have a better idea of the least squares estimate. Figure 8.15 shows a plot of the error mean squares for a selected set of values for the knot.

We can now see that for this sample, the least squares estimate of the knot is not near the true value of 5; it is actually around 7.15. In fact, the peculiar shape of the curve may be an indication of the problem; for linear models such a curve is always smooth. At this point we could readily use the estimates from the linear regression for that value. Instead, we will use these values as the ini-

TABLE 8.19

ESTIMATING THE KNOT IN A SEGMENTED REGRESSION

Non-Linear Least Squares Summary Statistics   Dependent Variable Y

Source	DF	Sum of Squares	Mean Square
Regression	6	202.35346116	33.72557686
Residual	35	1.44361369	0.04124611
Uncorrected Total	41	203.79707485	
(Corrected Total)	40	23.65025485	

Parameter	Estimate	Asymptotic Std. Error	Asymptotic 95 % Confidence Interval Lower	Upper
B0	0.000006779	0.1062984	-0.2157891	0.2158027
B1	0.999994860	0.0655954	0.8668299	1.1331598
B2	-0.099999442	0.0097208	-0.1197336	-0.0802653
B3	-0.000055793	559.9346085	-1136.7209967	1136.7208851
B4	0.099996471	0.0272701	0.0446355	0.1553574
KNOT	4.999616707	2799.5095454	-5678.2731726	5688.2724060

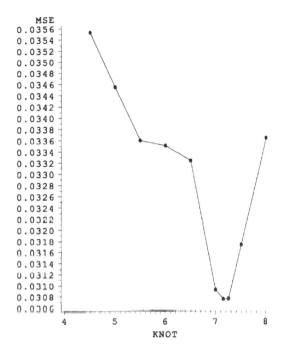

FIGURE 8.15 Plot of MSE and Knot

tial values for a nonlinear regression. The initial values from the linear regression (output not reproduced here) are as follows:

Parameter	Coefficient	Starting Value
INTERCEPT:	B0	−0.041
X1	B1	1.000
XSQ	B2	−0.092
X2	B3	0.535
X2SQ	B4	0.023
KNOT	KNOT	5.0

The results of the nonlinear regression are shown in Table 8.20.

The estimated parameters are quite similar to the initial values, and the confidence intervals now look reasonable. However, the error mean square is still somewhat larger than that obtained by the linear regression estimates. Note also that the estimated coefficients for X2 and X2SQ, as well as the knot, are quite different from the true values of the parameters. ◆

TABLE 8.20

SECOND NONLINEAR REGRESSION FOR SEGMENTED POLYNOMIAL

Non-Linear Least Squares Summary Statistics   Dependent Variable Y

Source	DF	Sum of Squares	Mean Square
Regression	6	202.68968437	33.78161406
Residual	35	1.10739048	0.03163973
Uncorrected Total	41	203.79707485	
(Corrected Total)	40	23.65025485	

Parameter	Estimate	Asymptotic Std. Error	Asymptotic 95 % Confidence Interval Lower	Upper
B0	-0.046178393	0.09263036506	-0.2342269149	0.1418701297
B1	1.009182408	0.06126312962	0.8848123810	1.1335524348
B2	-0.093408118	0.00845612190	-0.1105748566	-0.0762413801
B3	0.543113672	0.25310989219	0.0292763227	1.0569510204
B4	0.020336709	0.07835984316	-0.1387412855	0.1794147037
KNOT	7.166514178	0.35198257416	6.4519558046	7.8810725506

8.4 Summary

In general, nonlinear models fall into two categories. In the first category are those models that can be made linear through a transformation. This transformation may be made on the independent variables, the dependent variable, or a combination of the variables. The purpose of a transformation is to allow us to use the methods of linear regression to obtain estimates of the unknown parameters in the model. Unfortunately, this method only allows us to estimate the parameters of the relationship involving the transformed parameters. Of course, in some cases these relationships may be quite useful. Even so, it is as important to check assumptions when using intrinsically linear models as it is when using ordinary linear models.

The second category, intrinsically nonlinear models, cannot be made linear by transformations. This category of model can be handled using least squares methodology or maximum likelihood methodology. In either case, estimation of the unknown parameters is usually done by using iterative numerical methods. In previous chapters we have discussed the desirable properties of the regression model. For example, we have noted that the estimates of the coefficients are unbiased and have minimum variance. In this category of nonlinear models these properties of the estimators can only be verified in the limit. That is, the estimators are only unbiased and have minimum variance if the sample size is very large. As a result, for a specific nonlinear model and a specific sample size, very little can truly be stated regarding the properties of the estimates. Therefore, only asymptotic confidence intervals can be computed.

CHAPTER EXERCISES

1. Exercise 1 of Chapter 7 concerned the effect of drought conditions on the weight of pine seedlings. Six beds containing 24 pine seedlings of drought-resistant pines were subjected to drought over a period of 12 days. The average weight of the seedlings in each bed was recorded each day. The data are available in File REG07P01. Use a nonlinear regression to see if the decay models used in Example 8.4 are appropriate for these data.

2. In planning irrigation systems it is important to know how fast the water advances in a furrow. File REG08P02 gives the distance covered (in feet) at various times (in minutes). It may be appropriate to consider this as a growth curve, since eventually the water may all be absorbed and no further advance is possible. Fit a growth curve to these data. Compare the fit with a polynomial.

3. The data in Table 8.21 and in data file REG08P03 resulted from a kinetics study in which the velocity of a reaction (y) was expected to be related to the concentration (x) with the following equation:

$$y = \frac{\beta_0 x}{\beta_1 + x} + \epsilon.$$

(a) Obtain starting values for β_0 and β_1. To do this we can ignore the error term and note that we can transform the model into $z = \gamma_0 + \gamma_1 w$, where $z = 1/y$, $\gamma_0 = 1/\beta_0$, $\gamma_1 = \beta_1/\beta_0$, and $w = 1/x$. The initial values can be obtained from a linear regression of z on w and using $\beta_0 = 1/\gamma_0$ and $\beta_1 = \gamma_1/\gamma_0$.

(b) Using the starting values obtained in part (a), use nonlinear regression to estimate the parameters β_0 and β_1.

4. The data in Table 8.22 and in File REG08PO4 resulted from a study to determine the death rate of a certain strain of bacteria when exposed to air. The experimenter placed approximately equal samples of the bacteria on plates and exposed them to air. The measures taken were t = time in minutes, and y = percent of bacteria still viable. It is desired to fit the general form of the decay model,

$$y = \beta_0 + \beta_1 e^{-\beta_2 t} + \epsilon,$$

where the terms of the model are defined in Example 8.4.

TABLE 8.21

DATA FOR EXERCISE 3

Velocity (y)	Concentration (x)
1.92	1.0
2.13	1.0
2.77	1.5
2.48	1.5
4.63	2.0
5.05	2.0
5.50	3.0
5.46	3.0
7.30	4.0
6.34	4.0
8.23	5.0
8.56	5.0
9.59	6.0
9.62	6.0
12.15	10.0
12.60	10.0
16.78	20.0
17.91	20.0
19.55	30.0
19.70	30.0
21.71	40.0
21.6	40.0

(a) To obtain starting values for the coefficients, we note that as time increases, the percentage of deaths increases. It is logical, then, to use an estimate of $\beta_0 = 0$. To obtain starting values of the other parameters, note that if we ignore the error term in the model and make a logarithmic transformation, we obtain a simple linear model: $\log(y) = \log(\beta_1) + (-\beta_2)t$. Fit this simple linear regression model and solve for the initial estimates.

(b) Using the starting values obtained in (a), use nonlinear regression to fit the general decay model.

(c) Fit a polynomial model to the data and compare with (b).

5. To examine the growth rate of Atlantic salmon, a study was done in a controlled environment. The length of a sample of 10 fish was recorded every 7 days for a period of 21 weeks (starting with the second week). The average length of the 10 fish (LENGTH) and the age (AGE) of the fish are recorded in File REG08P05. Fit the logistic growth model illustrated in Example 8.6 to explain the growth of the Atlantic salmon.

TABLE 8.22

DATA FOR EXERCISE 4

Percent Viable (y)	Time (t)
0.92	1.0
0.93	1.0
0.77	2.5
0.80	2.5
0.63	5.0
0.65	5.0
0.50	10.0
0.46	10.0
0.30	20.0
0.34	20.0
0.23	30.0
0.26	30.0
0.19	40.0
0.17	40.0
0.15	50.0
0.12	50.0
0.06	75.0
0.08	75.0
0.04	100.0
0.05	100.0

The General Linear Model

9.1 Introduction

In Chapter 1 we showed how the various statistical procedures involving the estimation and comparison among means could be stated in terms of a linear model. In all subsequent chapters, all models have been "regression" models where the response variable is related to quantitative independent variables. In these models the parameters of the model are called regression coefficients, which measure the effect of independent variables on the values of the response variable.

In Example 6.3 we introduced the use of an indicator variable that allowed the response to differ from the regression line for a specific value of the independent variable. In this chapter we show how the use of several indicator variables, often called *dummy variables*, can be used in a regression model to perform the analysis of variance for comparing means. Although this method is more cumbersome to use than the standard analysis of variance procedure, it can be used where the standard method cannot be properly used. In addition, this method can be used for models that combine the features of the analysis of variance and regression.

As an introduction, we will show how the two-sample "pooled" t test can be performed as a regression analysis. Assume two samples of n each,[1] with means $\bar{y}_1$ and $\bar{y}_2$, respectively. The total sample size is $n. = 2n$. To describe this population we propose a regression model,

$$y = \beta_0 + \beta_1 x + \epsilon,$$

where the independent variable, x, takes on the following values:

$$x = 0 \text{ for observations in sample 1}$$

$$x = 1 \text{ for observations in sample 2.}$$

Substituting the values for x produces the model

$$y = \beta_0 + \epsilon, \text{ for observations in sample 1}$$

$$y = \beta_0 + \beta_1 + \epsilon, \text{ for sample 2.}$$

Define β_0 as μ_1 and $(\beta_0 + \beta_1)$ as μ_2, and the model describes two populations with means μ_1 and μ_2 which is appropriately analyzed by the two-sample t test.

Because of the nature of the independent variable, some of the elements of the formulas for the regression are simplified:

$$\Sigma x = n, \ \bar{x} = 1/2, \ \Sigma x^2 = n, \ \Sigma xy = n\,\bar{y}_2,$$

and we will define

$$\bar{y} = (\bar{y}_1 + \bar{y}_2)/2.$$

The quantities needed for the regression are

$$S_{xx} = \Sigma x^2 - 2n\bar{x}^2 = \frac{n}{2}$$

$$S_{xy} = \Sigma xy - 2n\bar{x}\bar{y} = \frac{n(\bar{y}_2 - \bar{y}_1)}{2}.$$

Then

$$\hat{\beta}_1 = \frac{S_{xy}}{S_{xx}} = (\bar{y}_2 - \bar{y}_1)$$

$$\hat{\beta}_0 = \bar{y}_1.$$

Note that for the sample from the first population where $x = 0$, $\hat{\mu}_{y|x} = \bar{y}_1$, and for the second population $\hat{\mu}_{y|x} = \bar{y}_2$, which are the values we would expect.

Using the relationships between the regression coefficients and the means of the two populations, we see that the test for $\beta_1 = 0$ is the test for $\mu_1 = \mu_2$. The value of MSE is given by

$$\text{MSE} = \frac{\text{SSE}}{2n - 2} = \frac{\Sigma(y - \mu_{y|x})^2}{2n - 2} = \frac{\Sigma(y_1 - \bar{y}_1)^2 + \Sigma(y_2 - \bar{y}_2)^2}{2n - 2},$$

[1] Equal sample sizes are used for algebraic simplicity.

which is identical to the pooled variance estimate for the pooled t test. Finally, the formula for the test statistic is

$$t = \frac{\hat{\beta}_1}{\sqrt{\dfrac{\text{MSE}}{S_{xx}}}} = \frac{\bar{y}_2 - \bar{y}_1}{\sqrt{\text{MSE}\dfrac{2}{n}}},$$

which is, of course, the formula for the t test.

Note that we have used a regression with an *indicator* variable: It *indicates* that when $x = 1$ the observation belongs in the second sample, while when it is 0 the observation is not, or equivalently, it is in sample 1. Indicator variables, which are also called *dummy* variables, are not restricted to having values of 0 or 1, although these values are most commonly used. The reader may wish to verify that using values of -1 and $+1$ will give the equivalent results for the t test. In fact, any two unique values will give equivalent results.

9.2 The Dummy Variable Model

We illustrate the dummy variable approach with the one-way analysis of variance or completely randomized design model.[2] Assuming data from independent samples of n_i from each of t populations or factor levels, the analysis of variance model is written

$$y_{ij} = \mu + \alpha_i + \epsilon_{ij}, i = 1, 2, \ldots, t, j = 1, 2, \ldots, n_i,$$

where

  n_i is the number of observations in each factor level
  t is the number of such factor levels
  μ is the overall mean
  α_i are the specific factor-level effects, subject to the restriction $\sum \alpha_i = 0$, which means that the "average" factor-level effect is zero

The hypotheses to be tested are

$$H_0: \alpha_i = 0, \text{ for all } i,$$

$$H_1: \alpha_i \neq 0 \text{ for one or more } i.$$

The obvious difference between this model and a regression model is the absence of *independent variables*. This is where indicator or dummy variables come in. The model using dummy variables is

$$y_{ij} = \mu z_0 + \alpha_1 z_1 + \alpha_2 z_2 + \ldots + \alpha_t z_t + \epsilon_{ij},$$

[2] For a more complete discussion see Section 1.5, or any statistical methods textbook.

where the z_i are the dummy variables indicating the presence or absence of certain conditions for observations as follows:

$z_0 = 1$ for all observations

$z_1 = 1$ for all observations occurring in factor level 1, and

$\quad = 0$ otherwise

$z_2 = 1$ for all observations occurring in factor level 2, and

$\quad = 0$ otherwise,

and so forth for all t factor levels. The definitions of μ, α_i, and ϵ_{ij} are as before. Note that substituting the actual values of the dummy variables for any observation does indeed produce the analysis of variance model:

$$y_{ij} = \mu + \alpha_i + \epsilon_{ij}.$$

In other words, the dummy variable model has the appearance of a regression model but, by using the actual values of the independent variables for any observation, exactly duplicates the analysis of variance model. Admittedly, these independent variables are not the usual quantitative variables that we have become accustomed to, but since the regression model makes no assumptions about any distribution of the independent variables, no assumptions are violated.

We now perform an analysis according to this model, for which we will use the procedures described in Chapter 3. The X and Y matrices[3] for a set of data described by this model are

$$
X =
\begin{bmatrix}
1 & 1 & 0 & \cdots & 0 \\
\cdot & \cdot & \cdot & \cdots & \cdot \\
\cdot & \cdot & \cdot & \cdots & \cdot \\
1 & 1 & 0 & \cdots & 0 \\
1 & 0 & 1 & \cdots & 0 \\
\cdot & \cdot & \cdot & \cdots & \cdot \\
\cdot & \cdot & \cdot & \cdots & \cdot \\
1 & 0 & 1 & \cdots & 0 \\
\cdot & \cdot & \cdot & \cdots & \cdot \\
\cdot & \cdot & \cdot & \cdots & \cdot \\
1 & 0 & 0 & \cdots & 1 \\
\cdot & \cdot & \cdot & \cdots & \cdot \\
\cdot & \cdot & \cdot & \cdots & \cdot \\
1 & 0 & 0 & \cdots & 1
\end{bmatrix},
\quad
Y =
\begin{bmatrix}
y_{11} \\
\cdot \\
\cdot \\
\cdot \\
y_{1n_1} \\
y_{21} \\
\cdot \\
\cdot \\
\cdot \\
y_{2n_2} \\
\cdot \\
\cdot \\
y_{t1} \\
\cdot \\
\cdot \\
\cdot \\
y_{tn_t}
\end{bmatrix}.
$$

[3] Although we have called the variables z_i, we will denote the corresponding matrix X to conform with the notation of Chapter 3.

It is not difficult to compute the $X'X$ and $X'Y$ matrices that specify the set of normal equations:

$$X'XB = X'Y.$$

The resulting matrices are

$$X'X = \begin{bmatrix} n. & n_1 & n_2 & \cdots & n_t \\ n_1 & n_1 & 0 & \cdots & 0 \\ n_2 & 0 & n_2 & \cdots & 0 \\ . & . & . & \cdots & . \\ . & . & . & \cdots & . \\ . & . & . & \cdots & . \\ n_t & 0 & 0 & \cdots & n_t \end{bmatrix}, \quad B = \begin{bmatrix} \mu \\ \alpha_1 \\ \alpha_2 \\ . \\ . \\ . \\ \alpha_t \end{bmatrix}, \quad X'Y = \begin{bmatrix} Y.. \\ Y_{1.} \\ Y_{2.} \\ . \\ . \\ . \\ Y_{t.} \end{bmatrix},$$

where $n. = \Sigma n_i$, $Y.. = \Sigma_{\text{all}} y_{ij}$, and $Y_{i.} = \Sigma_j y_{ij}$. The $X'X$ matrix is often referred to as the *incidence matrix*, as it is a matrix of frequencies or *incidences* of observations occurring in the various factor levels.

An inspection of $X'X$ and $X'Y$ shows that the sum of elements of rows 2 through $(t + 1)$ are equal to the elements of row 1. Remembering that each row corresponds to the coefficients of one equation, we see that the equation represented by the first row contributes no information over and above that provided by the other equations. For this reason, the $X'X$ matrix is singular; hence, it is not possible to immediately solve the set of normal equation to produce a set of unique parameter estimates.

The normal equations corresponding to the second and all subsequent rows represent equations of the form

$$\mu + \alpha_i - \bar{y}_{i.},$$

which reveal the obvious: Each factor level mean, $\bar{y}_{i.}$, estimates the mean, μ, plus the corresponding factor-level effect, α_i. We can solve each of these equations for α_i, producing the estimate

$$\hat{\alpha}_i = \bar{y}_{i.} - \mu.$$

However, if we now want a numerical estimate of $\hat{\alpha}_i$, we need to have a value for μ. It would appear reasonable to use the equation corresponding to the first row to estimate μ, but this equation requires values for the $\hat{\alpha}_i$, which we have not yet obtained. This is the result of the singularity of $X'X$: There are really only t equations for solving for the $(t + 1)$ parameters of the model.

Obviously, there must be a way to obtain estimates, because we do get estimates from the supposedly equivalent analysis of variance calculations. Now if we look closely at statements of the analysis of variance models we presented earlier, they were followed by the statement "subject to the restriction that $\Sigma \alpha_i = 0$." Note that implementing this restriction requires that only $(t - 1)$ of the $\hat{\alpha}_i$ need be estimated because any one parameter is simply the negative of the sum of all the others. As we will see, this is only one of a number of restrictions we can use, and because all will ultimately produce the same answers, the restrictions pose no loss of generality.

One way to implement this restriction is to omit α_t from the model, from which we obtain estimates of the first $(t - 1)$ parameters $(\alpha_i, i = 1, 2, \ldots, t - 1)$ and then

$$\hat{\alpha}_t = -\hat{\alpha}_1 - \hat{\alpha}_2 - \ldots - \hat{\alpha}_{t-1}.$$

Using this restriction in the X matrix, the dummy variables for $\hat{\alpha}_1, \hat{\alpha}_2, \ldots, \hat{\alpha}_{t-1}$ are defined as usual, but for observations in factor level t, all z_i are set to -1. Using these variables in the normal equations produces the following estimates:

$$\hat{\mu} = (1/t) \sum \bar{y}_{i.} = Y_{..}/t,$$

$$\hat{\alpha}_i = \bar{y}_{i.} - \hat{\mu}, i = 1, 2, \ldots, (t - 1),$$

and $\hat{\alpha}_t$ is computed by applying the restriction, that is,

$$\hat{\alpha}_t = -\hat{\alpha}_1 - \hat{\alpha}_2 - \ldots - \hat{\alpha}_{t-1}.$$

It is interesting to note that the resulting estimate of μ is not the weighted mean of factor-level means we would normally use when sample sizes are unequal.[4] Of course, the so-called overall mean is usually of little practical interest.

The inability to directly estimate all parameters and the necessity of applying restrictions is related to the degrees of freedom concept encountered in computing the sample variance in the first chapter of most introductory textbooks. There it is argued that, having to first obtain $\bar{y}$ to compute the sum of squared deviations for calculating the variance, we have lost one degree of freedom. The loss of that degree of freedom was supported by noting that $\sum(y - \bar{y}) = 0$, which is equivalent to the restriction we have just used. In the dummy variable model we start with t sample statistics, $\bar{y}_1, \bar{y}_2, \ldots, \bar{y}_t$. If we first estimate the overall mean (μ) from these statistics, there are only $(t - 1)$ degrees of freedom left for computing the estimates of the factor-level effect parameters (the t values of the α_i).

Other sets of restrictions may be used in the process of obtaining estimates, each of which may result in different numerical values of parameter estimates. For this reason, any set of estimates based on implementing a specific set of restrictions is said to be *biased*. However, the existence of this bias is not in itself a serious detriment to the use of this method, since these parameters are by themselves not overly useful. As we have seen, we are usually interested in functions of these parameters, such as contrasts or factor-level means, and numerical values for estimates of these functions, called *estimable functions*, are not affected by the specific restrictions applied.

A simple example illustrates this property. Assume a four-factor-level experiment with equal sample sizes for factor levels. The means are 4, 6, 7, and 7, respectively. Using the restriction $\sum \alpha_i = 0$, that is, the sum of factor-level effects is zero, provides factor-level effect estimates $\hat{\alpha}_i = -2, 0, 1, 1$, respectively, and $\hat{\mu} = 6$. Another popular restriction is to set the last factor level effect to 0, that is, $\hat{\alpha}_4 = 0$. The resulting estimates of factor-level effects are $-3, -1, 0, 0$, respectively, and $\hat{\mu} = 7$. One additional restriction lets $\hat{\mu} = 0$, and then the factor-level effects are 4, 6, 7, and 7, which are the factor-level means.

---

[4] The usual estimate of the mean is obtained by using the restriction $\sum n_i \alpha_i = 0$. However, we do not normally want sample frequencies to influence parameter estimates.

These sets of estimates are certainly not the same. However, for example, the estimate of the mean response for factor level 1, $(\hat{\mu} + \hat{\alpha}_1) = 4$, and likewise the estimate of the contrast $(\hat{\alpha}_1 - \hat{\alpha}_2) = -2$ for all sets of parameter estimates.

Another feature of the implementation of the dummy variable model is that numerical results for the partitioning of sums of squares are not affected by the particular restriction applied. This means that any hypothesis tests based on F ratios using the partitioning of sums of squares are valid tests for the associated hypotheses regardless of the specific restriction applied. We illustrate this method with a simple example.

EXAMPLE 9.1   Sleep-Inducing Drugs   In an experiment to determine the effectiveness of sleep-inducing drugs, eighteen insomniacs were randomly assigned to three factor levels:

1. Placebo (no drug)
2. Standard drug
3. New experimental drug

The response as shown in Table 9.1 is hours of sleep. We follow the procedure as implemented by PROC GLM of the SAS System, using selected portions of the output.

The first step is to construct the X and $X'X$ matrices. We will leave the construction of the X matrix to the reader, who can verify that the $X'X$ matrix as produced by PROC GLM of the SAS System is

	INTERCEPT	TRT 1	TRT 2	TRT 3	HOURS
INTERCEPT	18	6	6	6	134.2
TRT 1	6	6	0	0	29.9
TRT 2	6	0	6	0	47.7
TRT 3	6	0	0	6	46.6
HOURS	134.2	29.2	47.7	46.6	906.68

TABLE 9.1

DATA FOR EXAMPLE 9.1

Factor Level		
1	2	3
5.6	8.4	10.6
5.7	8.2	6.6
5.1	8.8	8.0
3.8	7.1	8.0
4.6	7.2	6.8
5.1	8.0	6.6

The output uses the mnemonic computer names INTERCEPT for μ and TRT 1, 2, and 3 for what we have called α_i. The response variable is HOURS. Note that the column labeled with the response variable is $X'Y$ and, equivalently, the corresponding row is $Y'X$. The last element in each of these is $Y'Y = \sum y^2$. The next step is to obtain estimates of the parameters, which are normally obtained from the inverse.

Now we know the $X'X$ matrix is singular and therefore no unique inverse can be obtained. However, because problems with singular matrices occur quite frequently, mathematicians have come up with so-called *generalized* or *pseudo* inverses to deal with singular matrices. But because there is no unique generalized inverse, any generalized inverse for our linear model will correspond to a particular restriction on the resulting parameter estimates. And just as there is no "correct" set of restrictions, there is no universally "correct" generalized inverse. The particular one employed by PROC GLM, sometimes called a G2 inverse, uses a sequential scheme for obtaining a solution: It essentially obtains rows (and corresponding columns) of the inverse, until it comes to one that, because of a singularity, it cannot obtain, and then it arbitrarily assigns a zero value to all elements of that row (and column) as well as to the corresponding parameter estimate. Thus, the solution obtained at this stage corresponds to the restriction that the last factor-level effect is 0. The generalized inverse and solution obtained by PROC GLM are as follows:

	INTERCEPT	TRT 1	TRT 2	TRT 3	HOURS
INTERCEPT	0.166666667	-0.166666667	-0.166666667	0	7.766666667
TRT 1	-0.166666667	0.333333333	0.166666667	0	-2.783333333
TRT 2	-0.166666667	0.166666667	0.333333333	0	0.183333333
TRT 3	0	0	0	0	0
HOURS	7.766666667	-2.783333333	0.183333333	0	16.536666667

As in the output of the $X'X$ matrix, the row and column labeled with the name of the response variable contain the parameter estimates; the very last element is the residual sum of squares. We can see that the last row and column of the generalized inverse as well as the corresponding parameter estimate all have zero values. However, the reader may verify that multiplying the inverse by the $X'X$ matrix will produce the identity matrix, except for zeroes in the last row and column. Also, multiplying the inverse by $X'Y$ will result in the given parameter estimates. However, as we have noted, these parameter values have very little practical use, and therefore the GLM procedure will not print them unless specifically requested.

The generalized inverse can, however, be used to compute sums of squares. Remember, the regression sum of squares is $\hat{B}' X'Y$, which in this example is

$$\text{SSR} = (7.76667)(124.2) + (-2.78333)(29.2) + (0.18333)(47.7) + (0)(46.6)$$

$$= 890.14.$$

Then

$$SSE = \Sigma y^2 - SSR = 906.68 - 890.14 = 16.54,$$

which is indeed what is given in the output (except for roundoff). The estimates of the factor-level means, defined as $\hat{\mu}_i = \hat{\mu} + \hat{\alpha}_i$, are obtained directly from the solution of the normal equations. Thus, $\hat{\mu}_1 = 7.7667 - 2.7833 = 4.9834$, and so forth. Note that $\hat{\mu}_3$ is the "intercept." Contrasts are computed using these estimates. ◆

The variances of these estimates are somewhat more difficult to compute because the components of these estimates are not uncorrelated. Hence, we need a formula for the variance of a sum of correlated variables.

Mean and Variance of a Linear Function of Correlated Variables

We have a set of random variables $y_1, y_2, \ldots, y_n$, with means $\mu_1, \mu_2, \ldots, \mu_n$, and variances and covariances given by the so-called variance–covariance matrix Σ, which has the variances (denoted by σ_{ii}) on the diagonal and the covariances between y_i and y_j (denoted by σ_{ij}) in the ith row and jth column.

Define

$$L = \Sigma a_i y_i = A'Y.$$

Then

$$\text{the mean of } L \text{ is } \Sigma a_i \mu_i = A'M,$$

and

$$\text{the variance } \Sigma\Sigma a_i a_j \sigma_{ij} = A'\Sigma A.$$

For the special case of $n = 2$, the variance of $(y_i + y_j)$ is

$$\text{Var}(y_i + y_j) = \text{Var}(y_i) + \text{Var}(y_j) + 2\text{Cov}(y_i, y_j),$$

where $\text{Cov}(y_i, y_j)$ is the covariance between y_i and y_j. Now in Section 3.5 we showed that

$$\text{Mean}(\hat{\beta}_j) = \beta_j$$

$$\text{Variance}(\hat{\beta}_j) = \sigma^2 c_{jj}$$

$$\text{Covariance}(\hat{\beta}_i, \hat{\beta}_j) = \sigma^2 c_{ij}.$$

Then

$$\text{Variance}(\beta_i + \beta_j) = (c_{ii} + c_{jj} + 2c_{ij})\sigma^2.$$

For any inferences, the appropriate error mean square is substituted for σ^2. In the computer output for the dummy variable model the coefficients are labeled differently; however, the same principle applies. For example, using the elements of the inverse given earlier, the variance of $\hat{\mu}_1 = \hat{\mu} + \hat{\alpha}_1$ is $\text{Var}(\mu + \alpha_1) =$

(0.16667 + 0.3333 + 2(–0.1667)) 1.10244 = 0.18377; hence, the standard error is 0.4287. The output from PROC GLM gives

TRT	HOURS LSMEAN	Std Err LSMEAN	Pr > |T| H0:LSMEAN=0
1	4.98333333	0.42864990	0.0001
2	7.95000000	0.42864990	0.0001
3	7.76666667	0.42864990	0.0001

Except for roundoff, the results agree. PROC GLM also provides the test of the hypothesis that the least squares means are zero; in most cases this is not a useful test. Options do exist for more meaningful statistics, such as contrasts or other estimates.

We have illustrated the implementation of the dummy variable model for the analysis of variance with output from PROC GLM, which uses one particular method for dealing with the singular $X'X$ matrix. However, other methods will provide the same results. Rare exceptions occur in nonstandard cases that are of little interest at this stage. The GLM procedure does, however, have features that allow for the study of special data structures. A special case is given in Section 9.3. For additional information, see Littell *et al.* (1991).

It is quite obvious that the dummy variable method is considerably more difficult to implement than the standard analysis of variance. However, the standard method cannot be used in all applications, and it becomes necessary to analyze the data using the dummy variable or general linear model analysis. Such applications include the following:

- The analysis of data involving two or more factors in a factorial structure with unequal frequencies in the cells.
- Models that include both qualitative and quantitative factors. An important model of this type is called the analysis of covariance.

These applications are presented in the remainder of this chapter.

9.3 Unequal Cell Frequencies

The standard analysis of variance calculations for multifactor designs can only be used if we have *balanced data,* which occurs if the number of observations for all factor-level combinations, usually called *cells*, are all equal We now show why this is so, and subsequently show how the dummy variable approach can be used to provide correct answers.

EXAMPLE 9.2   Unequal Cell Frequencies   We begin by showing why the "usual" formulas for the analysis of variance produce incorrect results in the case of unequal cell frequencies. Table 9.2 gives data for a 2×3 factorial with unequal cell frequencies. The factors are labeled A (row) and B (column). The cell entries are the responses and cell means. The marginal and grand means are shown in the last row and column. The lack of balance is obvious.

We will concentrate on the effect of factor A. If we examine the cell means, we see that there appears to be no effect due to factor A. That is, for level 1 of factor B, the cell means for the two levels of factor A are both 3.0; there is no difference. The same result occurs for levels 2 and 3 of factor B. However, the marginal means for the two levels of factor A are 5.857 and 4.143, implying a difference due to factor A. Using the means as calculated in Table 9.2, the usual calculations for the sum of squares due to factor A would be

$$\sum n_i (\bar{y}_{i.} - \bar{y}_{..})^2 - 7(5.86 - 5.00)^2 + 7(4.14 - 5.00)^2 = 10.282,$$

which is certainly not 0. In fact, using the usual analysis of variance calculations, presented in Table 9.3, produces a significant ($\alpha = 0.05$) effect due to factor A.[5] Further, notice that using this method the total of the A and B sum of squares is greater than the model sum of squares, implying a negative sum of squares for the interaction (which PROC ANOVA converts to 0!).

A closer examination of the data shows a possible reason for the apparently contradictory results. There does seem to an effect due to factor B: The means increase from level 1 to level 2 and again to level 3. Now the marginal mean for level 1 of factor A is heavily influenced by the four observations having large values resulting from level 3 of factor B, while the marginal mean for level 2 of factor A is heavily influenced by the low values from the four observations for level 1 of factor B. In other words, the apparent differences between the marginal

TABLE 9.2

UNBALANCED DATA

A Factor Levels	B Factor Levels			Marginal Means
	1	2	3	
1	3 Mean = 3.0	6, 4 Mean = 5.0	7, 8, 6, 7 Mean = 7.0	5.857
2	2, 3, 4, 3 Mean = 3.0	3, 7 Mean = 5.0	7 Mean = 7.0	4.143
Marginal Means	3.0	5.0	7.0	5.00

[5] The log for PROC ANOVA does note that the data are unbalanced and suggests the use of PROC GLM. Not all computer programs have this feature.

TABLE 9.3

ANALYSIS OF VARIANCE (PROC ANOVA)

Dependent Variable: Y

Source	DF	Sum of Squares	Mean Square	F Value	Pr > F
Model	5	40.00000000	8.00000000	4.57	0.0288
Error	8	14.00000000	1.75000000		
Corrected Total	13	54.00000000			

Source	DF	Anova SS	Mean Square	F Value	Pr > F
A	1	10.28571429	10.28571429	5.88	0.0416
B	2	40.00000000	20.00000000	11.43	0.0045
A*B	2	0.00000000	0.00000000	0.00	1.0000

means of A are apparently due to factor B, which is, of course not what we want. We can understand this effect more precisely by showing the marginal means estimate in terms of the model parameters.

For simplicity, we will use the model without interaction, that is,

$$y_{ijk} = \mu + \alpha_i + \beta_j + \epsilon_{ijk}.$$

Using this model we know, for example, that $\bar{y}_{11} = 3$, the mean of the A1, B1 cell, is an estimate of $\mu + \alpha_1 + \beta_1$. The "usual" formulas give us the following marginal means for factor A in terms of the model parameters:

$$\bar{y}_{1.} = \frac{\bar{y}_{11} + 2(\bar{y}_{12}) + 4(\bar{y}_{13})}{7}$$

$$\bar{y}_{1.} \doteq \frac{1}{7}[(\mu + \alpha_1 + \beta_1) + 2(\mu + \alpha_1 + \beta_2) + 4(\mu + \alpha_1 + \beta_3)]$$

$$\doteq \frac{1}{7}(7\mu + 7\alpha_1 + \beta_1 + 2\beta_2 + 4\beta_3)$$

$$\doteq (\mu + \alpha_1) + \frac{1}{7}(\beta_1 + 2\beta_2 + 4\beta_3)$$

and

$$\bar{y}_{2.} \doteq (\mu + \alpha_2) + \frac{1}{7}(4\beta_1 + 2\beta_2 + \beta_3).$$

The difference between these two means is:

$$\bar{y}_{1.} - \bar{y}_{2.} \doteq (\alpha_1 - \alpha_2) + \frac{3}{7}(\beta_3 - \beta_1).$$

Normally we would expect this difference to estimate $(\alpha_1 - \alpha_2)$, but because of the unequal cell frequencies it additionally estimates $\frac{1}{2}(\beta_3 - \beta_1)$. This supports the argument we noted previously.

Now if we had estimates of β_1 and β_3, we could obtain the needed estimate of $(\alpha_1 - \alpha_2)$. However, if the data are unbalanced, obtaining these estimates would require the estimates of the α_i. This means that we need to simultaneously estimate all of the parameters. This is, of course, what is accomplished by solving the normal equations for a regression model to obtain the partial regression coefficients. Therefore, using the dummy variable model in a regression setting provides the correct estimates. In other words, the estimates of the partial regression coefficients provide for the appropriate estimates as well as inferences.

Actually, unbalanced data are somewhat similar to having correlated variables in a regression. Remember that if the independent variables in a regression are uncorrelated, we can separately estimate the individual regression coefficients as if they were simple linear regressions. Similarly, in the balanced analysis of variance case we can compute the means and sums of squares independently for each factor. And just as correlations among independent variables (multicollinearity) reduce the effectiveness of individual partial regression coefficients, unbalanced data do not provide as efficient estimates and tests as are provided by balanced data. In other words: Good experimental designs are indeed optimal.

We illustrate by showing the analysis of variance, including the interaction, produced by PROC GLM for this example (Table 9.4).

For reasons that do not concern us, PROC GLM calls the partial sums of squares Type III sums of squares. We can see that the sum of squares due to factor A is indeed 0 and that the least squares means (LSMEAN) for factor A are also equal. Note also that the total of the factor sums of squares do not add to the model sum of squares, as is usual for most regression analyses. Also, the interaction sum of squares is indeed 0, as the data were constructed. Thus, this analysis of variance based on the dummy variable model agrees with the apparent nonexistence of the effect due to factor A. ◆

EXAMPLE 9.3   Senility and Brain Characteristics   In this example we will investigate how the size of the ventricle (a component of the brain) is related to senility and EEG. Senility is coded as follows:

     1: Not senile
     2: Mildly senile
     3: Moderately senile
     4: Severely senile

The EEG (electroencephalogram) readings, which measure electrical activity, are coded as follows:

     1: Normal
     2: Abnormal

TABLE 9.4

ANALYSIS OF VARIANCE BY GLM

Dependent Variable: Y

Source	DF	Sum of Squares	Mean Square	F Value	Pr > F
Model	5	40.00000000	8.00000000	4.57	0.0288
Error	8	14.00000000	1.75000000		
Corrected Total	13	54.00000000			

Source	DF	Type III SS	Mean Square	F Value	Pr > F
A	1	0.00000000	0.00000000	0.00	1.0000
B	2	25.60000000	12.80000000	7.31	0.0156
A*B	2	0.00000000	0.00000000	0.00	1.0000

Least Squares Means

A	Y LSMEAN	Std Err LSMEAN	Pr > |T| H0:LSMEAN=0
1	5.00000000	0.58333333	0.0001
2	5.00000000	0.58333333	0.0001

B	Y LSMEAN	Std Err LSMEAN	Pr > |T| H0:LSMEAN=0
1	3.00000000	0.73950997	0.0036
2	5.00000000	0.66143783	0.0001
3	7.00000000	0.73950997	0.0001

The data, available as File REG09X03, provide measurements of these variables on 88 elderly patients. The results of the analysis, using PROC GLM, are shown in Table 9.5. Here we see that although the model is significant ($p < 0.05$), none of the factors are. Hence, we may conclude that ventricle size is not related to senility or the EEG reading.

However, if we had performed the analysis using the "usual" analysis of variance calculations (results not shown), we would conclude that both main effects are significant ($p < 0.05$). The reasons for the different results can be explained by examining various statistics based on the cell frequencies (first value) and means (second value in parentheses) shown in Table 9.6.

We concentrate on differences due to senility. Here we see that the differences among the "ordinary" means are somewhat larger that those among the least squares means. The largest discrepancy occurs for the senility = 2 class, where the ordinary mean is dominated by the relatively large number of observations in the EEG = 1 cell, which has a low cell mean, while the least squares

TABLE 9.5

ANALYSIS OF SENILITY DATA

```
Dependent Variable: VENTRIC
                              Sum of          Mean
Source            DF         Squares        Square    F Value    Pr > F
Model             11      3428.692852     311.699350     2.25     0.0198
Error             76     10530.761693     138.562654

Corrected Total   87     13959.454545


Source            DF      Type III SS    Mean Square   F Value    Pr > F

EEG                2       530.958274     265.479137     1.92     0.1542
SENILITY           3       686.825037     228.941679     1.65     0.1844
EEG*SENILITY       6      1585.321680     264.220280     1.91     0.0905
```

TABLE 9.6

FREQUENCIES AND MEANS

EEG		SENILITY		
Frequency	1	2	3	4
1	23(57)	11(55)	5(64)	6(60)
2	5(59)	4(76)	5(61)	12(67)
3	2(57)	4(53)	3(65)	8(72)
MEAN	57.3	58.7	63.2	66.8
LSMEAN	57.5	61.0	63.5	66.3

mean is not affected by the cell frequencies. Similar but not so large differences occur in the senility = 1 and senility = 4 classes. The differences we see are not very large, but do account for the differences in the p-values. ◆

Large differences in inferences do not always occur for the two methods. However, since the dummy variable approach is the correct one, and since computer programs for this correct method are readily available, it should be used when data are unbalanced. Admittedly, the computer output from programs using the dummy variable approach is often more difficult to interpret,[6] and some other inference procedures, such as multiple comparisons, are not as easily performed (e.g., Montgomery, 1997). However, difficulties in execution should not affect the decision to use the correct method.

Although we have illustrated the dummy variable approach for a two-factor analysis, it can be used to analyze any data structure properly analyzed by the

[6] We have abbreviated the output from PROC GLM to avoid confusion.

analysis of variance. This includes special designs such as split plots and models having nested effects (hierarchial structure). Models can, of course, become unwieldy in terms of the number of parameters, but with the computing power available today, most can be handled without much difficulty. The method is not, however, a panacea that provides results that are not supported with data. In other words, the method will not rescue a poorly executed data-gathering effort, whether experiment, survey, or use of secondary data. And, as we will see in the next section, it cannot obtain estimates for data that do not exist.

Finally, since the method is a regression analysis, virtually all of the analytic procedures presented in this book may be applicable. Multicollinearity (extremely unbalanced data) and influential observations are, however, not very common phenomena, but outliers and nonnormal distribution of residuals may occur. Transformations of the response variable may be used, and the logarithmic transformation can be very useful. Of course, no transformation is made on the dummy variables; hence, the exponentiated parameter estimates become multiplier effects due to factor levels.

9.4 Empty Cells

As we have seen, the dummy variable approach allows us to perform the analysis of variance for unbalanced data. Unfortunately, there are some special cases of unbalanced data where even this method fails. One such situation occurs when there are empty or missing cells; that is, there are some factor-level combinations that contain no observations.

The problem with empty cells is that the model contains more parameters than there are observations to provide estimates. We already know that the dummy variable formulation produces more parameters than equations; hence, the $X'X$ matrix is singular and we have to impose restrictions on the parameters in order to provide useful estimates. And, because the reason for the singularity is known, it is possible to specify restrictions that will provide useful estimates and inferences. However, the additional singularities produced by the existence of empty cells are a different matter. Because empty cells can occur anywhere, the resulting singularities cannot be specified; hence, there are no universally acceptable restrictions that provide useful estimates. That is, any attempt to provide parameter estimates must impose arbitrary restrictions, and different restrictions may provide different results!

Computer programs for the general linear model are constructed to deal with the singularities that normally occur from the formulation of the model. Unfortunately, these programs cannot generally distinguish between the normally expected singularities and those that occur due to empty cells. We have seen that using different restrictions for dealing with the normal singularities does not affect useful estimates and effects. However, when there are empty cells, these different restrictions may affect estimates and effects. In other words, because various computer programs may implement different restrictions, they may provide different results, and furthermore, there is often little or no indication of what the resulting answers may mean. For a more extensive discussion, see Freund (1980).

EXAMPLE 9.4   **Example 9.2 Revisited (optional)**   We will illustrate the problem of empty cells using the data from Example 9.2 where we delete the single observation in the A = 1, B = 1 cell. We now include the interaction term, and use PROC GLM in the SAS System, requesting some options specifically available for this type of problem. The results are shown in Table 9.7.

TABLE 9.7

ANALYSIS FOR EMPTY CELL DATA

Source	DF	Sum of Squares	Mean Square	F Value	Pr > F
Model	4	35.69230769	8.92307692	5.10	0.0244
Error	8	14.00000000	1.75000000		
Corrected Total	12	49.69230769			

Source	DF	Type III SS	Mean Square	F Value	Pr > F
A	1	0.00000000	0.00000000	0.00	1.0000
B	2	18.04651163	9.02325581	5.16	0.0364
A*B	1	0.00000000	0.00000000	0.00	1.0000

Source	DF	Type IV SS	Mean Square	F Value	Pr > F
A	1*	0.00000000	0.00000000	0.00	1.0000
B	2*	13.24137931	6.62068966	3.78	0.0698
A*B	1	0.00000000	0.00000000	0.00	1.0000

*NOTE: Other Type IV Testable Hypotheses exist which may yield different SS.

Least Squares Means

A	Y LSMEAN	Std Err LSMEAN	Pr > |T| H0:LSMEAN=0
1	Non-est	.	.
2	5.00000000	0.58333333	0.0001

B	Y LSMEAN	Std Err LSMEAN	Pr > |T| H0:LSMEAN=0
1	Non-est	.	.
2	5.00000000	0.66143783	0.0001
3	7.00000000	0.73950997	0.0001

The analysis of variance shows only four degrees of freedom for the model. If there were no empty cell, there would be five degrees of freedom. When we turn to the Type III sum of squares, which we have seen to be the same as the partial sums of squares, we see that the interaction now has only one degree of freedom. This result is, in fact, the only sign that we have an empty cell in this example, because the sums of squares for A and A*B are 0, as they were with the complete data. However, this will not usually occur.[7] It is therefore very important to check that the degrees of freedom conform to expectations to ascertain the possibility of a potential empty cell problem.

We now turn to the TYPE IV sums of squares. These are calculated in a different manner and were developed by the author of PROC GLM for this type of situation. For "normal" situations, the Type III and Type IV sums of squares will be identical. However, as we can see in this instance, the results are different. Now there is no claim that the Type IV sums of squares are more "correct" than any other, and, in fact, many authorities prefer Type III. The only reason the Type IV sums of squares are calculated in PROC GLM is to demonstrate that there may be more than one solution, and no one set of estimates can be considered to be better than the other. This is the reason for the footnote "Other Type IV Testable Hypotheses exist which may yield different SS," and the existence of this footnote is a good reason for requesting the Type IV sums of squares if empty cells are suspected to exist.

Finally, the listing of the least squares means also shows that we have a problem. Remember that so-called estimable functions provide estimates that are not affected by the specific restrictions applied to solve the normal equations. Note that least squares means in Table 9.7 that involve the empty cell give the notation "Non-est," meaning they are not estimable. That is, the mathematical requirements for an "estimable function" do not exist for these estimates. In other words, unique estimates cannot be computed for these means. These statements will be printed by PROC GLM whether or not the Type IV sums of squares have been requested.   ◆

The question is what to do if we have empty cells? As we have noted, there is no unique correct answer. Omitting the interaction from the model is one restriction and will generally eliminate the problem, but omitting the interaction implies a possibly unwarranted assumption. Other restrictions may be applied, but are usually no less arbitrary and equally difficult to justify. Another possibility is to restrict the scope of the model by omitting or combining factor levels involved in empty cells. None of these alternatives are attractive, but the problem is that there is simply insufficient data to perform the desired analysis.

When there are more than two factors, the empty cell problem gets more complicated. For example, it may happen that there are complete data for all two-factor interactions, and if the higher-order interactions are considered of no interest, they can be omitted and the remaining results used. Of course, we must remember that this course of action involves an arbitrary restriction.

---

[7] The reason for this result is that there is no A or interaction effect in this example.

9.5 Models with Dummy and Continuous Variables

We consider in this section linear models that include parameters describing effects due to factor levels, as well as others describing regression relationships. In other words, these models include dummy variables representing factor levels, as well as quantitative variables associated with regression analyses. We illustrate with the simplest of these models, which has parameters representing levels of a single factor and a regression coefficient for one independent interval variable. The model is

$$y_{ij} = \beta_0 + \alpha_i + \beta_1 x_{ij} + \epsilon_{ij},$$

where

y_{ij}, $i = 1, 2, \ldots, t$, and $j = 1, 2, \ldots, n_i$, are values of the response variable for the jth observation of factor level i

x_{ij}, $i = 1, 2, \ldots, t$, and $j = 1, 2, \ldots, n_i$, are values of the independent variable for the jth observation of factor level i

α_i, $i = 1, 2, \ldots, t$, are the parameters for factor-level effects

β_0 and β_1 are the parameters of the regression relationship

ϵ_{ij} are the random error values.

If in this model we delete the term $\beta_1 x_{ij}$, the model is

$$y_{ij} = \beta_0 + \alpha_i + \epsilon_{ij},$$

which describes the one-way analysis of variance model (replacing β_0 with μ). On the other hand, if we delete the term α_i, the model is

$$y_{ij} = \beta_0 + \beta_1 x_{ij} + \epsilon_{ij},$$

which is that for a simple linear (one-variable) regression. Thus, the entire model describes a set of data consisting of pairs of values of variables x and y, arranged in a one-way structure or completely randomized design. The interpretation of the model may be aided by redefining parameters:

$$\beta_{0i} = \beta_0 + \alpha_i, \, i = 1, 2, \ldots, t,$$

which produces the model

$$y_{ij} = \beta_{0i} + \beta_1 x_{ij} + \epsilon_{ij}.$$

This model describes a set of t parallel regression lines, one for each factor level. Each has the same slope (β_1), but a different intercept (β_{0i}). A plot of a typical data set and estimated response lines with three factor levels is given in Figure 9.1, where the data points are identified by the factor levels (1, 2, or 3) and the three lines are the three parallel regression lines.

Of interest in this model are

1. The regression coefficient
2. Differences due to the factor levels

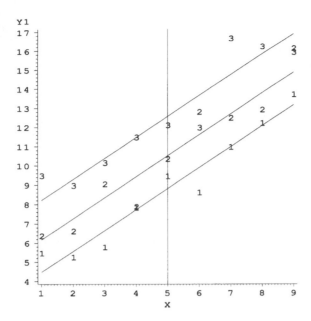

FIGURE 9.1 Data and Model Estimates

The interpretation of the regression coefficient is the same as in ordinary regression. Differences due to factor levels show in the degree of separation among the regression lines and, because they are parallel, are the same for any value of the independent variable. As a matter of convenience, the effects of the factor levels are usually given by the so-called adjusted or least squares means. These are defined as the points on the estimated regression lines ($\hat{\mu}_{y|x}$) at the overall mean of the independent variable, that is, at $\bar{x}$. The least squares mean may therefore be denoted by ($\hat{\mu}_{y|x}$). In Figure 9.1, $\bar{x} = 5$, which is represented by the vertical line, and the least squares means (from computer output, not reproduced here) are 8.8, 10.5, and 12.6.

The statistical analysis of this model starts with the dummy variable model:

$$y_{ij} = \mu z_0 + \alpha_1 z_1 + \alpha_2 z_2 + \ldots + \alpha_t z_t + \beta_1 x + \epsilon_{ij}.$$

This produces an X matrix that contains columns for the dummy variables for the factor levels and a column of values of the independent variable. The $X'X$ matrix is singular and standard restrictions must be used to solve the normal equations. However, the singularity does not affect the estimate of the regression coefficient.

As was the case for models with only dummy variables, models with quantitative and qualitative independent variables can take virtually any form, including dummy variables for design factors such as blocks, and linear and polynomial terms for one or more quantitative variables. As we will see later, we may have interactions between factor-level effects and interval independent variables. Problems of multicollinearity and influential observations may, of course, occur with the interval independent variables, and may be more difficult to detect and remedy because of the complexity of the overall model. Furthermore, computer programs for such models often do not have extensive diagnostic tools for some of these data problems. Therefore, it is of utmost importance to become familiar with the computer program used and thoroughly understand what a particular program does and does not do.

EXAMPLE 9.5   Counting Grubs   Grubs are larval stages of beetles and often cause injury to crops. In a study of the distribution of grubs, a random location was picked 24 times during a 2-month period in a city park known to be infested with grubs. In each location a pit was dug in 4 separate 3-inch depth increments, and the number of grubs of 2 species counted. Also measured for each sample were soil temperature and moisture content. We want to relate grub count to time of day and soil conditions. The data are available as REG09X05. The model is

$$y_{ij} - \mu + \delta_i + \lambda_j + (\delta\lambda)_{ij} + \beta_1(\text{DEPTH}) + \beta_2(\text{TEMP}) + \beta_3(\text{MOIST}) + \epsilon_{ij},$$

where

y_{ij} is the response (COUNT) in the jth species, $j = 1, 2$, in the ith time, $i = 1, 2, \ldots, 12$

μ is the mean (or intercept)

δ_i is the effect of the ith time

λ_j is the effect of the jth species[8]

$(\delta\lambda)_{ij}$ is the interaction between time and species[9]

$\beta_1, \beta_2, \beta_3$ are the regression coefficients for DEPTH, TEMP, and MOIST, respectively (we should note that DEPTH is not strictly an interval variable, although it does represent one that is roughly measured)

ϵ_{ij} is the random error, normally distributed with mean zero and variance σ^2.

Note that μ, δ_i, and λ_j are parameters describing factor levels, whereas the β_i are regression coefficients.

We will use PROC GLM. There are, however, some uncertainties about this model:

- The response variable is a frequency or count variable, which may have a distinctly nonnormal distribution.
- As we noted, depth is not strictly an interval variable.

For this reason we first show the residual plot from the preceding model in Figure 9.2. Here we can see that we do indeed have a nonnormal distribution and there seems to be evidence of a possible curvilinear effect. The complete absence of residuals in the lower left is due to the fact that there cannot be negative counts, which restricts residuals from that area.

[8] Instead of using species as a factor, one could specify a separate analysis for the two species. This is left as an exercise for the reader.

[9] For those familiar with experimental design, time may be considered a block effect and this interaction is the error for testing the species effect. Because some students may not be aware of this distinction, we will ignore it in the discussion of results.

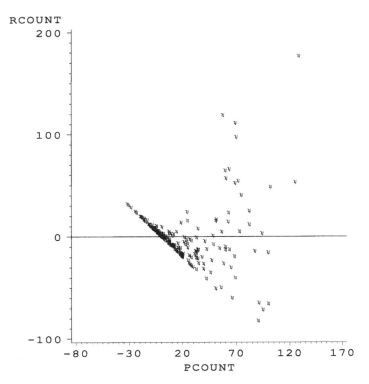

FIGURE 9.2 Residuals from Initial Model

Count data are known to have a Poisson distribution, for which the square root transformation is considered useful. We therefore perform this transformation on the response variable denoted by SQCOUNT and also add a quadratic term to DEPTH, which is denoted by DEPTH*DEPTH. The results of this analysis are shown in Table 9.8.

The model is obviously significant. The times or species appear to have no effect, but there is an interaction between time and species. Among the regression coefficients, only depth and the square of depth are significant. A plot of the least squares means for the time–species interactions shows no discernible pattern and is therefore not reproduced here.

The coefficients for the independent variables are shown in the output. We can use the coefficients to show the nature of the response curve for a typical pit, whose shape is shown in Figure 9.3. Of course, these are responses for the square root of counts, but the general characteristic of the response should be valid for counts.

Another method for plotting the response to depth is to declare DEPTH as a factor and obtain the least squares means, which may then be plotted. A lack of fit test may be used to see if the quadratic curve is adequate. This is left as an exercise for the reader. ◆

TABLE 9.8

ANALYSIS OF GRUB DATA

Dependent Variable: SQCOUNT

Source	DF	Sum of Squares	Mean Square	F Value	Pr > F
Model	51	1974.955210	38.724612	13.97	0.0001
Error	140	388.139445	2.772425		
Corrected Total	191	2363.094655			

Source	DF	Type III SS	Mean Square	F Value	Pr > F
TIME	23	70.9201718	3.0834857	1.11	0.3397
SPEC	1	3.8211817	3.8211817	1.38	0.2424
TIME*SPEC	23	174.5548913	7.5893431	2.74	0.0002
DEPTH	1	254.2976701	254.2976701	91.72	0.0001
DEPTH*DEPTH	1	149.0124852	149.0124852	53.75	0.0001
TEMP	1	6.0954685	6.0954685	2.20	0.1404
MOIST	1	3.9673560	3.9673560	1.43	0.2336

Parameter	Estimate	T for H0: Parameter=0	Pr > |T|	Std Error of Estimate
DEPTH	-7.52190840	-9.58	0.0001	0.78539245
DEPTH*DEPTH	1.04155075	7.33	0.0001	0.14206889
TEMP	-0.21253537	-1.48	0.1404	0.14333675
MOIST	0.09796414	1.20	0.2336	0.08189293

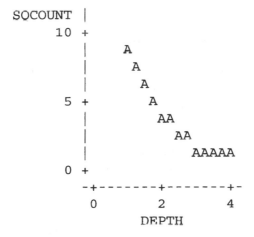

FIGURE 9.3 Response to Depth

9.6 A Special Application: The Analysis of Covariance

A general principle in any data-collecting effort is to minimize the error variance, which will, in turn, provide for higher power for hypothesis tests and narrower confidence intervals. This is usually accomplished by identifying and accounting for known sources of variation. For example, in experimental design, blocking is used to obtain more homogeneous experimental units, which in turn provides for a smaller error variance.

In some cases, a response variable may be affected by measured variables that have nothing to do with the factors in an experiment. For example, in an experiment on methods for inducing weight loss, the final weight of subjects will be affected by their initial weight, as well as the effect of the weight reduction method. Now if an analysis is based on only the final weights of the subjects, the error variance will include the possible effect of the initial weights. On the other hand, if we can somehow "adjust" for the initial weights, it follows that the resulting error variance will only measure the variation in weight *losses*.

One way to do this is to simply analyze the weight losses, and this is indeed an acceptable method. However, this simple subtraction only works if the two variables are measured in the same scale. For example, if we want to adjust the results of a chemical experiment for variations in ambient temperature, no simple subtraction is possible. In other words, we need an analysis procedure that will account for variation due to factors that are not part of the experimental factors.

The analysis of covariance is such a method. The model for the analysis of covariance is indeed the one we have been discussing. That is, for a one-factor experiment and one variable, the model is

$$y_{ij} = \beta_0 + \alpha_i + \beta_1 x_{ij} + \epsilon_{ij},$$

where the parameters and variables are as previously described. However, in the analysis of covariance the independent (regression) variable is known as the *covariate*. Furthermore, in the analysis of covariance the focus of inference is on the least squares or adjusted factor *means*, whereas the nature of the effect of the covariate is of secondary importance.

Two assumptions for the model are critical to ensure the proper inferences:

1. The covariate is not affected by the experimental factors. If this is not true, then the inferences of the factor effects are compromised because they must take into account the values of the covariate. Therefore, covariates are often measures of conditions that exist prior to the conduct of an experiment.
2. The regression relationship as measured by β_1 must be the same for all factor levels. If this assumption does not hold, the least squares means would depend on the value of the covariate. In other words, any inference on differences due to the factor will only be valid for a specific value of x. This would certainly not be a useful inference. A test for the existence of unequal slopes is given in the next section.

Note that in the preceding example of weight losses, analyzing the weight losses is equivalent to the analysis of covariance with β_1 given the value of unity. Hence, the analysis of covariance provides a more flexible model. It could be, for example, that individuals with heavier weights lose more, resulting in $\beta_1 > 1$.

EXAMPLE 9.6   Teaching Methods   The data result from an experiment to determine the effect of three methods of teaching history. Method 1 uses the standard lecture format, method 2 uses short movie clips at the beginning of each period, and method 3 uses a short interactive computer module at the end of the period. Three classes of 20 students are randomly assigned to the methods.[10] The response variable is the students' scores on a uniform final exam.

It is, of course, well known that not all students learn at the same rate: Some students learn better than others, regardless of teaching method. An intelligence test, such as the standard IQ test, may be used as a predictor of learning ability. For these students, this IQ test was administered before the experiment, hence the IQ scores make an ideal covariate. The data are shown in Table 9.9.

TABLE 9.9

TEACHING METHODS DATA

Method 1		Method 2		Method 3	
IQ	Score	IQ	Score	IQ	Score
91	76	102	75	103	91
90	75	91	78	110	89
102	75	90	79	91	89
102	73	80	72	96	94
98	77	94	78	114	91
94	71	104	76	100	94
105	73	107	81	112	95
102	77	96	79	94	90
89	69	109	82	92	85
88	71	100	76	93	90
96	78	105	84	93	92
89	71	112	86	100	94
122	86	94	81	114	95
101	73	97	79	107	92
123	88	97	76	89	87
109	74	80	71	112	100
103	80	101	73	111	95
92	67	97	78	89	85
86	71	101	84	82	82
102	74	94	76	98	90

[10] A preferred design would have at least two sections per method, since classes rather than students are appropriate experimental units for such an experiment.

The model is

$$y_{ij} = \beta_0 + \alpha_i + \beta_1 x_{ij} + \epsilon_{ij},$$

where

y_{ij}, $i = 1, 2, 3$, and $j = 1, 2, \ldots, 20$, are scores on the final exam
x_{ij}, $i = 1, 2, 3$, and $j = 1, 2, \ldots, 20$, are scores of the IQ test
α_i, $i, = 1, 2, 3$, are the parameters for the factor teaching method
β_0 and β_1 are the parameters of the regression relationship
ϵ_{ij} are the random error values

The output from PROC GLM from the SAS System is shown in Table 9.10.

The model is obviously significant. We first look at the effect of the covariate because if it is not significant, the analysis of variance would suffice and the results of that analysis are easier to interpret. The sum of squares due to IQ is significant, and the coefficient indicates a 0.35 unit increase in final exam score associated with a unit increase in IQ. The method is also significant, and the least squares means of 74.85, 78.68, and 90.62, with standard errors of about 0.68, obviously all differ. Tests for paired differences may be made, but do not adjust for the experimentwise error. Contrasts may be computed but, even if they are con-

TABLE 9.10

ANALYSIS OF COVARIANCE

Source	DF	Sum of Squares	Mean Square	F Value	Pr > F
Model	3	3512.745262	1170.915087	125.27	0.0001
Error	56	523.438072	9.347108		
Corrected Total	59	4036.183333			

Source	DF	Type III SS	Mean Square	F Value	Pr > F
METHOD	2	2695.816947	1347.908474	144.21	0.0001
IQ	1	632.711928	632.711928	67.69	0.0001

Parameter	Estimate	T for H0: Parameter=0	Pr > |T|	Std Error of Estimate
IQ	0.34975784	8.23	0.0001	0.04251117

Least Squares Means

METHOD	SCORE LSMEAN	Std Err LSMEAN	Pr > |T| H0:LSMEAN=0
1	74.8509019	0.6837401	0.0001
2	78.6780024	0.6860983	0.0001
3	90.6210957	0.6851835	0.0001

structed to be orthogonal, they are also somewhat correlated. Paired comparisons (such as Duncan's or Tukey's) are difficult to perform because the estimated means are correlated and have different standard errors. Be sure to check program specifications and instructions.

The data and regression lines are shown in Figure 9.4, where the plotting symbol indicates the teaching method. The least squares means occur at the intersection of the vertical line at $\bar{x}$ and agree with the printed results.

It is of interest to see the results of an analysis of variance without the covariate. The major difference is that the error standard deviation is 4.58, compared to 3.06 for the analysis of covariance. In other words, confidence intervals for means are reduced about one-third by using the covariate. Because the differences among means of the teaching methods are quite large, significances are only minimally affected. The means and least squares means are

	Mean	LS Mean
Method 1	74.95	74.85
Method 2	78.20	78.68
Method 3	91.00	90.62

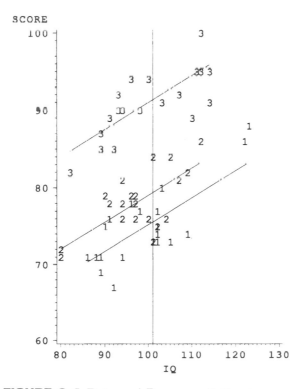

FIGURE 9.4 Data and Response Estimates

We can see that the differences are minor. This is because the means of the covariate differ very little among the three classes. If the mean of the covariate differs among factor levels, least squares means will differ from the ordinary means. ◆

The use of the analysis of covariance is not restricted to the completely randomized design, or to a single covariate. For complex designs, such as split plots, care must be taken to use appropriate error terms. And if there are several covariates, multicollinearity may be a problem, although the fact that we are not usually interested in the coefficients themselves alleviates difficulties.

9.7 Heterogeneous Slopes in the Analysis of Covariance

In all analysis of covariance models we have presented thus far, the regression coefficient is common for all factor levels. This condition is indeed necessary for the validity of the analysis of covariance. Therefore, if we are using the analysis of covariance, we need a test to ascertain that this condition holds. Of course, other models where regression coefficients vary among factor levels may occur, and it is therefore useful to be able to implement analyses for such models.

The existence of variability of the regression coefficients among factor levels is, in fact, an *interaction* between factors and the regression variable(s). That is, the effect of one factor, say the regression coefficient, is different across levels of the other factor levels.

The dummy variable model for a single factor and a single regression variable is

$$y_{ij} = \mu z_0 + \alpha_1 z_1 + \alpha_2 z_2 + \ldots + \alpha_t z_t + \beta_m x + \epsilon_{ij},$$

where the z_i are the dummy variables as previously defined. We have added an "m" subscript to the regression coefficient to distinguish it from those we will need to describe the coefficients for the individual factor levels. Remember that in factor level i, for example, $z_i = 1$ and all other z_i are 0, resulting in the model

$$y_{ij} = \mu + \alpha_i + \beta_m x + \epsilon_{ij}.$$

Now interactions are constructed as products of the main effect variables. Thus, the model that includes the interactions is

$$y_{ij} = \mu z_0 + \alpha_1 z_1 + \alpha_2 z_2 + \ldots + \alpha_t z_t + \beta_m x + \beta_1 z_1 x + \beta_2 z_2 x + \ldots + \beta_t z_t x + \epsilon_{ij}.$$

Using the definition of the dummy variables, the model becomes

$$y_{ij} = \mu + \alpha_i + \beta_m x + \beta_i x + \epsilon_{ij}$$
$$= \mu + \alpha_i + (\beta_m + \beta_i)x + \epsilon_{ij},$$

which defines a model with different intercepts, α_i, and slopes, $(\beta_m + \beta_i)$, for each factor level. Note that as was the case for the dummy variables, there are $(t + 1)$ regression coefficients to be estimated from t factor levels. This introduces another singularity into the model; however, the same principles used for the solution with the dummy variables will also work here.

The test for equality of regression coefficients is now the test for

$$H_0: \beta_i = 0, \text{ for all } i,$$

which is the test for the interaction coefficients.

Some computer programs, such as PROC GLM of the SAS System, have provisions for this test, as we will illustrate next. However, if such a program is not available, the test is readily performed as a restricted/unrestricted model test. The unrestricted model simply estimates a separate regression for each factor level and the error sum of squares is simply the sum of error SS for all models. The restricted model is the analysis of covariance that is restricted to having one regression coefficient. Subtracting the error sum of squares and degrees of freedom as outlined in Chapter 1 provides for the test.

EXAMPLE 9.7   Livestock Prices   From a larger data set, we have extracted data on sales of heifers at an auction market. The response variable is price (PRICE) in dollars per hundred weight. The factors are

GRADE: Coded PRIME, CHOICE, and GOOD
WGT: Weight in hundreds of pounds

The data are shown in Table 9.11 and are available as File REG09X07.

Because weight effect may not be the same for all grades, we propose a model that allows the weight coefficient to vary among grades. This model is implemented with PROC GLM of the SAS System with a model that includes the interaction of weight and class. The results are shown in Table 9.12 on page 369.

The model has five degrees of freedom: two for GRADE, one for WGT, and two for the interaction that allows for the different slopes. The interaction is significant ($p = 0.0028$). The estimated coefficients for weight, labeled wgt/[grade], are shown at the bottom of the table. The outstanding feature is that for the prime grade: Increased weight has a much more negative effect on price than it does for the other grades. Therefore, as in any factorial structure, the main effects of grade and weight may not have a useful interpretation. The plot of the data, with points labeled by the first letter of GRADE is shown in Figure 9.5 on page 369 and clearly demonstrates the different slopes and reinforces the result that the main effects are not readily meaningful.   ◆

EXAMPLE 9.8   Example 9.6 Revisited   In Example 9.6 the IQ scores had approximately the same effect on test scores for all three methods and, in fact, the test for heterogeneous slopes (not shown) is not rejected. We have altered the data so that the effect of IQ increases from method 1 to method 2 and again for method 3. This would be the result if methods 2 and 3 appealed more to students with higher aptitudes. The data are not shown but are available as File REG09X08.

We implement PROC GLM, including in the model statement the interaction between METHOD and IQ. We do not request the printing of the least squares means, as they are not useful, but do request the printing of the estimated

TABLE 9.11

LIVESTOCK MARKETING DATA

OBS	GRADE	WGT	PRICE
1	PRIME	2.55	58.00
2	PRIME	2.55	57.75
3	PRIME	2.70	42.00
4	PRIME	2.90	42.25
5	PRIME	2.65	60.00
6	PRIME	2.90	48.75
7	PRIME	2.50	63.00
8	PRIME	2.50	62.25
9	PRIME	2.50	56.50
10	CHOICE	2.55	48.00
11	CHOICE	3.05	38.25
12	CHOICE	2.60	40.50
13	CHOICE	3.35	40.75
14	CHOICE	4.23	32.25
15	CHOICE	3.10	37.75
16	CHOICE	3.75	36.75
17	CHOICE	3.60	37.00
18	CHOICE	2.70	44.25
19	CHOICE	2.70	40.50
20	CHOICE	3.05	39.75
21	CHOICE	3.65	34.50
22	GOOD	2.50	39.00
23	GOOD	2.55	44.00
24	GOOD	2.60	45.00
25	GOOD	2.55	44.00
26	GOOD	2.90	41.25
27	GOOD	3.40	34.25
28	GOOD	2.02	33.25
29	GOOD	3.95	33.00

coefficients (beta1, beta2, and beta3) for the three methods. The results are shown in Table 9.13 on page 370. The model now has five parameters (plus the intercept): two for the factors, one for the overall regression, and two for the additional two regressions. Again, we first check the interaction; it is significant ($p = 0.0074$), and hence, we conclude that the effect of the covariate is not the same for the three methods. An analysis of covariance is not appropriate. At this point, none of the other tests are useful, as they represent parameters that are not meaningful. That is, if regressions are different, the overall or "mean" coefficient has no meaning, and differences among response means depend on specific values of the covariate (IQ score). The last portion of the output shows the estimated coefficients and their standard errors. We see that it indeed appears that $\beta_1 < \beta_2 < \beta_3$. The data and estimated lines are shown in Figure 9.6 on page 370.

Remember that in a factorial experiment the effect of an interaction was that it prevented making useful inferences on main effects. This is exactly what happens here: The effect of the teaching methods depends on the IQ scores of

TABLE 9.12

ANALYSIS OF LIVESTOCK MARKETING DATA

Source	DF	Sum of Squares	Mean Square	F Value	Pr > F
Model	5	1963.156520	392.631304	22.91	0.0001
Error	23	394.140894	17.136561		
Corrected Total	28	2357.297414			

R-Square	C.V.	Root MSE	PRICE Mean
0.832800	9.419329	4.139633	43.94828

Source	DF	Type III SS	Mean Square	F Value	Pr > F
GRADE	2	343.8817604	171.9408802	10.03	0.0007
WGT	1	464.9348084	464.9348084	27.13	0.0001
WGT*GRADE	2	263.1206149	131.5603074	7.68	0.0028

Parameter	Estimate	T for H0: Parameter=0	Pr > |T|	Std Error of Estimate
wgt/choice	-6.7215787	-2.85	0.0090	2.35667471
wgt/good	-3.4263512	-1.32	0.1988	2.58965521
wgt/prime	-39.9155844	-4.46	0.0002	8.95092088

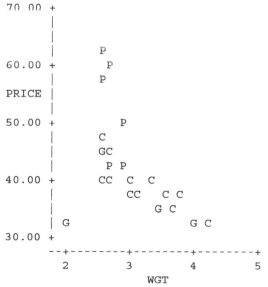

Plot of PRICE*WGT=GRADE.

NOTE: 7 obs hidden.

FIGURE 9.5 Plot of Livestock Marketing Data

TABLE 9.13

ANALYSIS WITH DIFFERENT SLOPES

Source	DF	Sum of Squares	Mean Square	F Value	Pr > F
Model	5	3581.530397	716.306079	75.42	0.0001
Error	54	512.869603	9.497585		
Corrected Total	59	4094.4000000			

Source	DF	Type III SS	Mean Square	F Value	Pr > F
METHOD	2	26.6406241	13.3203121	1.40	0.2548
IQ	1	574.5357250	574.5357250	60.49	0.0001
IQ*METHOD	2	102.2348437	51.1174219	5.38	0.0074

Parameter	Estimate	T for H0: Parameter=0	Pr > |T|	Std Error of Estimate
beta1	0.18618499	2.71	0.0090	0.06865110
beta2	0.31719119	3.76	0.0004	0.08441112
beta3	0.51206140	7.10	0.0001	0.07215961

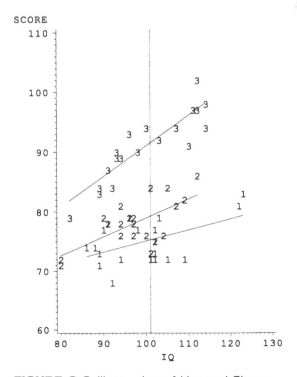

FIGURE 9.6 Illustration of Unequal Slopes

students. As we can see in Figure 9.6, method 3 is indeed superior to the others for all students, although the difference is most marked for students with higher IQs. Method 2 is virtually no better that method 1 for students with lower IQs. ◆

As before, differences in slopes may occur with other data structures and/or several covariates. Of course, interpretations become more complicated. For example, if we have a factorial experiment, the regressions may differ across levels of any one or more main effects, or even across all factor-level combinations. For such situations a sequential analysis procedure must be used, starting with the most complicated (unrestricted) model and reducing the scope (adding restrictions) when nonsignificances are found. Thus, for a two-factor factorial, the model will start with different coefficients for all cells; if these are found to differ, no simplification is possible. However, if these are found nonsignificant, continue with testing for differences among levels of factor B, and so forth.

For models with several independent variables, there will be a sum of squares for interaction with each variable, and variable selection may be used. However, programs for models with dummy and interval variables usually do not provide for variable selection. Hence, such a program must be rerun after any variable is deleted.

9.8 Summary

In this chapter we have introduced the use of "dummy" variables to perform analysis of variance using regression methodology. The method is more cumbersome than the usual analysis of variance calculations, but must be used when analyzing factorial data with unequal cell frequencies. However, even the dummy variable method fails if there are empty cells.

Models including both dummy and interval independent variables are a simple extension. The analysis of covariance is a special application where the focus is on the analysis of the effect of the factor levels holding constant the independent interval variable, called the covariate. However, if the effect due to the covariate is not the same for all levels of the factors, the analysis of covariance may not be appropriate.

CHAPTER EXERCISES

1. A psychology student obtained data on a study of aggressive behavior in nursery-school children, shown in Table 9.14. His analysis used the cell and marginal means shown in the table. The p-values of that analysis, using standard ANOVA computations gave STATUS, $p = 0.0225$; GENDER, $p = 0.0001$; and interaction, $p = 0.0450$. His analysis was incorrect in *two* ways.

Perform the correct analysis and see if results are different. The data are available as File REG09P01.

2. It is of interest to determine if the existence of a particular gene, called the TG gene, affects the weaning weight of mice. A sample of 97 mice of two strains (A and B) are randomly assigned to five cages. Variables recorded are the response, weight at weaning (WGT, in grams), the presence of TG (TG: coded Y or N), and sex (SEX: coded M or F). Because the age at weaning also affects weight, this variable was also recorded (AGE, in days).The data are available in File REG09P02. Perform an analysis to determine if the existence of the gene affects weaning weight. Discuss and, if possible, analyze for violations of assumptions.

3. The pay of basketball players is obviously related to performance, but it may also be a function of the position they play. Data for the 1984–1985 season on pay (SAL, in thousands of dollars) and performance, as measured by scoring average (AVG, in points per game), are obtained for eight randomly selected players from each of the following positions (POS): (1) scoring forward, (2) power forward, (3) center, (4) off guard, and (5) point guard. The data are given in File REG09P03. Perform the analysis to ascertain whether position affects pay over and above the effect of scoring. Plotting the data may be useful.

4. It is desired to estimate the weights of five items using a scale that is not completely accurate. Obviously, randomly replicated weighings are needed, but that would require too much time. Instead, all 10 combinations of three items are weighed. The results are shown in Table 9.15. Construct the data and set up the model to estimate the individual weights. (Hint: Use a model without intercept.)

5. We have quarterly data for the years 1955–1968 on the number of feeder cattle (PLACE) in feedlots for fattening. We want to estimate this number as a function of the price of range cattle (PRANGE), which are the cattle that enter the feedlots for fattening; the price of slaughter cattle (PSLTR), which are the products of the fattening process; and the price of corn (PCORN), the main ingredient of the feed for the fattening process. It is well known that there is a seasonal pattern of feeder placement as well as a possibility of a long-term trend. The data are available in File REG09P05.

TABLE 9.14

NUMBER OF AGGRESSIVE BEHAVIORS BY SEX AND SOCIABILITY STATUS

	Sociable	Shy	Means
Female	0,1,2,1,0,1,2,4,0,0,1	0,1,2,1,0,0,1	0.9444
Male	3,7,8,6,6,7,2,0	2,1,3,1,2	3.692
Means	2.6842	1.1667	2.097

TABLE 9.15

WEIGHTS

Combination	Weight
123	5
124	7
125	8
134	9
135	9
145	12
234	8
235	8
245	11
345	13

Perform an analysis for estimating feeder placement. (Hint: There are a number of issues to be faced in developing this model.)

6. Exercise 5 of Chapter 3 (data in File REG03P05) concerned the relationship of the sales of three types of oranges to their prices. Because purchasing patterns may differ among the days of the week, the variable DAY is the day of the week (Sunday was not included). Reanalyze the data to see the effect of the day of the week. Check assumptions.

7. From the *Statistical Abstract of the United States, 1995*, we have data on college enrollment by sex from 1975 through 1993. The data are available in File REG09P07, which identifies the sex and enrollment by year. Perform a regression to estimate the trend in enrollment for both males and females and perform a test to see if the trends differ

Categorical Response Variables

10.1 Introduction

The primary emphasis of this text up to this point has been that of modeling a continuous response variable. We have seen how this response can be modeled using continuous or categorical independent or factor variables, or even a combination of both. Obviously, situations arise where it is desirable to be able to construct a statistical model using a categorical response variable. Basic courses in statistical methods do present methodology for analyzing relationships involving categorical variables; however, the analyses usually involve relationships between only two categorical variables, which are analyzed with the use of contingency tables and the chi-square test for independence. This analysis very rarely involves the construction of a model. In this chapter we will consider analyses of categorical response variables in the form of regression models. We first examine models with a binary response variable and continuous independent variable(s), followed by the more general case of response variables with any number of categories. We then consider models where a categorical response variable is related to any number of categorical independent variables.

10.2 Binary Response Variables

In a variety of applications we may have a response variable that has only two possible outcomes. As in the case of a dichotomous independent variable, we can represent such a variable by a dummy variable. In this context, such a variable is often called a **quantal** or **binary** response. It is often useful to study the behavior of such a variable as related to one or more numeric independent or factor variables. In other words, we may want to do a regression analysis where the dependent variable is a dummy variable and the independent variable or variables may be interval variables.

For example:

- An economist may investigate the incidence of failure of savings and loan banks as related to the size of their deposits. The independent variable is the average size of deposits at the end of the first year of business and the dependent variable can be coded as

  $y = 1$ if the bank succeeded for 5 years
  $y = 0$ if it failed within the 5-year period

- A biologist is investigating the effect of pollution on the survival of a certain species of organism. The independent variable is the level of pollution as measured in the habitat of this particular species, and the dependent variable is

  $y = 1$ if an individual of the species survived to adulthood
  $y = 0$ if it died prior to adulthood

- A study to determine the effect of an insecticide on insects will use as the independent variable the strength of the insecticide, and a dependent variable defined as

  $y = 1$ if an individual insect exposed to the insecticide dies
  $y = 0$ if the individual does not die

Because many applications of such models are concerned with response to medical drugs, the independent variable is often called the "dose" and the dependent variable the "response." In fact, this approach to modeling furnishes the foundation for a branch of statistics called **bioassay**. We will briefly discuss some methods used in bioassay later in this section. The reader is referred to Finney (1971) for a complete discussion of this subject.

A number of statistical methods have been developed for analyzing models with a dichotomous response variable. We will present two such methods in some detail:

1. The standard linear regression model,

$$y = \beta_0 + \beta_1 x + \epsilon$$

2. The logistic regression model,

$$y = \frac{\exp(\beta_0 + \beta_1 x)}{1 + \exp(\beta_0 + \beta_1 x)} + \epsilon$$

The first model is a straight-line fit of the data, whereas the second model provides a special curved line. Both have practical applications and have been found appropriate in a wide variety of situations. Both models may also be used with more than one independent variable.

Before discussing the procedures for using sample data to estimate the regression coefficients for either model, we will examine the effect of using a dummy response variable.

The Linear Model with a Dichotomous Dependent Variable

To illustrate a linear model with a response variable that only has values of 0 or 1, consider the following example. A medical researcher is interested in determining whether the amount of a certain antibiotic given to mothers after caesarean delivery affects the incidence of infection. The researcher proposes a simple linear regression model,

$$y = \beta_0 + \beta_1 x + \epsilon$$

where

$y = 1$ if infection occurs within 2 weeks
$y = 0$ if not
$x =$ amount of the antibiotic in ml/hr
$\epsilon =$ random error, a random variable with mean 0 and variance σ^2

The researcher is to control values of x at specified levels for a sample of patients.

In this model, the expected response has a special meaning. Since the error term has mean 0, the expected response is

$$\mu_{y|x} = \beta_0 + \beta_1 x$$

The response variable has the properties of a binomial random variable with the following discrete probability distribution.

y	$p(y)$
0	$1 - p$
1	p

where p is the probability that y takes the value 1. What this is saying is that the regression model actually provides a mechanism for estimating the probability that $y = 1$, that is, the probability of a patient suffering a postoperative infection. In other words, the researcher is modeling how the probability of postoperative infection is affected by different strengths of the antibiotic.

Unfortunately, special problems arise with the regression process when the response variable is dichotomous. Recall that the error terms in a regression model are assumed to have a normal distribution with a constant variance for all observations. In the model that uses a dummy variable for a dependent variable, the error terms are not normal, nor do they have a constant variance.

According to the definition of the dependent variable, the error terms will have the values

$$\epsilon = 1 - \beta_0 - \beta_1 x, \text{ when } y = 1,$$

and

$$\epsilon = -\beta_0 - \beta_1 x, \text{ when } y = 0.$$

Obviously, the assumption of normality does not hold for this model. Additionally, since y is a binomial variable, the variance of y is

$$\sigma^2 = p(1 - p).$$

But $p = \mu_{y|x} = \beta_0 + \beta_1 x$; hence,

$$\sigma^2 = (\beta_0 + \beta_1 x)(1 - \beta_0 - \beta_1 x).$$

Clearly, the variance depends on x, which is a violation of the equal variance assumptions.

Finally, since $\mu_{y|x}$ is really a probability, its values are bounded by 0 and 1. This imposes a constraint on the regression model that limits the estimation of the regression parameters. In fact, ordinary least squares may predict values for the dependent variable that are negative or larger than 1 even for values of the independent variable that are within the range of the sample data.

Although these violations of the assumptions cause a certain amount of difficulty, solutions are available:

- The problem of nonnormality is mitigated by recalling that the central limit theorem indicates that for most distributions, the sampling distribution of the mean will be approximately normal for reasonably large samples. Furthermore, even in the case of a small sample, the estimates of the regression coefficients, and consequently the estimated responses, are unbiased point estimates.
- The problem of unequal variances is solved by the use of **weighted least squares**, which was presented in Section 4.3.
- If the linear model predicts values for μ_x that are outside the interval, we choose a curvilinear model that does not. The logistic regression model is one such choice.

10.3 Weighted Least Squares

In Section 4.3 we noted that in the case of nonconstant variances, the appropriate weight to be assigned to the ith observation is

$$w_i = 1/\sigma_i^2,$$

where σ_i^2 is the variance of the ith observation. This procedure gives smaller weights to observations with large variances and vice versa. In other words, more "reliable" observations provide more information and vice versa. After weighting, all other estimation and inference procedures are performed in the usual manner, except that the actual values of sums of squares as well as mean squares reflect the numerical values of the weights.

In a model with a dichotomous response variable, σ_i^2 is equal to $p_i(1 - p_i)$, where p_i is the probability that the ith observation is 1. We do not know this probability, but according to our model,

$$p_i = \beta_0 + \beta_1 x_i .$$

Therefore, a logical procedure for doing weighted least squares to obtain estimates of the regression coefficients is as follows:

1. Use the desired model and perform an ordinary least squares regression to compute the predicted value of y for all x_i. Call these $\hat{\mu}_i$.
2. Estimate the weights by

$$\hat{w}_i = \frac{1}{\hat{\mu}_i(1 - \hat{\mu}_i)} .$$

3. Use these weights in a weighted least squares and obtain estimates of the regression coefficients.
4. This procedure may be iterated until the estimates of the coefficients stabilize. That is, repetition is stopped when estimates change very little from iteration to iteration.

Usually the estimates obtained in this way will stabilize very quickly, making step 4 unnecessary. In fact, in many cases, the estimates obtained from the first weighted least squares will differ very little from those obtained from the ordinary least squares procedure. Thus, ordinary least squares does give satisfactory results in many cases

As we noted in Section 4.3, the estimates of coefficients usually change little due to weighting, but the confidence and prediction intervals for the response will reflect the relative degrees of precision based on the appropriate variances. That is, intervals for observations having small variances will be smaller than those for observations with large variances. However, even here the differences due to weighting may not be very large.

EXAMPLE 10.1   In a recent study of urban planning in Florida, a survey was taken of 50 cities, 24 of which used tax increment funding (TIF) and 26 of which did not. One part of the study was to investigate the relationship between the presence or absence of TIF and the median family income of the city. The data are given in Table 10.1, which is described next.

The linear model is

$$y = \beta_0 + \beta_1 x + \epsilon,$$

TABLE 10.1

DATA ON URBAN PLANNING STUDY

y	Income	y	Income
0	9.2	0	12.9
0	9.2	1	9.6
0	9.3	1	10.1
0	9.4	1	10.3
0	9.5	1	10.9
0	9.5	1	10.9
0	9.5	1	11.1
0	9.6	1	11.1
0	9.7	1	11.1
0	9.7	1	11.5
0	9.8	1	11.8
0	9.8	1	11.9
0	9.9	1	12.1
0	10.5	1	12.2
0	10.5	1	12.5
0	10.9	1	12.6
0	11.0	1	12.6
0	11.2	1	12.6
0	11.2	1	12.9
0	11.5	1	12.9
0	11.7	1	12.9
0	11.8	1	12.9
0	12.1	1	13.1
0	12.3	1	13.2
0	12.5	1	13.5

where

$y = 0$ if the city did not use TIF
$y = 1$ if it did
x = median income of the city
ϵ = random error

The first step in obtaining the desired estimates of the regression coefficients is to perform an ordinary least squares regression. The results are given in Table 10.2. The values of the estimated coefficients are used to obtain the estimated values, $\hat{\mu}_i$, of y for each x, which are then used to calculate weights for estimation by weighted least squares. **Caution:** The linear model can produce $\hat{\mu}_i$ values less than 0 or greater than 1. If this has occurred, the weights will be undefined and an alternative model, such as the logistic model (described later in this chapter), must be considered. The predicted values and weights are given in Table 10.3.

The computer output of the weighted least squares regression is given in Table 10.4. Note that these estimates differ very little from the ordinary least squares estimates in Table 10.2. Rounding the parameter estimates in the output, we get the desired regression equation:

$$\hat{\mu}_{y|x} = -1.980 + 0.21913(\text{INCOME}).$$

The data and estimated line are shown in Figure 10.1. The plot suggests a rather poor fit, which is supported by an R-square value of 0.45, but the p-value of 0.0001 suggests that median income does have some bearing on the participation in TIF. Thus, for example, the estimated probability of a city with median income of \$10,000 using TIF is $-1.980 + 0.21913(10) = 0.2113$. That is, there is about a 21% chance that a city with median income of \$10,000 is participating in tax increment funding.

To illustrate the fact that the weighted least squares estimate stabilizes quite rapidly, two more iterations were performed. The results are

$$\text{Iteration 2: } \hat{\mu}_{y|x} = -1.992 + 0.2200(\text{INCOME})$$

and

$$\text{Iteration 3: } \hat{\mu}_{y|x} = -2.015 + 0.2218(\text{INCOME}).$$

The regression estimates change very little, and virtually no benefit in the standard error of the estimates is realized by the additional iterations.

Notice that the regression equation does not predict negative values or values greater than 1 as long as we consider only median incomes within the range of the data. Thus, the equation satisfies the constraints discussed previously. In addition, the sample size of 50 is sufficiently large to overcome the nonnormality of the distribution of the residuals. ◆

TABLE 10.2

REGRESSION OF INCOME ON TIF

Analysis of Variance

Source	DF	Sum of Squares	Mean Square	F Value	Prob>F
Model	1	3.53957	3.53957	19.003	0.0001
Error	48	8.94043	0.18626		
C Total	49	12.48000			

Parameter Estimates

Variable	DF	Parameter Estimate	Standard Error	T for H0: Parameter=0	Prob > |T|
INTERCEP	1	-1.818872	0.53086972	-3.426	0.0013
INCOME	1	0.205073	0.04704277	4.359	0.0001

TABLE 10.3

ESTIMATION OF WEIGHTS

y	Income	Predicted Value	Weight	y	Income	Predicted Value	Weight
0	9.2	0.068	15.821	0	12.9	0.827	6.976
0	9.2	0.068	15.821	1	9.6	0.150	7.850
0	9.3	0.088	12.421	1	10.1	0.252	5.300
0	9.4	0.109	10.312	1	10.3	0.293	4.824
0	9.5	0.129	8.881	1	10.9	0.416	4.115
0	9.5	0.129	8.881	1	10.9	0.416	4.115
0	9.5	0.129	8.881	1	11.1	0.457	4.029
0	9.6	0.150	7.850	1	11.1	0.457	4.029
0	9.7	0.170	7.076	1	11.1	0.457	4.029
0	9.7	0.170	7.076	1	11.5	0.539	4.025
0	9.8	0.191	6.476	1	11.8	0.601	4.170
0	9.8	0.191	6.476	1	11.9	0.622	4.251
0	9.9	0.211	5.999	1	12.1	0.663	4.472
0	10.5	0.334	4.493	1	12.2	0.683	4.619
0	10.5	0.334	4.493	1	12.5	0.745	5.258
0	10.9	0.416	4.115	1	12.6	0.765	5.563
0	11.0	0.437	4.065	1	12.6	0.765	5.563
0	11.2	0.478	4.008	1	12.6	0.765	5.563
0	11.2	0.478	4.008	1	12.9	0.827	6.976
0	11.5	0.539	4.025	1	12.9	0.827	6.976
0	11.7	0.580	4.106	1	12.9	0.827	6.976
0	11.8	0.601	4.170	1	12.9	0.827	6.976
0	12.1	0.663	4.472	1	13.1	0.868	8.705
0	12.3	0.704	4.794	1	13.2	0.888	10.062
0	12.5	0.745	5.258	1	13.5	0.950	20.901

TABLE 10.4

WEIGHTED REGRESSION

Analysis of Variance

Source	DF	Sum of Squares	Mean Square	F Value	Prob>F
Model	1	36.49604	36.49604	38.651	0.0001
Error	48	45.32389	0.94425		
C Total	49	81.81993			

Parameter Estimates

Variable	DF	Parameter Estimate	Standard Error	T for H0: Parameter=0	Prob > |T|
INTERCEP	1	-1.979665	0.39479503	-5.014	0.0001
INCOME	1	0.219126	0.03524632	6.217	0.0001

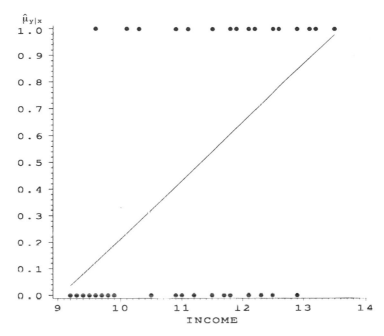

FIGURE 10.1 Linear Regression

10.4 Simple Logistic Regression

If a simple linear regression equation model using weighted least squares violates the constraints on the model or does not properly fit the data, we may need to use a curvilinear model. One such model with a wide range of applicability is the logistic regression model:

$$\mu_{y|x} = \frac{\exp(\beta_0 + \beta_1 x)}{1 + \exp(\beta_0 + \beta_1 x)}.$$

The curve described by the logistic model has the following properties:

- As x becomes large, $\mu_{y|x}$ approaches 1 if $\beta_1 > 0$, or approaches 0 if $\beta_1 < 0$. Similarly, as x becomes small, $\mu_{y|x}$ approaches 0 if $\beta_1 > 0$, or approaches 1 if $\beta_1 < 0$.
- $\mu_{y|x} = \frac{1}{2}$ when $x = -(\beta_0/\beta_1)$.
- The curve describing $\mu_{y|x}$ is monotone, that is, it either increases (or decreases) everywhere.

A typical simple logistic regression function for $\beta_1 > 0$ is shown in Figure 10.2. Notice that the graph is sigmoidal or "S"-shaped. This feature makes it more useful when there are observations for which the response probability is near 0 or 1, since the curve can never go below 0 or above 1, which is not true of the strictly linear model.

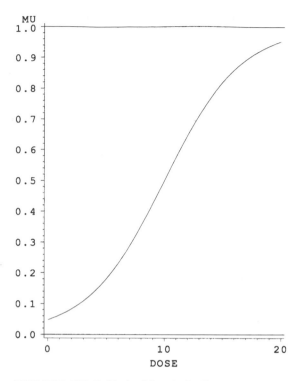

FIGURE 10.2 Typical Logistic Curve

Although the function itself is certainly not linear and appears very complex, it is, in fact, relatively easy to use. The model has two unknown parameters, β_0 and β_1. It is not coincidental that these parameters have the same symbols as those of the simple linear regression model. Estimating the two parameters from sample data is reasonably straightforward. We first make a **logit** transformation of the form

$$\mu_p = \log\left[\frac{\mu_{y|x}}{1 - \mu_{y|x}}\right],$$

where log is the natural logarithm. Substituting this transformation for $\mu_{y|x}$ in the logistic model results in a model of the form

$$\mu_p = \beta_0 + \beta_1 x + \epsilon,$$

which is a simple linear regression model. Of course, the values of the μ_p are usually not known; hence, preliminary estimates must be used. If multiple observations exist for each x, preliminary estimates of the $\mu_{y|x}$ are simply the sample proportions. If such multiples are not available, an alternative procedure using the maximum likelihood method is recommended and discussed later in this section.

The logit transformation linearizes the model, but does not eliminate the problem of nonconstant variance. Therefore, the regression coefficients in this simple linear regression model should be estimated using weighted least squares. We will illustrate the procedure with an example where multiple observations for each value of x are used as preliminary estimates of μ_p.

EXAMPLE 10.2   A toxicologist is interested in the effect of a toxic substance on tumor incidence in laboratory animals. A sample of animals is exposed to various concentrations of the substance and subsequently examined for the presence or absence of tumors. The response variable for an individual animal is then either 1 if a tumor is present, or 0 if not. The independent variable is the concentration of the toxic substance (CONC). The number of animals at each concentration (N) and the number of individuals with the value 1, that is, the number having tumors (NUMBER), make up the results, which are shown in Table 10.5.

The first step is to use the logit transformation to "linearize" the model. The second step consists of the use of weighted least squares to obtain estimates of the unknown parameters. Because the experiment was conducted at only six distinct values of the independent variable, concentration of the substance, the task is not difficult.

We calculate $\hat{p}$, the proportion of 1's at each value of CONC. These are given in Table 10.6 under the column PHAT. We then make the logit transformation on the resulting values:

$$\hat{\mu}_p = \ln [\hat{p}/(1 - \hat{p})].$$

These are given in Table 10.6 under the column LOG.

TABLE 10.5

DATA FOR TOXICOLOGY STUDY

CONC	N	NUMBER
0.0	50	2
2.1	54	5
5.4	46	5
8.0	51	10
15.0	50	40
19.5	52	42

TABLE 10.6

CALCULATIONS FOR LOGISTIC REGRESSION

CONC	N	NUMBER	PHAT	LOG	W
0.0	50	2	0.04000	-3.17805	1.92000
2.1	54	5	0.09259	-2.28238	4.53704
5.4	46	5	0.10870	-2.10413	4.45652
8.0	51	10	0.19608	-1.41099	8.03922
15.0	50	40	0.80000	1.38629	8.00000
19.5	52	42	0.80769	1.43508	8.07692

Because the variances are still not constant, we have to use weighted regression. The weights are computed as

$$\hat{w}_i = n_i \hat{p}_i (1 - \hat{p}_i),$$

where

n_i = total number of animals at concentration x_i

$\hat{p}_i$ = sample proportion of animals with tumors at concentration x_i

These values are listed in Table 10.6 under the column W. We now perform the weighted least squares regression, using LOG as the dependent variable and concentration as the independent variable.

The results of the weighted least squares estimation are given in Table 10.7. The model is certainly significant, with a *p*-value of 0.0017, and the coefficient of determination is a respectable 0.93. The residual variation is somewhat difficult to interpret since we are using the log scale.

The coefficients of the estimated simple linear regression model are rounded to give:

$$\hat{LOG} = -3.139 + 0.254(CONC).$$

This can be transformed back into the original units using the transformation:

$$ESTPROP = \exp(\hat{LOG})/\{1 + \exp(\hat{LOG})\},$$

which for CONC = 10 gives the value 0.355. This means that, on the average, there is a 35.5% chance that exposure to concentrations of 10 units results in tumors in laboratory animals.

The response curve for this example is shown in Figure 10.3, which also shows the original values. From this plot we can verify that the estimated probability of a tumor when the concentration is 10 units is approximately 0.35. ◆

TABLE 10.7

LOGISTIC REGRESSION ESTIMATES

Analysis of Variance

Source	DF	Sum of Squares	Mean Square	F Value	Prob>F
Model	1	97.79495	97.79495	56.063	0.0017
Error	4	6.97750	1.74437		
C Total	5	104.77245			

Parameter Estimates

Variable	DF	Parameter Estimate	Standard Error	T for H0: Parameter=0	Prob > |T|
INTERCEP	1	-3.138831	0.42690670	-7.352	0.0018
CONC	1	0.254274	0.03395972	7.488	0.0017

Another feature of the simple logistic regression function is the interpretation of the coefficient β_1. Recall that we defined μ_p as

$$\mu_p = \log\left[\frac{\mu_{y|x}}{1 - \mu_{y|x}}\right].$$

The quantity $\{\mu_{y|x}/(1 - \mu_{y|x})\}$ is called the odds in favor of the event, in this case, having a tumor. Then μ_p, the log of the odds at x, is denoted as $\log\{$Odds at $x\}$. Suppose we consider the same value at $(x + 1)$. Then

$$\mu_p = \log\left[\frac{\mu_{y|x+1}}{1 - \mu_{y|x+1}}\right],$$

would be $\log\{$Odds at $(x + 1)\}$. According to the linear model, $\log\{$Odds at $x\}$ $= \beta_0 + \beta_1 x$ and $\log\{$Odds at $\{x + 1\}$ $= \beta_0 + \beta_1(x + 1)$. It can be shown that the difference between the odds at $(x + 1)$ and at x is

$$\log\{\text{Odds at } (x + 1)\} - \log\{\text{Odds at } x\} = \beta_1,$$

which is equivalent to

$$\log\{(\text{Odds at } x + 1)/(\text{Odds at } x)\} = \beta_1.$$

Taking antilogs of both sides gives the relationship

$$\frac{\text{Odds at } x + 1}{\text{Odds at } x} = e^{\beta_1}.$$

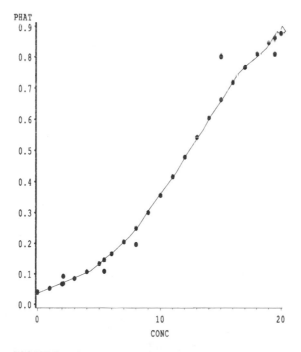

FIGURE 10.3 Plot of Logistics Curve

The estimate of this quantity is known as the **odds ratio**, and is interpreted as the increase in the odds, or the proportional increase in the response proportion, for a unit increase in the independent variable. In our example, $\hat{\beta}_1 = 0.25$; hence, the estimated odds ratio is $e^{0.25} = 1.28$. Therefore, the odds of getting a tumor are estimated to increase by 28% with a unit increase in concentration of the toxin.

The logistic model can also be used to find certain critical values of the independent variable. For example, suppose that the toxicologist in Example 10.2 wants to estimate the concentration of the substance at which 75% of the animals exposed would be expected to develop a tumor. In other words, we are looking for a value of the independent variable for a given value of the response. A rough approximation can be obtained from Figure 10.3 by locating the value of CONC corresponding to a PHAT of 0.75. From that graph, the value would appear to be approximately 17. We can use the estimated logistic regression to solve for this value.

We start with the assumption that $\mu_{y|x} = 0.75$, and then:

$$\mu_p = \log\left[\frac{\mu_{y|x}}{1 - \mu_{y|x}}\right] = \log\left[\frac{0.75}{1 - 0.75}\right] = 1.099.$$

Using the estimated coefficients from Table 4.7 provides the equation

$$1.099 = -3.139 + 0.254x,$$

which is solved for x to provide the estimate of 16.69. This agrees with the approximation found from the graph.

The procedure presented in this section will not work for data in which one or more of the distinct x values has a $\hat{p}$ of 0 or 1, because the logit is undefined for these values. Modifications in the definition of these extreme values can be made that remedy this problem. One procedure is to define $\hat{p}_i$ to be $1/2n_i$ if the sample proportion is 0 and $\hat{p}_i$ to be $(1 - 1/2n_i)$ if the sample proportion is 1, where n_i are the number of observations in each factor level.

This procedure for calculating estimates of the regression coefficients can be very cumbersome, and in fact cannot be done if multiple observations are not available at all values of the independent variable. Therefore, most logistic regression is performed by estimating the regression coefficients using the method known as **maximum likelihood estimation**. This method uses the logistic function and an assumed distribution of y to obtain estimates for the coefficients that are most consistent with the sample data. A discussion of the maximum likelihood method of estimation is given in Appendix C. The procedure is complex and usually requires numerical search methods; hence, maximum likelihood estimation of a logistic regression is done on a computer. Most computer packages equipped to do logistic regression offer this option. The result of using the maximum likelihood method using PROC CATMOD in SAS on the data from Example 10.2 is given in Table 10.8. Notice that the estimates are very similar to those given in Table 10.7.

As an example of a logistic regression where multiple observations are not available, let us return to Example 10.1. Recall that the response variable, y, was denoted as 0 if the city did not use tax increment funding and 1 if it did, and the independent variable was the median family income of the city. Because there are no multiple observations, we will use the maximum likelihood method of estimation. Table 10.9 gives a portion of the output from PROC LOGISTIC in SAS.[1] Notice that the table lists the parameter estimates and a test on the parameters called the Wald chi-square. This test plays the part of the t test for regression coefficients in the standard regression model.

Standard errors of the estimates and the p-values associated with the tests for significance are also presented. Notice that both coefficients are significant. The odds ratio is given as 2.723. Recall that the odds increase multiplicatively by the value of the odds ratio for a unit increase in the independent variable, x. In other words, the odds will increase by a multiple of 2.723 for an increase in median income of $1000. This means that the odds of a city participating in TIF increase by about 172% for every increase in median income of $1000. Further, using the estimates of the coefficients in the logistic model, we can determine that for a city with median income of $10,000 (INCOME=10), the estimated probability is about 0.22. This compares favorably with the estimate of 21% we got using the weighted regression in Example 10.1.

Figure 10.4 shows the graph of the estimated logistic regression model. Compare this graph with the one in Figure 10.1.

TABLE 10.8

MAXIMUM LIKELIHOOD ESTIMATES

EFFECT	PARAMETER	ESTIMATE	STANDARD ERROR
INTERCEPT	1	− 3.20423	0.33125
X	2	.262767	0.0273256

TABLE 10.9

LOGISTIC REGRESSION FOR EXAMPLE 10.1

Variable	DF	Parameter Estimate	Standard Error	Wald Chi-Square	Pr > Chi-Square	Odds Ratio
INTERCPT	1	-11.3487	3.3513	11.4673	0.0007	0.000
INCOME	1	1.0019	0.2954	11.5027	0.0007	2.723

[1] Because PROC LOGISTIC uses $y - 1$ if the characteristic of interest is present and $y = 2$ otherwise, the data were recoded prior to running the program to ensure that the signs of the coefficients fit the problem. As always, it is recommended that documentation relating to any computer program used be consulted prior to doing any analysis.

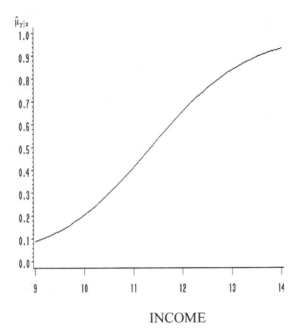

INCOME

FIGURE 10.4 Logistic Regression for Example 10.1

10.5 Multiple Logistic Regression

The simple logistic regression model can easily be extended to two or more independent variables. Of course, the more variables, the harder it is to get multiple observations at all levels of all variables. Therefore, most logistic regressions with more than one independent variable are done using the maximum likelihood method. The extension from a single independent variable to m independent variables simply involves replacing $\beta_0 + \beta_1 x$ with $\beta_0 + \beta_1 x_1 + \beta_2 x_2 + \ldots + \beta_m x_m$ in the simple logistic regression equation given in Section 10.4. The corresponding logistic regression equation then becomes

$$\mu_{y|x} = \frac{\exp(\beta_0 + \beta_1 x_1 + \beta_2 x_2 + \ldots \beta_m x_m)}{1 + \exp(\beta_0 + \beta_1 x_1 + \beta_2 x_2 + \ldots \beta_m x_m)}.$$

Making the same logit transformation as before,

$$\mu_p = \log\left[\frac{\mu_{y|x}}{1 - \mu_{y|x}}\right],$$

we obtain the multiple linear regression model:

$$\mu_p = \beta_0 + \beta_1 x_1 + \beta_2 x_2 + \ldots + \beta_m x_m.$$

We then estimate the coefficients of this model using maximum likelihood methods, similar to those used in the simple logistic regression problem.

EXAMPLE 10.3 As an illustration of the multiple logistic regression model, suppose that the toxicology study of Example 10.2 involved two types of substances. The logistic regression model used to analyze the effect of concentration now involves a second independent variable, type of substance. The data given in Table 10.10 show the results. Again, the response variable is either 1 if a tumor is present, or 0 if not. The concentration of toxic substance is again CONC, and the type of substance (TYPE) is either 1 or 2. The number of animals at each combination of concentration and type is N, and the number of animals having tumors is labeled NUMBER.

To analyze the data, we will use the multiple logistic regression model, with two independent variables, CONC and TYPE. Even though we have multiple observations at each combination of levels of the independent variables, we will use the maximum likelihood method to do the analysis. Using PROC LOGISTIC in SAS, we obtain the results given in Table 10.11. The presented results are only a portion of the output.

This output resembles that of a multiple linear regression analysis using ordinary least squares. The differences lie in the test statistic used to evaluate the significance of the coefficients. The maximum likelihood method uses the Wald chi-square statistic rather than the t distribution. The output also gives us standardized estimates and the odds ratio.

TABLE 10.10

DATA FOR TOXICOLOGY STUDY

ODS	CONC	TYPE	N	NUMBER
1	0.0	1	25	2
2	0.0	2	25	0
3	2.1	1	27	4
4	2.1	2	27	1
5	5.4	1	23	3
6	5.4	2	23	2
7	8.0	1	26	6
8	8.0	2	25	4
9	15.0	1	25	25
10	15.0	2	25	15
11	19.5	1	27	25
12	19.5	2	25	17

TABLE 10.11

MAXIMUM LIKELIHOOD ESTIMATES

Variable	DF	Parameter Estimate	Standard Error	Wald Chi-Square	Pr > Chi-Square	Standardized Estimate	Odds Ratio
INTERCPT	1	-1.3856	0.5346	6.7176	0.0095		0.250
CONC	1	0.2853	0.0305	87.2364	0.0001	1.096480	1.330
TYPE	1	-1.3974	0.3697	14.2823	0.0002	-0.385820	0.247

The interpretation of the estimated regression coefficients in the multiple logistic regression model parallels that for the simple logistic regression, with the exception that the coefficients are the partial coefficients of the multiple linear regression model (see Section 3.4). From Table 10.11 we can see that both the independent variables are significant; therefore, there is an effect due to the concentration of the toxic substance on incidence of tumors with type fixed, and there is a difference in type of toxic substance with concentration fixed. The interpretation of the estimated odds ratio for one independent variable assumes all other independent variables are held constant. From Table 10.11 we see that the odds ratio for concentration is 1.33. Therefore, we can say that the odds of getting a tumor increase by 33% for a unit increase in concentration of the toxin for a fixed type. That is, the risk increases by approximately 33% as long as the type of toxin does not change. Further, we can see from the table that the estimated odds ratio for type is 0.247. From this we can conclude that the risk of tumors for type 1 toxin is about 1/4 or 25% that of type 2.   ◆

As in all regression analyses, it is important to justify the necessary assumptions on the model. In the case of logistic regression, we need to be sure that the estimated response function, $\mu_{y|x}$, is monotonic and sigmoidal in shape. This can usually be determined by plotting the estimated response function. The detection of outliers and influential observations and determining whether the logistic regression is appropriate are much more difficult to do for binary response variables. Some procedures for this are given in Neter *et. al.* (1996).

There are several other curvilinear models that can be used to model binary response variables. Long (1997) discusses four such models, one of which is known as the **probit model**. The probit model has almost the same shape as the logistic model and is obtained by transforming the $\mu_{y|x}$ by means of the cumulative normal distribution. The probit transformation is less flexible than the logistic regression model because it cannot be readily extended to more than one predictor variable. Also, formal inference procedures are more difficult to carry out with the probit regression model. In many cases, the two models agree closely except near the endpoints. Long (1997) refers to both the probit and logit jointly as the binary response model.

To demonstrate the use of the probit model, we reanalyze the data from Table 10.5 using PROC PROBIT in SAS. The results are shown in Table 10.12, along with comparative results from the logistic regression shown in Table 10.8. Notice that there is very little difference in the predicted values.

On occasion we may have a response variable that has more than two levels. For example, in the toxin study described earlier, we may have animals that are unaffected, have pretumor lesions, or have defined tumors. Therefore, we would have a response variable that had three categories. Logistic regression can still be employed to analyze this type of data by means of a **polytomous logistic regression model**.

Polytomous logistic regression is simply an extension of the binary logistic regression model. Various complexities arise from this extension, but the basic ideas used are the same. Hosmer and Lemeshow (1989) provide details for the polytomous logistic regression analysis.

TABLE 10.12

COMPARISON OF OBSERVED, LOGISTIC, AND
PROBITMODELS FOR EXAMPLE 10.2

OBS	CONC	PHAT	LOGISTIC	PROBIT
1	0.0	0.04000	0.03333	0.03901
2	2.1	0.09259	0.06451	0.06584
3	5.4	0.10870	0.15351	0.14365
4	8.0	0.19608	0.26423	0.24935
5	15.0	0.80000	0.66377	0.67640
6	19.5	0.80769	0.86429	0.87211

An approximate method of handling three or more response categories in a logistic regression is to carry out the analysis using several individual binary logistic regression models. For example, if the toxin study had three outcomes, we could construct three separate binary logistic models. One would use two categories: no tumor and pretumor lesion; the second would use no tumor and tumor; and the third would use pretumor lesion and tumor. This type of analysis is easier to do than a single polytomous logistic regression and often results in only a moderate loss of efficiency. See Begg and Gray (1984) for a discussion of the two methods.

10.6 Loglinear Model

When both the response variable and the independent variables are categorical, the logit model becomes very cumbersome to use. Instead of using logistic regression to analyze such a process, we usually use what is known as the **loglinear model**,[2] which is designed for categorical data analysis. A complete discussion of this model and its wide range of applications can be found in Agresti (1990). We will discuss the use of the loglinear model to describe the relationship between a categorical response variable and one or more categorical independent variables.

A convenient way to present data collected on two or more categorical variables simultaneously is in the form of a contingency table. If the data are measured only on two variables, one independent variable and the response variable, the contingency table is simply a two-way frequency table. If the study involves more than one independent variable, the contingency table takes the form of a multiway frequency table. Further, since the categorical variables may not have any ordering (relative magnitude) of the levels, the sequencing of the levels is often arbitrary, so there is not one unique table.

A general strategy for the analysis of contingency tables involves testing several models, including models that represent various associations or interactions among the variables. Each model generates expected cell frequencies that are com-

---

[2] Notice that we use the terminology "loglinear" to describe this model. This is to differentiate it from the "linear in log" model of Chapter 8.

pared with the observed frequencies. The model that best fits the observed data is chosen. This allows for the analysis of problems with more than two variables and for identification of simple and complex associations among these variables.

One such way of analyzing contingency tables is known as **loglinear modeling**. In the loglinear modeling approach, the expected frequencies are computed under the assumption that a certain specified model is appropriate to explain the relationship among variables. The complexity of this model usually results in computational problems obtaining the expected frequencies. These problems can be resolved only through the use of iterative methods. As a consequence of this, most analyses are done with computers.

As an example of a loglinear model, consider the following example.

EXAMPLE 10.4   A random sample of 102 registered voters was taken from the Supervisor of Elections' roll. Each of the registered voters was asked the following two questions:

1. What is your political party affiliation?
2. Are you in favor of increased arms spending?

The results are given in Table 10.13.

The variables are "party affiliation" and "opinion." We will designate the probability of an individual belonging to the ijth cell as p_{ij}, the marginal probability of belonging to the ith row (opinion) as p_i, and the marginal probability of belonging to the jth column (party) as p_j. If the two variables are statistically independent, then

$$p_{ij} = p_i p_j.$$

Under this condition the expected frequencies are

$$E_{ij} = np_{ij} = np_i p_j.$$

Taking natural logs of both sides results in the relationship

$$\log(E_{ij}) = \log(n) + \log(p_i) + \log(p_j).$$

Therefore, if the two variables are independent, the log of the expected frequencies is a linear function of the marginal probabilities. We turn this around and see that a test for independence is really a test to see if the log of the expected frequencies is a linear function of the marginal probabilities.

TABLE 10.13

FREQUENCIES OF OPINION BY PARTY

OPINION	PARTY			
	DEM	REP	NONE	TOTAL
FAVOR	16	21	11	48
NOFAVOR	24	17	13	54
TOTAL	40	38	24	102

Define $\mu_{ij} = \log(E_{ij})$, $\log(n) = \mu$, $\log(p_i) = \lambda_i^A$, and $\log(p_j) = \lambda_j^B$. Then the model[3] can be written as

$$\mu_{ij} = \mu + \lambda_i^A + \lambda_j^B.$$

This model closely resembles a linear model with two categorical independent variables, which is the two-factor ANOVA model. In fact, the analysis very closely resembles that of a two-way analysis of variance model. The terms λ^A represent the effects of variable A designated as "rows" (opinion), and the terms λ^B represent the effects of the variable B, or "columns" (party affiliation).

Notice that the model is constructed under the assumption that rows and columns of the contingency table are independent. If they are not independent, this model requires an additional term, which can be called an "association" or interaction factor. Using consistent notation, this term may be designated λ_{ij}^{AB}. This term is analogous to the interaction term in the ANOVA model and has a similar interpretation. The test for independence then becomes one of determining whether the association factor should be in the model or not. This is done by what is called a "lack of fit" test, usually using the likelihood ratio statistic.

This test follows the same pattern as the test for interaction in the factorial ANOVA model, and the results are usually displayed in a table very similar to the ANOVA table. Instead of using sums of squares and the F distribution to test hypotheses about the parameters in the model, we use the likelihood ratio statistic and the chi-square distribution. The likelihood ratio test statistic is used because it can be subdivided corresponding to the various terms in the model.

We first perform the test of independence using a loglinear model. If we specify the model as outlined previously, the hypothesis of independence becomes:

$$H_0: \lambda_{ij}^{AB} = 0, \text{ for all } i \text{ and } j$$

$$H_1: \lambda_{ij}^{AB} \neq 0, \text{ for some } i \text{ and } j.$$

The analysis is performed by PROC CATMOD from the SAS System with results shown in Table 10.14.

As in the analysis of a factorial experiment, we start by examining the interaction, here called association. The last item in that output is the likelihood ratio

TABLE 10.14

LOGLINEAR ANALYSIS FOR EXAMPLE 10.4

SOURCE	DF	CHI-SQUARE	PROB
PARTY	2	4.38	0.1117
OPINION	1	0.35	0.5527
LIKELIHOOD RATIO	2	1.85	0.3972

[3] A and B are not exponents; they are identifiers and are used in a superscript mode to avoid complicated subscripts.

test for goodness of fit and has a value of 1.85 and a p-value of 0.3972. Thus, we cannot reject H_0, and we conclude the independence model fits.

The other items in the printout are the tests on the "main effects," which are a feature of the use of this type of analysis. It is interesting to note that both the opinion and the party likelihood ratio statistics are not significant. Although the exact hypotheses tested by these statistics are expressed in terms of means of logarithms of expected frequencies, the general interpretation is that there is no difference in the marginal values for opinion nor in party. By looking at the data in Table 10.13 we see that the total favoring the issue is 48, whereas the total not favoring it is 54. Further, the proportions of the number of Democrats, Republicans, and "none" listed in the margin of the table are quite close. In conclusion, there is nothing about this table that differs significantly![4] ◆

EXAMPLE 10.5 A study by Aylward *et al.* (1984) and reported in Green (1988) examines the relationship between neurological status and gestational age. The researchers were interested in determining whether knowing an infant's gestational age can provide additional information regarding the infant's neurological status. For this study a total of 505 newborn infants were cross-classified on two variables: overall neurological status, as measured by the Prechtl examination, and gestational age. The data are shown in Table 10.15. Notice that the age of the infant is recorded by intervals and can therefore be considered a categorical variable.

We will analyze these data using the loglinear modeling approach. That is, we will develop a set of hierarchical models, starting with the simplest, which may be of little interest, and going to the most complex, testing each model for goodness of fit. The model that best fits the data will be adopted. Some of the computations will be done by hand for illustrative purposes only, but the resulting statistics were provided by computer output.

We start with the simplest model, one that contains only the overall mean. This model has the form

$$\log(E_{ij}) = \mu_{ij} = \mu.$$

TABLE 10.15

NUMBER OF INFANTS

	Gestational Age (in weeks)				
	31 or less	32–33	34–36	37 or More	All Infants
Prechtl Status					
Normal	46	111	169	103	429
Dubious	11	15	19	11	56
Abnormal	8	5	4	3	20
All Infants	65	131	192	117	505

[4] In some applications these main effects may not be of interest.

The expected frequencies under this model are given in Table 10.16.

Notice that all the expected frequencies are the same, 42. This is because the model assumes all the cells have the same value, μ. The expected frequencies are then the total divided by the number of cells, or $505/12 = 42$ (rounded to integers). The likelihood ratio statistic for testing the lack of fit of this model, obtained by PROC CATMOD from the SAS System, has a huge value of 252.7. This value obviously exceeds the 0.05 tabled value of 19.675 for the χ^2 distribution with eleven degrees of freedom; hence, we readily reject the model and go to the next.

The next model has only one term in addition to the mean. We can choose a model that has only the grand mean and a row effect, or we can choose a model with only the grand mean and a column effect. For the purposes of this example, we choose the model with a grand mean and a row effect. This model is

$$\log(E_{ij}) = \mu_{ij} = \mu + \lambda_i^A.$$

The term λ_i^A represents the effect due to Prechtl scores. Note that there is no effect due to age groups. The expected frequencies are listed in Table 10.17. They are obtained by dividing each row total by 4, the number of columns.

For example, the first row is obtained by dividing 429 by 4 (rounded to integers). The likelihood ratio test for lack of fit has a value of 80.85, which is compared to the value $\chi^2_{0.05}(9) = 16.919$. Again, the model does not fit, so we must go to the next model. The next model has both age and Prechtl as factors. That is, the model is

$$\ln(E_{ij}) = \mu_{ij} = \lambda_i^P + \lambda_i^A.$$

TABLE 10.16

EXPECTED FREQUENCIES, NO EFFECTS

Prechtl Status	Age Group 1	2	3	4	Total
Normal	42	42	42	42	168
Dubious	42	42	42	42	168
Abnormal	42	42	42	42	168

TABLE 10.17

EXPECTED FREQUENCIES WITH ROW EFFECT

Prechtl Status	Age Group 1	2	3	4	Total
Normal	107	107	107	107	429
Dubious	14	14	14	14	56
Abnormal	5	5	5	5	20

We will be testing the goodness of fit of the model, but really we will be testing for independence. This is because this is a lack of fit test against the hierarchical scheme which uses the "saturated" model, or the model that contained the terms above as well as the "interaction" term, λ_{ij}^{AB}.

The expected frequencies are given in Table 10.18. The values are calculated by multiplying row totals by column totals and dividing by the total. The likelihood ratio test statistic for testing the goodness of fit of this model has a value of 14.30. This exceeds the critical value of $X_{0.05}^2(6) = 12.592$, so this model does not fit, either. That is, there is a significant relationship between the gestational age of newborn infants and their neurological status. Examination of Table 10.15 indicates that 40% of abnormal infants were less than 31 weeks of age, and that the percentage of abnormal infants decreases across age. ◆

The extension of the loglinear model to more than two categorical variables is relatively straightforward, and most computer packages offer this option. The procedure for extending this type of analysis to three categorical variables simply follows the preceding pattern. As an illustration of the procedure, consider the following example.

EXAMPLE 10.6   A school psychologist was interested in determining a relationship between socioeconomic status, race, and the ability to pass a standardized reading exam of students in the sixth-grade. The data in Table 10.19 resulted from a review of one sixth-grade class. The variable Race has two levels, White and Nonwhite. The variable School Lunch, a measure of socioeconomic status, also has two levels,

TABLE 10.18

EXPECTED FREQUENCIES, ROW AND COLUMN EFFECT

Prechtl Status	Age Group 1	2	3	4	Total
Normal	55	111	163	99	429
Dubious	7	15	21	13	56
Abnormal	3	5	8	5	20

TABLE 10.19

STUDENT DATA

Race	School Lunch	Passed Test No	Yes
White	No	25	150
	Yes	43	143
Nonwhite	No	23	29
	Yes	36	22

Yes and No. The variable Passed Test indicates whether the student passed or did not pass the standardized test. The table lists the frequency of occurrence of each combination of the three variables. The total sample size is 471 students.

We want to obtain the best model to fit this data, and consequently explain the relationship between these three variables. To do so we will employ a hierarchical approach to loglinear modeling.

Starting with the model with no interactions,

$$\mu_{ijk} = \lambda_i^R + \lambda_j^L + \lambda_k^P,$$

we perform the analysis using PROC CATMOD from the SAS System, with the results shown in Table 10.20. Notice that the likelihood ratio test for lack of fit is significant, indicating that the model with no interactions is not sufficient to explain the relationship between the three variables. We next try a model with only the two-way interactions. That is, we fit the model

$$\mu_{ijk} = \lambda_i^R + \lambda_j^L + \lambda_k^P + \lambda_{ij}^{RL} + \lambda_{ik}^{RP} + \lambda_{jk}^{LP}.$$

Again, the model is fit using PROC CATMOD, and the results are presented in Table 10.21.

TABLE 10.20

MODEL WITH NO INTERACTION

MAXIMUM-LIKELIHOOD ANALYSIS-OF-VARIANCE TABLE

Source	DF	Chi-Square	Prob
RACE	1	119.07	0.0000
LUNCH	1	0.61	0.4335
PASS	1	92.10	0.0000
LIKELIHOOD RATIO	4	56.07	0.0000

TABLE 10.21

ANALYSIS WITH INTERACTIONS

MAXIMUM-LIKELIHOOD ANALYSIS-OF-VARIANCE TABLE

Source	DF	Chi-Square	Prob
RACE	1	63.94	0.0000
LUNCH	1	2.31	0.1283
RACE*LUNCH	1	0.55	0.4596
PASS	1	33.07	0.0000
RACE*PASS	1	47.35	0.0000
LUNCH*PASS	1	7.90	0.0049
LIKELIHOOD RATIO	1	0.08	0.7787

Now the likelihood ratio test for lack of fit is not significant, indicating a reasonable fit of the model. No three-way interaction is present. Notice that the two-way interaction between race and lunch is not significant. Therefore, we may try a model without that term. Even though the "main effect" Lunch is not significant, its interaction with Pass is, so we will use the convention that main effects involved in significant interactions remain in the model. The model without the interaction between Race and Lunch is then tested, with the results given in Table 10.22.

The model fits very well. The likelihood ratio test for the goodness of fit indicates the model fits adequately. The individual terms are all significant at the 0.05 level.

To interpret the results, Table 10.23 gives the proportion of students of each race in the other two categories. For example, 18% of the White students did not pass the test, whereas 54% of the Nonwhite students did not pass.   ◆

TABLE 10.22

ANALYSIS WITHOUT RACE–LUNCH INTERACTIONS

MAXIMUM-LIKELIHOOD ANALYSIS-OF-VARIANCE TABLE

Source	DF	Chi-Square	Prob
RACE	1	65.35	0.0000
LUNCH	1	3.85	0.0498
PASS	1	33.34	0.0000
LUNCH*PASS	1	7.44	0.0064
RACE*PASS	1	47.20	0.0000
LIKELIHOOD RATIO	2	0.63	0.7308

TABLE 10.23

PROPORTIONS

Race	School Lunch	Passed Test	
		No	**Yes**
White	Yes	0.07	0.42
	No	0.11	0.40
		0.18	0.82
Nonwhite	Yes	0.21	0.26
	No	0.33	0.20
		0.54	0.46

10.7 Summary

In this chapter we briefly examined the problems of modeling a categorical response variable. The use of a binary response variable led to yet another nonlinear model, the logistic regression model. We examined two strategies to handle responses that had more than two categories but had continuous independent variables. We then briefly looked at how we could model categorical responses with categorical independent variables through the use of the loglinear model.

There are many variations of the modeling approach to the analysis of categorical data. These topics are discussed in various texts, including Bishop, Feinberg, and Holland (1975) and Upton (1978). A discussion of categorical data with ordered categories is given in Agresti (1984). A methodology that clearly distinguishes between independent and dependent variables is given in Grizzle, Starmer, and Koch (1969). This methodology is often called the "linear model" approach and emphasizes estimation and hypothesis testing of the model parameters. Therefore, it is easily used to test for differences among probabilities, but is awkward to use for tests of independence. Conversely, the loglinear model is relatively easy to use to test independence, but not so easily used to test for differences among probabilities. Most computer packages offer the user a choice of approaches. As in all methodology that relies heavily on computer calculations, the user should make sure that the analysis is really what is expected by carefully reading documentation on the particular program used.

CHAPTER EXERCISES

1. In a study to determine the effectiveness of a new insecticide on common cockroaches, samples of 100 roaches were exposed to five levels of the insecticide. After 20 minutes the number of dead roaches was counted. Table 10.24 gives the results.

    (a) Calculate the estimated logistic response curve.
    (b) Find the estimated probability of death when the concentration is 17%.

TABLE 10.24

DATA FOR EXERCISE 1

Level (% concentration)	Number of Roaches	Number of Dead Roaches
5	100	15
10	100	27
15	100	35
20	100	50
30	100	69

(c) Find the odds ratio.

(d) Estimate the concentration for which 50% of the roaches treated are expected to die.

2. Using the results of Exercise 1, plot the estimated logistic curve and the observed values. Does the regression appear to fit?

3. A recent heart disease study examined the effect of blood pressure on the incident of heart disease. The average blood pressure of a sample of adult males was taken over a 6-year period. At the end of the period the subjects were classified as having coronary heart disease or not having it. The results are in Table 10.25.

(a) Calculate the estimated logistic response curve

(b) What is the probability of heart disease for an adult male with average blood pressure of 150?

(c) At what value of the average blood pressure would we expect the chance of heart disease to be 75%?

4. Reaven and Miller (1979) examined the relationship between chemical subclinical and overt nonketotic diabetes in nonobese adult subjects. The three primary variables used in the analysis are glucose intolerance (GLUCOS), insulin response to oral glucose (RESP), and insulin resistance (RESIST). The patients were then classified as "normal" (N), "chemical diabetic" (C), or "overt diabetic" (O). Table 10.26 and File REG10P04 give the results for a sample of 50 patients from the study.

Use the classification as a response variable and the other three as independent variables and perform three separate binary logistic regressions. Explain the results.

5. Using the data in Table 10.26 or File REG10P04, do the following:

(a) Use the classification as a response variable and GLUCOS as an independent variable and perform three separate binary logistic regressions.

(b) Use the classification as a response variable and RESP as an independent variable and perform three separate binary logistic regressions.

TABLE 10.25

DATA FOR EXERCISE 3

Average Blood Pressure	Number of Subjects	Number with Heart Disease
117	156	3
126	252	17
136	285	13
146	271	16
156	140	13
166	85	8
176	100	17
186	43	8

TABLE 10.26

DATA FOR EXERCISE 4

SUBJ	GLUCOS	RESP	RESIST	CLASS
1	56	24	55	N
2	289	117	76	N
3	319	143	105	N
4	356	199	108	N
5	323	240	143	N
6	381	157	165	N
7	350	221	119	N
8	301	186	105	N
9	379	142	98	N
10	296	131	94	N
11	353	221	53	N
12	306	178	66	N
13	290	136	142	N
14	371	200	93	N
15	312	208	68	N
16	393	202	102	N
17	425	143	204	C
18	465	237	111	C
19	558	748	122	C
20	503	320	253	C
21	540	188	211	C
22	469	607	271	C
23	486	297	220	C
24	568	232	276	C
25	527	480	233	C
26	537	622	264	C
27	466	287	231	C
28	599	266	268	C
29	477	124	60	C
30	472	297	272	C
31	456	326	235	C
32	517	564	206	C
33	503	408	300	C
34	522	325	286	C
35	1468	28	455	O
36	1487	23	327	O
37	714	232	279	O
38	1470	54	382	O
39	1113	81	378	O
40	972	87	374	O
41	854	76	260	O
42	1364	42	346	O
43	832	102	319	O
44	967	138	351	O
45	920	160	357	O
46	613	131	248	O
47	857	145	324	O
48	1373	45	300	O
49	1133	118	300	O
50	849	159	310	O

(c) Use the classification as a response variable and RESIST as an independent variable and perform three separate binary logistic regressions.

(d) Compare the results in (a) through (c) with the results in Exercise 4.

6. The market research department for a large department store conducted a survey of credit card customers to determine if they thought that buying with a credit card was quicker than paying cash. The customers were from three different metropolitan areas. The results are given in Table 10.27. Use the hierarchical approach to loglinear modeling to determine which model best fits the data. Explain the results.

7. Table 10.28 gives the results of a political poll of registered voters in Florida that indicated the relationship between political party, race, and support for a tax on sugar to be used in restoration of the Everglades in South Florida. Use the hierarchical approach to loglinear modeling to determine which model best fits the data. Explain the results.

8. Miller and Halpern (1982) report data from the Stanford Heart Transplant Program that began in 1967. Table 10.29 gives a sample of the data. The variable STATUS is coded 1 if the patient was reported dead by the end of the study or 0 if still alive; the variable AGE is the age of the patient at transplant. Do a logistic regression to determine the relationship between age and status. Calculate the odds ratio and explain it. If the appropriate computer program is available, fit the probit model and compare the two.

TABLE 10.27

DATA FOR EXERCISE 6

Rating	City 1	City 2	City 3
Easier	62	51	45
Same	28	30	35
Harder	10	19	20

TABLE 10.28

DATA FOR EXERCISE 7

Race	Political Party	Support the Sugar Tax Yes	No
White	Republican	15	125
	Democrat	75	40
	Independent	44	26
Nonwhite	Republican	21	32
	Democrat	66	28
	Independent	36	22

TABLE 10.29

DATA FOR EXERCISE 8

Status	Age	Status	Age
1	54	1	40
1	42	1	51
1	52	1	44
0	50	1	32
1	46	1	41
1	18	1	42
0	46	1	38
0	41	0	41
1	31	1	33
0	50	1	19
0	52	0	34
0	47	1	36
0	24	1	53
0	14	0	18
0	39	1	39
1	34	0	43
0	30	0	46
1	49	1	45
1	48	0	48
0	49	0	19
0	20	0	43
0	41	0	20
1	51	1	51
0	24	0	38
0	27	1	50

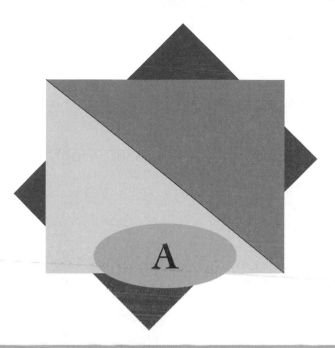

Statistical Tables

TABLE A.1

THE NORMAL DISTRIBUTION—PROBABILITIES EXCEEDING Z

Z	PROB > Z	Z	PROB > Z	Z	PROB > Z	Z	PROB > Z
−3.99	1.0000	−3.49	0.9998	−2.99	0.9986	−2.49	0.9936
−3.98	1.0000	−3.48	0.9997	−2.98	0.9986	−2.48	0.9934
−3.97	1.0000	−3.47	0.9997	−2.97	0.9985	−2.47	0.9932
−3.96	1.0000	−3.46	0.9997	−2.96	0.9985	−2.46	0.9931
−3.95	1.0000	−3.45	0.9997	−2.95	0.9984	−2.45	0.9929
−3.94	1.0000	−3.44	0.9997	−2.94	0.9984	−2.44	0.9927
−3.93	1.0000	−3.43	0.9997	−2.93	0.9983	−2.43	0.9925
−3.92	1.0000	−3.42	0.9997	−2.92	0.9982	−2.42	0.9922
−3.91	1.0000	−3.41	0.9997	−2.91	0.9982	−2.41	0.9920
−3.90	1.0000	−3.40	0.9997	−2.90	0.9981	−2.40	0.9918
−3.89	0.9999	−3.39	0.9997	−2.89	0.9981	−2.39	0.9916
−3.88	0.9999	−3.38	0.9996	−2.88	0.9980	−2.38	0.9913
−3.87	0.9999	−3.37	0.9996	−2.87	0.9979	−2.37	0.9911
−3.86	0.9999	−3.36	0.9996	−2.86	0.9979	−2.36	0.9909
−3.85	0.9999	−3.35	0.9996	−2.85	0.9978	−2.35	0.9906
−3.84	0.9999	−3.34	0.9996	−2.84	0.9977	−2.34	0.9904
−3.83	0.9999	−3.33	0.9996	−2.83	0.9977	−2.33	0.9901
−3.82	0.9999	−3.32	0.9995	−2.82	0.9976	−2.32	0.9898
−3.81	0.9999	−3.31	0.9995	−2.81	0.9975	−2.31	0.9896
−3.80	0.9999	−3.30	0.9995	−2.80	0.9974	−2.30	0.9893
−3.79	0.9999	−3.29	0.9995	−2.79	0.9974	−2.29	0.9890
−3.78	0.9999	−3.28	0.9995	−2.78	0.9973	−2.28	0.9887
−3.77	0.9999	−3.27	0.9995	−2.77	0.9972	−2.27	0.9884
−3.76	0.9999	−3.26	0.9994	−2.76	0.9971	−2.26	0.9881
−3.75	0.9999	−3.25	0.9994	−2.75	0.9970	−2.25	0.9878
−3.74	0.9999	−3.24	0.9994	−2.74	0.9969	−2.24	0.9875
−3.73	0.9999	−3.23	0.9994	−2.73	0.9968	−2.23	0.9871
−3.72	0.9999	−3.22	0.9994	−2.72	0.9967	−2.22	0.9868
−3.71	0.9999	−3.21	0.9993	−2.71	0.9966	−2.21	0.9864
−3.70	0.9999	−3.20	0.9993	−2.70	0.9965	−2.20	0.9861
−3.69	0.9999	−3.19	0.9993	−2.69	0.9964	−2.19	0.9857
−3.68	0.9999	−3.18	0.9993	−2.68	0.9963	−2.18	0.9854
−3.67	0.9999	−3.17	0.9992	−2.67	0.9962	−2.17	0.9850
−3.66	0.9999	−3.16	0.9992	−2.66	0.9961	−2.16	0.9846
−3.65	0.9999	−3.15	0.9992	−2.65	0.9960	−2.15	0.9842
−3.64	0.9999	−3.14	0.9992	−2.64	0.9959	−2.14	0.9838
−3.63	0.9999	−3.13	0.9991	−2.63	0.9957	−2.13	0.9834
−3.62	0.9999	−3.12	0.9991	−2.62	0.9956	−2.12	0.9830
−3.61	0.9998	−3.11	0.9991	−2.61	0.9955	−2.11	0.9826
−3.60	0.9998	−3.10	0.9990	−2.60	0.9953	−2.10	0.9821
−3.59	0.9998	−3.09	0.9990	−2.59	0.9952	−2.09	0.9817
−3.58	0.9998	−3.08	0.9990	−2.58	0.9951	−2.08	0.9812
−3.57	0.9998	−3.07	0.9989	−2.57	0.9949	−2.07	0.9808
−3.56	0.9998	−3.06	0.9989	−2.56	0.9948	−2.06	0.9803
−3.55	0.9998	−3.05	0.9989	−2.55	0.9946	−2.05	0.9798
−3.54	0.9998	−3.04	0.9988	−2.54	0.9945	−2.04	0.9793
−3.53	0.9998	−3.03	0.9988	−2.53	0.9943	−2.03	0.9788
−3.52	0.9998	−3.02	0.9987	−2.52	0.9941	−2.02	0.9783
−3.51	0.9998	−3.01	0.9987	−2.51	0.9940	−2.01	0.9778
−3.50	0.9998	−3.00	0.9987	−2.50	0.9938	−2.00	0.9772

TABLE A.1—Continued

Z	PROB > Z	Z	PROB > Z	Z	PROB > Z	Z	PROB > Z
−1.99	0.9767	−1.49	0.9319	−0.99	0.8389	−0.49	0.6879
−1.98	0.9761	−1.48	0.9306	−0.98	0.8365	−0.48	0.6844
−1.97	0.9756	−1.47	0.9292	−0.97	0.8340	−0.47	0.6808
−1.96	0.9750	−1.46	0.9279	−0.96	0.8315	−0.46	0.6772
−1.95	0.9744	−1.45	0.9265	−0.95	0.8289	−0.45	0.6736
−1.94	0.9738	−1.44	0.9251	−0.94	0.8264	−0.44	0.6700
−1.93	0.9732	−1.43	0.9236	−0.93	0.8238	−0.43	0.6664
−1.92	0.9726	−1.42	0.9222	−0.92	0.8212	−0.42	0.6628
−1.91	0.9719	−1.41	0.9207	−0.91	0.8186	−0.41	0.6591
−1.90	0.9713	−1.40	0.9192	−0.90	0.8159	−0.40	0.6554
−1.89	0.9706	−1.39	0.9177	−0.89	0.8133	−0.39	0.6517
−1.88	0.9699	−1.38	0.9162	−0.88	0.8106	−0.38	0.6480
−1.87	0.9693	−1.37	0.9147	−0.87	0.8078	−0.37	0.6443
−1.86	0.9686	−1.36	0.9131	−0.86	0.8051	−0.36	0.6406
−1.85	0.9678	−1.35	0.9115	−0.85	0.8023	−0.35	0.6368
−1.84	0.9671	−1.34	0.9099	−0.84	0.7995	−0.34	0.6331
−1.83	0.9664	−1.33	0.9082	−0.83	0.7967	−0.33	0.6293
−1.82	0.9656	−1.32	0.9066	−0.82	0.7939	−0.32	0.6255
−1.81	0.9649	−1.31	0.9049	−0.81	0.7910	−0.31	0.6217
−1.80	0.9641	−1.30	0.9032	−0.80	0.7881	−0.30	0.6179
−1.79	0.9633	−1.29	0.9015	−0.79	0.7852	−0.29	0.6141
−1.78	0.9625	−1.28	0.8997	−0.78	0.7823	−0.28	0.6103
−1.77	0.9616	−1.27	0.8980	−0.77	0.7794	−0.27	0.6064
−1.76	0.9608	−1.26	0.8962	−0.76	0.7764	−0.26	0.6026
−1.75	0.9599	−1.25	0.8944	−0.75	0.7734	−0.25	0.5987
−1.74	0.9591	−1.24	0.8925	−0.74	0.7704	−0.24	0.5948
−1.73	0.9582	−1.23	0.8907	−0.73	0.7673	−0.23	0.5910
−1.72	0.9573	−1.22	0.8888	−0.72	0.7642	−0.22	0.5871
−1.71	0.9564	−1.21	0.8869	−0.71	0.7611	−0.21	0.5832
−1.70	0.9554	−1.20	0.8849	−0.70	0.7580	−0.20	0.5793
−1.69	0.9545	−1.19	0.8830	−0.69	0.7549	0.19	0.5753
−1.68	0.9535	−1.18	0.8810	−0.68	0.7517	−0.18	0.5714
−1.67	0.9525	−1.17	0.8790	−0.67	0.7486	−0.17	0.5675
−1.66	0.9515	−1.16	0.8770	−0.66	0.7454	−0.16	0.5636
−1.65	0.9505	−1.15	0.8749	−0.65	0.7422	−0.15	0.5596
−1.64	0.9495	−1.14	0.8729	−0.64	0.7389	−0.14	0.5557
−1.63	0.9484	−1.13	0.8708	−0.63	0.7357	−0.13	0.5517
−1.62	0.9474	−1.12	0.8686	−0.62	0.7324	−0.12	0.5478
−1.61	0.9463	−1.11	0.8665	−0.61	0.7291	−0.11	0.5438
−1.60	0.9452	−1.10	0.8643	−0.60	0.7257	−0.10	0.5398
−1.59	0.9441	−1.09	0.8621	−0.59	0.7224	−0.09	0.5359
−1.58	0.9429	−1.08	0.8599	−0.58	0.7190	−0.08	0.5319
−1.57	0.9418	−1.07	0.8577	−0.57	0.7157	−0.07	0.5279
−1.56	0.9406	−1.06	0.8554	−0.56	0.7123	−0.06	0.5239
−1.55	0.9394	−1.05	0.8531	−0.55	0.7088	−0.05	0.5199
−1.54	0.9382	−1.04	0.8508	−0.54	0.7054	−0.04	0.5160
−1.53	0.9370	−1.03	0.8485	−0.53	0.7019	−0.03	0.5120
−1.52	0.9357	−1.02	0.8461	−0.52	0.6985	−0.02	0.5080
−1.51	0.9345	−1.01	0.8438	−0.51	0.6950	−0.01	0.5040
−1.50	0.9332	−1.00	0.8413	−0.50	0.6915	0.00	0.5000

TABLE A.1—Continued

Z	PROB > Z	Z	PROB > Z	Z	PROB > Z	Z	PROB > Z
0.01	0.4960	0.51	0.3050	1.01	0.1562	1.51	0.0655
0.02	0.4920	0.52	0.3015	1.02	0.1539	1.52	0.0643
0.03	0.4880	0.53	0.2981	1.03	0.1515	1.53	0.0630
0.04	0.4840	0.54	0.2946	1.04	0.1492	1.54	0.0618
0.05	0.4801	0.55	0.2912	1.05	0.1469	1.55	0.0606
0.06	0.4761	0.56	0.2877	1.06	0.1446	1.56	0.0594
0.07	0.4721	0.57	0.2843	1.07	0.1423	1.57	0.0582
0.08	0.4681	0.58	0.2810	1.08	0.1401	1.58	0.0571
0.09	0.4641	0.59	0.2776	1.09	0.1379	1.59	0.0559
0.10	0.4602	0.60	0.2743	1.10	0.1357	1.60	0.0548
0.11	0.4562	0.61	0.2709	1.11	0.1335	1.61	0.0537
0.12	0.4522	0.62	0.2676	1.12	0.1314	1.62	0.0526
0.13	0.4483	0.63	0.2643	1.13	0.1292	1.63	0.0516
0.14	0.4443	0.64	0.2611	1.14	0.1271	1.64	0.0505
0.15	0.4404	0.65	0.2578	1.15	0.1251	1.65	0.0495
0.16	0.4364	0.66	0.2546	1.16	0.1230	1.66	0.0485
0.17	0.4325	0.67	0.2514	1.17	0.1210	1.67	0.0475
0.18	0.4286	0.68	0.2483	1.18	0.1190	1.68	0.0465
0.19	0.4247	0.69	0.2451	1.19	0.1170	1.69	0.0455
0.20	0.4207	0.70	0.2420	1.20	0.1151	1.70	0.0446
0.21	0.4168	0.71	0.2389	1.21	0.1131	1.71	0.0436
0.22	0.4129	0.72	0.2358	1.22	0.1112	1.72	0.0427
0.23	0.4090	0.73	0.2327	1.23	0.1093	1.73	0.0418
0.24	0.4052	0.74	0.2296	1.24	0.1075	1.74	0.0409
0.25	0.4013	0.75	0.2266	1.25	0.1056	1.75	0.0401
0.26	0.3974	0.76	0.2236	1.26	0.1038	1.76	0.0392
0.27	0.3936	0.77	0.2206	1.27	0.1020	1.77	0.0384
0.28	0.3897	0.78	0.2177	1.28	0.1003	1.78	0.0375
0.29	0.3859	0.79	0.2148	1.29	0.0985	1.79	0.0367
0.30	0.3821	0.80	0.2119	1.30	0.0968	1.80	0.0359
0.31	0.3783	0.81	0.2090	1.31	0.0951	1.81	0.0351
0.32	0.3745	0.82	0.2061	1.32	0.0934	1.82	0.0344
0.33	0.3707	0.83	0.2033	1.33	0.0918	1.83	0.0336
0.34	0.3669	0.84	0.2005	1.34	0.0901	1.84	0.0329
0.35	0.3632	0.85	0.1977	1.35	0.0885	1.85	0.0322
0.36	0.3594	0.86	0.1949	1.36	0.0869	1.86	0.0314
0.37	0.3557	0.87	0.1922	1.37	0.0853	1.87	0.0307
0.38	0.3520	0.88	0.1894	1.38	0.0838	1.88	0.0301
0.39	0.3483	0.89	0.1867	1.39	0.0823	1.89	0.0294
0.40	0.3446	0.90	0.1841	1.40	0.0808	1.90	0.0287
0.41	0.3409	0.91	0.1814	1.41	0.0793	1.91	0.0281
0.42	0.3372	0.92	0.1788	1.42	0.0778	1.92	0.0274
0.43	0.3336	0.93	0.1762	1.43	0.0764	1.93	0.0268
0.44	0.3300	0.94	0.1736	1.44	0.0749	1.94	0.0262
0.45	0.3264	0.95	0.1711	1.45	0.0735	1.95	0.0256
0.46	0.3228	0.96	0.1685	1.46	0.0721	1.96	0.0250
0.47	0.3192	0.97	0.1660	1.47	0.0708	1.97	0.0244
0.48	0.3156	0.98	0.1635	1.48	0.0694	1.98	0.0239
0.49	0.3121	0.99	0.1611	1.49	0.0681	1.99	0.0233
0.50	0.3085	1.00	0.1587	1.50	0.0668	2.00	0.0228

TABLE A.1—Continued

Z	PROB > Z	Z	PROB > Z	Z	PROB > Z	Z	PROB > Z
2.01	0.0222	2.51	0.0060	3.01	0.0013	3.51	0.0002
2.02	0.0217	2.52	0.0059	3.02	0.0013	3.52	0.0002
2.03	0.0212	2.53	0.0057	3.03	0.0012	3.53	0.0002
2.04	0.0207	2.54	0.0055	3.04	0.0012	3.54	0.0002
2.05	0.0202	2.55	0.0054	3.05	0.0011	3.55	0.0002
2.06	0.0197	2.56	0.0052	3.06	0.0011	3.56	0.0002
2.07	0.0192	2.57	0.0051	3.07	0.0011	3.57	0.0002
2.08	0.0188	2.58	0.0049	3.08	0.0010	3.58	0.0002
2.09	0.0183	2.59	0.0048	3.09	0.0010	3.59	0.0002
2.10	0.0179	2.60	0.0047	3.10	0.0010	3.60	0.0002
2.11	0.0174	2.61	0.0045	3.11	0.0009	3.61	0.0002
2.12	0.0170	2.62	0.0044	3.12	0.0009	3.62	0.0001
2.13	0.0166	2.63	0.0043	3.13	0.0009	3.63	0.0001
2.14	0.0162	2.64	0.0041	3.14	0.0008	3.64	0.0001
2.15	0.0158	2.65	0.0040	3.15	0.0008	3.65	0.0001
2.16	0.0154	2.66	0.0039	3.16	0.0008	3.66	0.0001
2.17	0.0150	2.67	0.0038	3.17	0.0008	3.67	0.0001
2.18	0.0146	2.68	0.0037	3.18	0.0007	3.68	0.0001
2.19	0.0143	2.69	0.0036	3.19	0.0007	3.69	0.0001
2.20	0.0139	2.70	0.0035	3.20	0.0007	3.70	0.0001
2.21	0.0136	2.71	0.0034	3.21	0.0007	3.71	0.0001
2.22	0.0132	2.72	0.0033	3.22	0.0006	3.72	0.0001
2.23	0.0129	2.73	0.0032	3.23	0.0006	3.73	0.0001
2.24	0.0125	2.74	0.0031	3.24	0.0006	3.74	0.0001
2.25	0.0122	2.75	0.0030	3.25	0.0006	3.75	0.0001
2.26	0.0119	2.76	0.0029	3.26	0.0006	3.76	0.0001
2.27	0.0116	2.77	0.0028	3.27	0.0005	3.77	0.0001
2.28	0.0113	2.78	0.0027	3.28	0.0005	3.78	0.0001
2.29	0.0110	2.79	0.0026	3.29	0.0005	3.79	0.0001
2.30	0.0107	2.80	0.0026	3.30	0.0005	3.80	0.0001
2.31	0.0104	2.81	0.0025	3.31	0.0005	3.81	0.0001
2.32	0.0102	2.82	0.0024	3.32	0.0005	3.82	0.0001
2.33	0.0099	2.83	0.0023	3.33	0.0004	3.83	0.0001
2.34	0.0096	2.84	0.0023	3.34	0.0004	3.84	0.0001
2.35	0.0094	2.85	0.0022	3.35	0.0004	3.85	0.0001
2.36	0.0091	2.86	0.0021	3.36	0.0004	3.86	0.0001
2.37	0.0089	2.87	0.0021	3.37	0.0004	3.87	0.0001
2.38	0.0087	2.88	0.0020	3.38	0.0004	3.88	0.0001
2.39	0.0084	2.89	0.0019	3.39	0.0003	3.89	0.0001
2.40	0.0082	2.90	0.0019	3.40	0.0003	3.90	0.0000
2.41	0.0080	2.91	0.0018	3.41	0.0003	3.91	0.0000
2.42	0.0078	2.92	0.0018	3.42	0.0003	3.92	0.0000
2.43	0.0075	2.93	0.0017	3.43	0.0003	3.93	0.0000
2.44	0.0073	2.94	0.0016	3.44	0.0003	3.94	0.0000
2.45	0.0071	2.95	0.0016	3.45	0.0003	3.95	0.0000
2.46	0.0069	2.96	0.0015	3.46	0.0003	3.96	0.0000
2.47	0.0068	2.97	0.0015	3.47	0.0003	3.97	0.0000
2.48	0.0066	2.98	0.0014	3.48	0.0003	3.98	0.0000
2.49	0.0064	2.99	0.0014	3.49	0.0002	3.99	0.0000
2.50	0.0062	3.00	0.0013	3.50	0.0002	4.00	0.0000

TABLE A.1A

SELECTED PROBABILITY VALUES FOR THE NORMAL DISTRIBUTION
VALUES OF Z EXCEEDED WITH GIVEN PROBABILITY

PROB	Z
0.5000	0.00000
0.4000	0.25335
0.3000	0.52440
0.2000	0.84162
0.1000	1.28155
0.0500	1.64485
0.0250	1.95996
0.0100	2.32635
0.0050	2.57583
0.0020	2.87816
0.0010	3.09023
0.0005	3.29053
0.0001	3.71902

TABLE A.2

THE T DISTRIBUTION—VALUES OF T EXCEEDED WITH GIVEN PROBABILITY

df	P = 0.25	P = 0.10	P = 0.05	P = 0.025	P = 0.01	P = 0.005	P = 0.001	P = 0.0005	df
1	1.0000	3.0777	6.3138	12.706	31.821	63.657	318.31	636.62	1
2	0.8165	1.8856	2.9200	4.3027	6.9646	9.9248	22.327	31.599	2
3	0.7649	1.6377	2.3534	3.1824	4.5407	5.8409	10.215	12.924	3
4	0.7407	1.5332	2.1318	2.7764	3.7469	4.6041	7.1732	8.6103	4
5	0.7267	1.4759	2.0150	2.5706	3.3649	4.0321	5.8934	6.8688	5
6	0.7176	1.4398	1.9432	2.4469	3.1427	3.7074	5.2076	5.9588	6
7	0.7111	1.4149	1.8946	2.3646	2.9980	3.4995	4.7853	5.4079	7
8	0.7064	1.3968	1.8595	2.3060	2.8965	3.3554	4.5008	5.0413	8
9	0.7027	1.3830	1.8331	2.2622	2.8214	3.2498	4.2968	4.7809	9
10	0.6998	1.3722	1.8125	2.2281	2.7638	3.1693	4.1437	4.5869	10
11	0.6974	1.3634	1.7959	2.2010	2.7181	3.1058	4.0247	4.4370	11
12	0.6955	1.3562	1.7823	2.1788	2.6810	3.0545	3.9296	4.3178	12
13	0.6938	1.3502	1.7709	2.1604	2.6503	3.0123	3.8520	4.2208	13
14	0.6924	1.3450	1.7613	2.1448	2.6245	2.9768	3.7874	4.1405	14
15	0.6912	1.3406	1.7531	2.1314	2.6025	2.9467	3.7329	4.0728	15
16	0.6901	1.3368	1.7459	2.1199	2.5835	2.9208	3.6862	4.0150	16
17	0.6892	1.3334	1.7396	2.1098	2.5669	2.8982	3.6458	3.9652	17
18	0.6884	1.3304	1.7341	2.1009	2.5524	2.8784	3.6105	3.9217	18
19	0.6876	1.3277	1.7291	2.0930	2.5395	2.8609	3.5794	3.8834	19
20	0.6870	1.3253	1.7247	2.0860	2.5280	2.8453	3.5518	3.8495	20
21	0.6864	1.3232	1.7207	2.0796	2.5176	2.8314	3.5272	3.8193	21
22	0.6858	1.3212	1.7171	2.0739	2.5083	2.8188	3.5050	3.7922	22
23	0.6853	1.3195	1.7139	2.0687	2.4999	2.8073	3.4850	3.7677	23
24	0.6848	1.3178	1.7109	2.0639	2.4922	2.7969	3.4668	3.7454	24
25	0.6844	1.3163	1.7081	2.0595	2.4851	2.7874	3.4502	3.7252	25
26	0.6840	1.3150	1.7056	2.0555	2.4786	2.7787	3.4350	3.7066	26
27	0.6837	1.3137	1.7033	2.0518	2.4727	2.7707	3.4210	3.6896	27
28	0.6834	1.3125	1.7011	2.0484	2.4671	2.7633	3.4082	3.6739	28
29	0.6830	1.3114	1.6991	2.0452	2.4620	2.7564	3.3963	3.6594	29
30	0.6828	1.3104	1.6973	2.0423	2.4573	2.7500	3.3852	3.6460	30
35	0.6816	1.3062	1.6896	2.0301	2.4377	2.7238	3.3401	3.5912	35
40	0.6807	1.3031	1.6839	2.0211	2.4233	2.7045	3.3069	3.5510	40
45	0.6800	1.3006	1.6794	2.0141	2.4121	2.6896	3.2815	3.5203	45
50	0.6794	1.2987	1.6759	2.0086	2.4033	2.6778	3.2614	3.4960	50
55	0.6790	1.2971	1.6730	2.0040	2.3961	2.6682	3.2452	3.4764	55
60	0.6786	1.2958	1.6706	2.0003	2.3901	2.6603	3.2317	3.4602	60
65	0.6783	1.2947	1.6686	1.9971	2.3851	2.6536	3.2204	3.4466	65
70	0.6780	1.2938	1.6669	1.9944	2.3808	2.6479	3.2108	3.4350	70
75	0.6778	1.2929	1.6654	1.9921	2.3771	2.6430	3.2025	3.4250	75
90	0.6772	1.2910	1.6620	1.9867	2.3685	2.6316	3.1833	3.4019	90
105	0.6768	1.2897	1.6595	1.9828	2.3624	2.6235	3.1697	3.3856	105
120	0.6765	1.2886	1.6577	1.9799	2.3578	2.6174	3.1595	3.3735	120
INF	0.6745	1.2816	1.6449	1.9600	2.3263	2.5758	3.0902	3.2905	INF

TABLE A.3

χ^2 DISTRIBUTION—χ^2 VALUES EXCEEDED WITH GIVEN PROBABILITY

df	0.995	0.99	0.975	0.95	0.90	0.75	0.50	0.25	0.10	0.05	0.025	0.01	0.005
1	0.000	0.000	0.001	0.004	0.016	0.102	0.455	1.323	2.706	3.841	5.024	6.635	7.879
2	0.010	0.020	0.051	0.103	0.211	0.575	1.386	2.773	4.605	5.991	7.378	9.210	10.579
3	0.072	0.115	0.216	0.352	0.584	1.213	2.366	4.108	6.251	7.815	9.348	11.345	12.838
4	0.207	0.297	0.484	0.711	1.064	1.923	3.357	5.385	7.779	9.488	11.143	13.277	14.860
5	0.412	0.554	0.831	1.145	1.610	2.675	4.351	6.626	9.236	11.070	12.833	15.086	16.750
6	0.676	0.872	1.237	1.635	2.204	3.455	5.348	7.841	10.645	12.592	14.449	16.812	18.548
7	0.989	1.239	1.690	2.167	2.833	4.255	6.346	9.037	12.017	14.067	16.013	18.475	20.278
8	1.344	1.646	2.180	2.733	3.490	5.071	7.344	10.219	13.362	15.507	17.535	20.090	21.955
9	1.735	2.088	2.700	3.325	4.168	5.899	8.343	11.389	14.684	16.919	19.023	21.666	23.589
10	2.156	2.558	3.247	3.940	4.865	6.737	9.342	12.549	15.987	18.307	20.483	23.209	25.188
11	2.603	3.053	3.816	4.575	5.578	7.584	10.341	13.701	17.275	19.675	21.920	24.725	26.757
12	3.074	3.571	4.404	5.226	6.304	8.438	11.340	14.845	18.549	21.026	23.337	26.217	28.300
13	3.565	4.107	5.009	5.892	7.042	9.299	12.340	15.984	19.812	22.362	24.736	27.688	29.819
14	4.075	4.660	5.629	6.571	7.790	10.165	13.339	17.117	21.064	23.685	26.119	29.141	31.319
15	4.601	5.229	6.262	7.261	8.547	11.037	14.339	18.245	22.307	24.996	27.488	30.578	32.801
16	5.142	5.812	6.908	7.962	9.312	11.912	15.338	19.369	23.542	26.296	28.845	32.000	34.267
17	5.697	6.408	7.564	8.672	10.085	12.792	16.338	20.489	24.769	27.587	30.191	33.409	35.718
18	6.265	7.015	8.231	9.390	10.865	13.675	17.338	21.605	25.989	28.869	31.526	34.805	37.156
19	6.844	7.633	8.907	10.117	11.651	14.562	18.338	22.718	27.204	30.144	32.852	36.191	38.582
20	7.434	8.260	9.591	10.851	12.443	15.452	19.337	23.828	28.412	31.410	34.170	37.566	39.997
21	8.034	8.897	10.283	11.591	13.240	16.344	20.337	24.935	29.615	32.671	35.479	38.932	41.401
22	8.643	9.542	10.982	12.338	14.041	17.240	21.337	26.039	30.813	33.924	36.781	40.289	42.796
23	9.260	10.196	11.689	13.091	14.848	18.137	22.337	27.141	32.007	35.172	38.076	41.638	44.181
24	9.886	10.856	12.401	13.848	15.659	19.037	23.337	28.241	33.196	36.415	39.364	42.980	45.559
25	10.520	11.524	13.120	14.611	16.473	19.939	24.337	29.339	34.382	37.652	40.646	44.314	46.928
26	11.160	12.198	13.844	15.379	17.292	20.843	25.336	30.435	35.563	38.885	41.923	45.642	48.290
27	11.808	12.879	14.573	16.151	18.114	21.749	26.336	31.528	36.741	40.113	43.195	46.963	49.645
28	12.461	13.565	15.308	16.928	18.939	22.657	27.336	32.620	37.916	41.337	44.461	48.278	50.993
29	13.121	14.256	16.047	17.708	19.768	23.567	28.336	33.711	39.087	42.557	45.722	49.588	52.336
30	13.787	14.953	16.791	18.493	20.599	24.478	29.336	34.800	40.256	43.773	46.979	50.892	53.672
35	17.192	18.509	20.569	22.465	24.797	29.054	34.336	40.223	46.059	49.802	53.203	57.342	60.275
40	20.707	22.164	24.433	26.509	29.051	33.660	39.335	45.616	51.805	55.758	59.342	63.691	66.766
45	24.311	25.901	28.366	30.612	33.350	38.291	44.335	50.985	57.505	61.656	65.410	69.957	73.166
50	27.991	29.707	32.357	34.764	37.689	42.942	49.335	56.334	63.167	67.505	71.420	76.154	79.490
55	31.735	33.570	36.398	38.958	42.060	47.610	54.335	61.665	68.796	73.311	77.380	82.292	85.749
60	35.534	37.485	40.482	43.188	46.459	52.294	59.335	66.981	74.397	79.082	83.298	88.379	91.952
65	39.383	41.444	44.603	47.450	50.883	56.990	64.335	72.285	79.973	84.821	89.177	94.422	98.105
70	43.275	45.442	48.758	51.739	55.329	61.698	69.334	77.577	85.527	90.531	95.023	100.425	104.215
75	47.206	49.475	52.942	56.054	59.795	66.417	74.334	82.858	91.061	96.217	100.839	106.393	110.286
80	51.172	53.540	57.153	60.391	64.278	71.145	79.334	88.130	96.578	101.879	106.629	112.329	116.321
85	55.170	57.634	61.389	64.749	68.777	75.881	84.334	93.394	102.079	107.522	112.393	118.236	122.325
90	59.196	61.754	65.647	69.126	73.291	80.625	89.334	98.650	107.565	113.145	118.136	124.116	128.299
95	63.250	65.898	69.925	73.520	77.818	85.376	94.334	103.899	113.038	118.752	123.858	129.973	134.247
100	67.328	70.065	74.222	77.929	82.358	90.133	99.334	109.141	118.498	124.342	129.561	135.807	140.169

TABLE A.4
THE F DISTRIBUTION p = 0.1

Denominator df	Numerator df										
	1	2	3	4	5	6	7	8	9	10	11
1	39.9	49.5	53.6	55.8	57.2	58.2	58.9	59.4	59.9	60.2	60.5
2	8.53	9.00	9.16	9.24	9.29	9.33	9.35	9.37	9.38	9.39	9.40
3	5.54	5.46	5.39	5.34	5.31	5.28	5.27	5.25	5.24	5.23	5.22
4	4.54	4.32	4.19	4.11	4.05	4.01	3.98	3.95	3.94	3.92	3.91
5	4.06	3.78	3.62	3.52	3.45	3.40	3.37	3.34	3.32	3.30	3.28
6	3.78	3.46	3.29	3.18	3.11	3.05	3.01	2.98	2.96	2.94	2.92
7	3.59	3.26	3.07	2.96	2.88	2.83	2.78	2.75	2.72	2.70	2.68
8	3.46	3.11	2.92	2.81	2.73	2.67	2.62	2.59	2.56	2.54	2.52
9	3.36	3.01	2.81	2.69	2.61	2.55	2.51	2.47	2.44	2.42	2.40
10	3.29	2.92	2.73	2.61	2.52	2.46	2.41	2.38	2.35	2.32	2.30
11	3.23	2.86	2.66	2.54	2.45	2.39	2.34	2.30	2.27	2.25	2.23
12	3.18	2.81	2.61	2.48	2.39	2.33	2.28	2.24	2.21	2.19	2.17
13	3.14	2.76	2.56	2.43	2.35	2.28	2.23	2.20	2.16	2.14	2.12
14	3.10	2.73	2.52	2.39	2.31	2.24	2.19	2.15	2.12	2.10	2.07
15	3.07	2.70	2.49	2.36	2.27	2.21	2.16	2.12	2.09	2.06	2.04
16	3.05	2.67	2.46	2.33	2.24	2.18	2.13	2.09	2.06	2.03	2.01
17	3.03	2.64	2.44	2.31	2.22	2.15	2.10	2.06	2.03	2.00	1.98
18	3.01	2.62	2.42	2.29	2.20	2.13	2.08	2.04	2.00	1.98	1.95
19	2.99	2.61	2.40	2.27	2.18	2.11	2.06	2.02	1.98	1.96	1.93
20	2.97	2.59	2.38	2.25	2.16	2.09	2.04	2.00	1.96	1.94	1.91
21	2.96	2.57	2.36	2.23	2.14	2.08	2.02	1.98	1.95	1.92	1.90
22	2.95	2.56	2.35	2.22	2.13	2.06	2.01	1.97	1.93	1.90	1.88
23	2.94	2.55	2.34	2.21	2.11	2.05	1.99	1.95	1.92	1.89	1.87
24	2.93	2.54	2.33	2.19	2.10	2.04	1.98	1.94	1.91	1.88	1.85
25	2.92	2.53	2.32	2.18	2.09	2.02	1.97	1.93	1.89	1.87	1.84
30	2.88	2.49	2.28	2.14	2.05	1.98	1.93	1.88	1.85	1.82	1.79
35	2.85	2.46	2.25	2.11	2.02	1.95	1.90	1.85	1.82	1.79	1.76
40	2.84	2.44	2.23	2.09	2.00	1.93	1.87	1.83	1.79	1.76	1.74
45	2.82	2.42	2.21	2.07	1.98	1.91	1.85	1.81	1.77	1.74	1.72
50	2.81	2.41	2.20	2.06	1.97	1.90	1.84	1.80	1.76	1.73	1.70
55	2.80	2.40	2.19	2.05	1.95	1.88	1.83	1.78	1.75	1.72	1.69
60	2.79	2.39	2.18	2.04	1.95	1.87	1.82	1.77	1.74	1.71	1.68
75	2.77	2.37	2.16	2.02	1.93	1.85	1.80	1.75	1.72	1.69	1.66
100	2.76	2.36	2.14	2.00	1.91	1.83	1.78	1.73	1.69	1.66	1.64
INF	2.71	2.30	2.08	1.94	1.85	1.77	1.72	1.67	1.63	1.60	1.57

TABLE A.4—Continued

Denominator df	Numerator df										
	12	13	14	15	16	20	24	30	45	60	120
1	60.7	60.9	61.1	61.2	61.3	61.7	62	62.3	62.6	62.8	63.1
2	9.41	9.41	9.42	9.42	9.43	9.44	9.45	9.46	9.47	9.47	9.48
3	5.22	5.21	5.20	5.20	5.20	5.18	5.18	5.17	5.16	5.15	5.14
4	3.90	3.89	3.88	3.87	3.86	3.84	3.83	3.82	3.80	3.79	3.78
5	3.27	3.26	3.25	3.24	3.23	3.21	3.19	3.17	3.15	3.14	3.12
6	2.90	2.89	2.88	2.87	2.86	2.84	2.82	2.80	2.77	2.76	2.74
7	2.67	2.65	2.64	2.63	2.62	2.59	2.58	2.56	2.53	2.51	2.49
8	2.50	2.49	2.48	2.46	2.45	2.42	2.40	2.38	2.35	2.34	2.32
9	2.38	2.36	2.35	2.34	2.33	2.30	2.28	2.25	2.22	2.21	2.18
10	2.28	2.27	2.26	2.24	2.23	2.20	2.18	2.16	2.12	2.11	2.08
11	2.21	2.19	2.18	2.17	2.16	2.12	2.10	2.08	2.04	2.03	2.00
12	2.15	2.13	2.12	2.10	2.09	2.06	2.04	2.01	1.98	1.96	1.93
13	2.10	2.08	2.07	2.05	2.04	2.01	1.98	1.96	1.92	1.90	1.88
14	2.05	2.04	2.02	2.01	2.00	1.96	1.94	1.91	1.88	1.86	1.83
15	2.02	2.00	1.99	1.97	1.96	1.92	1.90	1.87	1.84	1.82	1.79
16	1.99	1.97	1.95	1.94	1.93	1.89	1.87	1.84	1.80	1.78	1.75
17	1.96	1.94	1.93	1.91	1.90	1.86	1.84	1.81	1.77	1.75	1.72
18	1.93	1.92	1.90	1.89	1.87	1.84	1.81	1.78	1.74	1.72	1.69
19	1.91	1.89	1.88	1.86	1.85	1.81	1.79	1.76	1.72	1.70	1.67
20	1.89	1.87	1.86	1.84	1.83	1.79	1.77	1.74	1.70	1.68	1.64
21	1.87	1.86	1.84	1.83	1.81	1.78	1.75	1.72	1.68	1.66	1.62
22	1.86	1.84	1.83	1.81	1.80	1.76	1.73	1.70	1.66	1.64	1.60
23	1.84	1.83	1.81	1.80	1.78	1.74	1.72	1.69	1.64	1.62	1.59
24	1.83	1.81	1.80	1.78	1.77	1.73	1.70	1.67	1.63	1.61	1.57
25	1.82	1.80	1.79	1.77	1.76	1.72	1.69	1.66	1.62	1.59	1.56
30	1.77	1.75	1.74	1.72	1.71	1.67	1.64	1.61	1.56	1.54	1.50
35	1.74	1.72	1.70	1.69	1.67	1.63	1.60	1.57	1.52	1.50	1.46
40	1.71	1.70	1.68	1.66	1.65	1.61	1.57	1.54	1.49	1.47	1.42
45	1.70	1.68	1.66	1.64	1.63	1.58	1.55	1.52	1.47	1.44	1.40
50	1.68	1.66	1.64	1.63	1.61	1.57	1.54	1.50	1.45	1.42	1.38
55	1.67	1.65	1.63	1.61	1.60	1.55	1.52	1.49	1.44	1.41	1.36
60	1.66	1.64	1.62	1.60	1.59	1.54	1.51	1.48	1.42	1.40	1.35
75	1.63	1.61	1.60	1.58	1.57	1.52	1.49	1.45	1.40	1.37	1.32
100	1.61	1.59	1.57	1.56	1.54	1.49	1.46	1.42	1.37	1.34	1.28
INF	1.55	1.52	1.50	1.49	1.47	1.42	1.38	1.34	1.28	1.24	1.17

TABLE A.4A

THE F DISTRIBUTION p = 0.05

Denominator df	Numerator df										
	1	2	3	4	5	6	7	8	9	10	11
1	161	199	216	225	230	234	237	239	241	242	243
2	18.5	19	19.2	19.2	19.3	19.3	19.4	19.4	19.4	19.4	19.4
3	10.1	9.55	9.28	9.12	9.01	8.94	8.89	8.85	8.81	8.79	8.76
4	7.71	6.94	6.59	6.39	6.26	6.16	6.09	6.04	6.00	5.96	5.94
5	6.61	5.79	5.41	5.19	5.05	4.95	4.88	4.82	4.77	4.74	4.70
6	5.99	5.14	4.76	4.53	4.39	4.28	4.21	4.15	4.10	4.06	4.03
7	5.59	4.74	4.35	4.12	3.97	3.87	3.79	3.73	3.68	3.64	3.60
8	5.32	4.46	4.07	3.84	3.69	3.58	3.50	3.44	3.39	3.35	3.31
9	5.12	4.26	3.86	3.63	3.48	3.37	3.29	3.23	3.18	3.14	3.10
10	4.96	4.10	3.71	3.48	3.33	3.22	3.14	3.07	3.02	2.98	2.94
11	4.84	3.98	3.59	3.36	3.20	3.09	3.01	2.95	2.90	2.85	2.82
12	4.75	3.89	3.49	3.26	3.11	3.00	2.91	2.85	2.80	2.75	2.72
13	4.67	3.81	3.41	3.18	3.03	2.92	2.83	2.77	2.71	2.67	2.63
14	4.60	3.74	3.34	3.11	2.96	2.85	2.76	2.70	2.65	2.60	2.57
15	4.54	3.68	3.29	3.06	2.90	2.79	2.71	2.64	2.59	2.54	2.51
16	4.49	3.63	3.24	3.01	2.85	2.74	2.66	2.59	2.54	2.49	2.46
17	4.45	3.59	3.20	2.96	2.81	2.70	2.61	2.55	2.49	2.45	2.41
18	4.41	3.55	3.16	2.93	2.77	2.66	2.58	2.51	2.46	2.41	2.37
19	4.38	3.52	3.13	2.90	2.74	2.63	2.54	2.48	2.42	2.38	2.34
20	4.35	3.49	3.10	2.87	2.71	2.60	2.51	2.45	2.39	2.35	2.31
21	4.32	3.47	3.07	2.84	2.68	2.57	2.49	2.42	2.37	2.32	2.28
22	4.30	3.44	3.05	2.82	2.66	2.55	2.46	2.40	2.34	2.30	2.26
23	4.28	3.42	3.03	2.80	2.64	2.53	2.44	2.37	2.32	2.27	2.24
24	4.26	3.40	3.01	2.78	2.62	2.51	2.42	2.36	2.30	2.25	2.22
25	4.24	3.39	2.99	2.76	2.60	2.49	2.40	2.34	2.28	2.24	2.20
30	4.17	3.32	2.92	2.69	2.53	2.42	2.33	2.27	2.21	2.16	2.13
35	4.12	3.27	2.87	2.64	2.49	2.37	2.29	2.22	2.16	2.11	2.07
40	4.08	3.23	2.84	2.61	2.45	2.34	2.25	2.18	2.12	2.08	2.04
45	4.06	3.20	2.81	2.58	2.42	2.31	2.22	2.15	2.10	2.05	2.01
50	4.03	3.18	2.79	2.56	2.40	2.29	2.20	2.13	2.07	2.03	1.99
55	4.02	3.16	2.77	2.54	2.38	2.27	2.18	2.11	2.06	2.01	1.97
60	4.00	3.15	2.76	2.53	2.37	2.25	2.17	2.10	2.04	1.99	1.95
75	3.97	3.12	2.73	2.49	2.34	2.22	2.13	2.06	2.01	1.96	1.92
100	3.94	3.09	2.70	2.46	2.31	2.19	2.10	2.03	1.97	1.93	1.89
INF	3.84	3.00	2.60	2.37	2.21	2.10	2.01	1.94	1.88	1.83	1.79

TABLE A.4A—Continued

Denominator df	Numerator df										
	12	13	14	15	16	20	24	30	45	60	120
1	244	245	245	246	246	248	249	250	251	252	253
2	19.4	19.4	19.4	19.4	19.4	19.4	19.5	19.5	19.5	19.5	19.5
3	8.74	8.73	8.71	8.70	8.69	8.66	8.64	8.62	8.59	8.57	8.55
4	5.91	5.89	5.87	5.86	5.84	5.80	5.77	5.75	5.71	5.69	5.66
5	4.68	4.66	4.64	4.62	4.60	4.56	4.53	4.50	4.45	4.43	4.40
6	4.00	3.98	3.96	3.94	3.92	3.87	3.84	3.81	3.76	3.74	3.70
7	3.57	3.55	3.53	3.51	3.49	3.44	3.41	3.38	3.33	3.30	3.27
8	3.28	3.26	3.24	3.22	3.20	3.15	3.12	3.08	3.03	3.01	2.97
9	3.07	3.05	3.03	3.01	2.99	2.94	2.90	2.86	2.81	2.79	2.75
10	2.91	2.89	2.86	2.85	2.83	2.77	2.74	2.70	2.65	2.62	2.58
11	2.79	2.76	2.74	2.72	2.70	2.65	2.61	2.57	2.52	2.49	2.45
12	2.69	2.66	2.64	2.62	2.60	2.54	2.51	2.47	2.41	2.38	2.34
13	2.60	2.58	2.55	2.53	2.51	2.46	2.42	2.38	2.33	2.30	2.25
14	2.53	2.51	2.48	2.46	2.44	2.39	2.35	2.31	2.25	2.22	2.18
15	2.48	2.45	2.42	2.40	2.38	2.33	2.29	2.25	2.19	2.16	2.11
16	2.42	2.40	2.37	2.35	2.33	2.28	2.24	2.19	2.14	2.11	2.06
17	2.38	2.35	2.33	2.31	2.29	2.23	2.19	2.15	2.09	2.06	2.01
18	2.34	2.31	2.29	2.27	2.25	2.19	2.15	2.11	2.05	2.02	1.97
19	2.31	2.28	2.26	2.23	2.21	2.16	2.11	2.07	2.01	1.98	1.93
20	2.28	2.25	2.22	2.20	2.18	2.12	2.08	2.04	1.98	1.95	1.90
21	2.25	2.22	2.20	2.18	2.16	2.10	2.05	2.01	1.95	1.92	1.87
22	2.23	2.20	2.17	2.15	2.13	2.07	2.03	1.98	1.92	1.89	1.84
23	2.20	2.18	2.15	2.13	2.11	2.05	2.01	1.96	1.90	1.86	1.81
24	2.18	2.15	2.13	2.11	2.09	2.03	1.98	1.94	1.88	1.84	1.79
25	2.16	2.14	2.11	2.09	2.07	2.01	1.96	1.92	1.86	1.82	1.77
30	2.09	2.06	2.04	2.01	1.99	1.93	1.89	1.84	1.77	1.74	1.68
35	2.04	2.01	1.99	1.96	1.94	1.88	1.83	1.79	1.72	1.68	1.62
40	2.00	1.97	1.95	1.92	1.90	1.84	1.79	1.74	1.67	1.64	1.58
45	1.97	1.94	1.92	1.89	1.87	1.81	1.76	1.71	1.64	1.60	1.54
50	1.95	1.92	1.89	1.87	1.85	1.78	1.74	1.69	1.61	1.58	1.51
55	1.93	1.90	1.88	1.85	1.83	1.76	1.72	1.67	1.59	1.55	1.49
60	1.92	1.89	1.86	1.84	1.82	1.75	1.70	1.65	1.57	1.53	1.47
75	1.88	1.85	1.83	1.80	1.78	1.71	1.66	1.61	1.53	1.49	1.42
100	1.85	1.82	1.79	1.77	1.75	1.68	1.63	1.57	1.49	1.45	1.38
INF	1.75	1.72	1.69	1.67	1.64	1.57	1.52	1.46	1.37	1.32	1.22

TABLE A.4B

THE F DISTRIBUTION p = 0.025

Denominator df	Numerator df										
	1	2	3	4	5	6	7	8	9	10	11
1	648	800	864	900	922	937	948	957	963	969	973
2	38.5	39	39.2	39.2	39.3	39.3	39.4	39.4	39.4	39.4	39.4
3	17.4	16	15.4	15.1	14.9	14.7	14.6	14.5	14.5	14.4	14.4
4	12.2	10.6	9.98	9.60	9.36	9.20	9.07	8.98	8.90	8.84	8.79
5	10	8.43	7.76	7.39	7.15	6.98	6.85	6.76	6.68	6.62	6.57
6	8.81	7.26	6.60	6.23	5.99	5.82	5.70	5.60	5.52	5.46	5.41
7	8.07	6.54	5.89	5.52	5.29	5.12	4.99	4.90	4.82	4.76	4.71
8	7.57	6.06	5.42	5.05	4.82	4.65	4.53	4.43	4.36	4.30	4.24
9	7.21	5.71	5.08	4.72	4.48	4.32	4.20	4.10	4.03	3.96	3.91
10	6.94	5.46	4.83	4.47	4.24	4.07	3.95	3.85	3.78	3.72	3.66
11	6.72	5.26	4.63	4.28	4.04	3.88	3.76	3.66	3.59	3.53	3.47
12	6.55	5.10	4.47	4.12	3.89	3.73	3.61	3.51	3.44	3.37	3.32
13	6.41	4.97	4.35	4.00	3.77	3.60	3.48	3.39	3.31	3.25	3.20
14	6.30	4.86	4.24	3.89	3.66	3.50	3.38	3.29	3.21	3.15	3.09
15	6.20	4.77	4.15	3.80	3.58	3.41	3.29	3.20	3.12	3.06	3.01
16	6.12	4.69	4.08	3.73	3.50	3.34	3.22	3.12	3.05	2.99	2.93
17	6.04	4.62	4.01	3.66	3.44	3.28	3.16	3.06	2.98	2.92	2.87
18	5.98	4.56	3.95	3.61	3.38	3.22	3.10	3.01	2.93	2.87	2.81
19	5.92	4.51	3.90	3.56	3.33	3.17	3.05	2.96	2.88	2.82	2.76
20	5.87	4.46	3.86	3.51	3.29	3.13	3.01	2.91	2.84	2.77	2.72
21	5.83	4.42	3.82	3.48	3.25	3.09	2.97	2.87	2.80	2.73	2.68
22	5.79	4.38	3.78	3.44	3.22	3.05	2.93	2.84	2.76	2.70	2.65
23	5.75	4.35	3.75	3.41	3.18	3.02	2.90	2.81	2.73	2.67	2.62
24	5.72	4.32	3.72	3.38	3.15	2.99	2.87	2.78	2.70	2.64	2.59
25	5.69	4.29	3.69	3.35	3.13	2.97	2.85	2.75	2.68	2.61	2.56
30	5.57	4.18	3.59	3.25	3.03	2.87	2.75	2.65	2.57	2.51	2.46
35	5.48	4.11	3.52	3.18	2.96	2.80	2.68	2.58	2.50	2.44	2.39
40	5.42	4.05	3.46	3.13	2.90	2.74	2.62	2.53	2.45	2.39	2.33
45	5.38	4.01	3.42	3.09	2.86	2.70	2.58	2.49	2.41	2.35	2.29
50	5.34	3.97	3.39	3.05	2.83	2.67	2.55	2.46	2.38	2.32	2.26
55	5.31	3.95	3.36	3.03	2.81	2.65	2.53	2.43	2.36	2.29	2.24
60	5.29	3.93	3.34	3.01	2.79	2.63	2.51	2.41	2.33	2.27	2.22
75	5.23	3.88	3.30	2.96	2.74	2.58	2.46	2.37	2.29	2.22	2.17
100	5.18	3.83	3.25	2.92	2.70	2.54	2.42	2.32	2.24	2.18	2.12
INF	5.02	3.69	3.12	2.79	2.57	2.41	2.29	2.19	2.11	2.05	1.99

TABLE A.4B—Continued

Denominator df	Numerator df										
	12	13	14	15	16	20	24	30	45	60	120
1	977	980	983	985	987	993	997	1001	1007	1010	1014
2	39.4	39.4	39.4	39.4	39.4	39.4	39.5	39.5	39.5	39.5	39.5
3	14.3	14.3	14.3	14.3	14.2	14.2	14.1	14.1	14	14	13.9
4	8.75	8.71	8.68	8.66	8.63	8.56	8.51	8.46	8.39	8.36	8.31
5	6.52	6.49	6.46	6.43	6.40	6.33	6.28	6.23	6.16	6.12	6.07
6	5.37	5.33	5.30	5.27	5.24	5.17	5.12	5.07	4.99	4.96	4.90
7	4.67	4.63	4.60	4.57	4.54	4.47	4.41	4.36	4.29	4.25	4.20
8	4.20	4.16	4.13	4.10	4.08	4.00	3.95	3.89	3.82	3.78	3.73
9	3.87	3.83	3.80	3.77	3.74	3.67	3.61	3.56	3.49	3.45	3.39
10	3.62	3.58	3.55	3.52	3.50	3.42	3.37	3.31	3.24	3.20	3.14
11	3.43	3.39	3.36	3.33	3.30	3.23	3.17	3.12	3.04	3.00	2.94
12	3.28	3.24	3.21	3.18	3.15	3.07	3.02	2.96	2.89	2.85	2.79
13	3.15	3.12	3.08	3.05	3.03	2.95	2.89	2.84	2.76	2.72	2.66
14	3.05	3.01	2.98	2.95	2.92	2.84	2.79	2.73	2.65	2.61	2.55
15	2.96	2.92	2.89	2.86	2.84	2.76	2.70	2.64	2.56	2.52	2.46
16	2.89	2.85	2.82	2.79	2.76	2.68	2.63	2.57	2.49	2.45	2.38
17	2.82	2.79	2.75	2.72	2.70	2.62	2.56	2.50	2.42	2.38	2.32
18	2.77	2.73	2.70	2.67	2.64	2.56	2.50	2.44	2.36	2.32	2.26
19	2.72	2.68	2.65	2.62	2.59	2.51	2.45	2.39	2.31	2.27	2.20
20	2.68	2.64	2.60	2.57	2.55	2.46	2.41	2.35	2.27	2.22	2.16
21	2.64	2.60	2.56	2.53	2.51	2.42	2.37	2.31	2.23	2.18	2.11
22	2.60	2.56	2.53	2.50	2.47	2.39	2.33	2.27	2.19	2.14	2.08
23	2.57	2.53	2.50	2.47	2.44	2.36	2.30	2.24	2.15	2.11	2.04
24	2.54	2.50	2.47	2.44	2.41	2.33	2.27	2.21	2.12	2.08	2.01
25	2.51	2.48	2.44	2.41	2.38	2.30	2.24	2.18	2.10	2.05	1.98
30	2.41	2.37	2.34	2.31	2.28	2.20	2.14	2.07	1.99	1.94	1.87
35	2.34	2.30	2.27	2.23	2.21	2.12	2.06	2.00	1.91	1.86	1.79
40	2.29	2.25	2.21	2.18	2.15	2.07	2.01	1.94	1.85	1.80	1.72
45	2.25	2.21	2.17	2.14	2.11	2.03	1.96	1.90	1.81	1.76	1.68
50	2.22	2.18	2.14	2.11	2.08	1.99	1.93	1.87	1.77	1.72	1.64
55	2.19	2.15	2.11	2.08	2.05	1.97	1.90	1.84	1.74	1.69	1.61
60	2.17	2.13	2.09	2.06	2.03	1.94	1.88	1.82	1.72	1.67	1.58
75	2.12	2.08	2.05	2.01	1.99	1.90	1.83	1.76	1.67	1.61	1.52
100	2.08	2.04	2.00	1.97	1.94	1.85	1.78	1.71	1.61	1.56	1.46
INF	1.94	1.90	1.87	1.83	1.80	1.71	1.64	1.57	1.45	1.39	1.27

TABLE A.4C

THE F DISTRIBUTION p = 0.01

Denominator df	Numerator df										
	1	2	3	4	5	6	7	8	9	10	11
1	4052	5000	5403	5625	5764	5859	5928	5981	6022	6056	6083
2	98.5	99	99.2	99.2	99.3	99.3	99.4	99.4	99.4	99.4	99.4
3	34.1	30.8	29.5	28.7	28.2	27.9	27.7	27.5	27.3	27.2	27.1
4	21.2	18	16.7	16	15.5	15.2	15	14.8	14.7	14.5	14.5
5	16.3	13.3	12.1	11.4	11	10.7	10.5	10.3	10.2	10.1	9.96
6	13.7	10.9	9.78	9.15	8.75	8.47	8.26	8.10	7.98	7.87	7.79
7	12.2	9.55	8.45	7.85	7.46	7.19	6.99	6.84	6.72	6.62	6.54
8	11.3	8.65	7.59	7.01	6.63	6.37	6.18	6.03	5.91	5.81	5.73
9	10.6	8.02	6.99	6.42	6.06	5.80	5.61	5.47	5.35	5.26	5.18
10	10	7.56	6.55	5.99	5.64	5.39	5.20	5.06	4.94	4.85	4.77
11	9.65	7.21	6.22	5.67	5.32	5.07	4.89	4.74	4.63	4.54	4.46
12	9.33	6.93	5.95	5.41	5.06	4.82	4.64	4.50	4.39	4.30	4.22
13	9.07	6.70	5.74	5.21	4.86	4.62	4.44	4.30	4.19	4.10	4.02
14	8.86	6.51	5.56	5.04	4.69	4.46	4.28	4.14	4.03	3.94	3.86
15	8.68	6.36	5.42	4.89	4.56	4.32	4.14	4.00	3.89	3.80	3.73
16	8.53	6.23	5.29	4.77	4.44	4.20	4.03	3.89	3.78	3.69	3.62
17	8.40	6.11	5.19	4.67	4.34	4.10	3.93	3.79	3.68	3.59	3.52
18	8.29	6.01	5.09	4.58	4.25	4.01	3.84	3.71	3.60	3.51	3.43
19	8.18	5.93	5.01	4.50	4.17	3.94	3.77	3.63	3.52	3.43	3.36
20	8.10	5.85	4.94	4.43	4.10	3.87	3.70	3.56	3.46	3.37	3.29
21	8.02	5.78	4.87	4.37	4.04	3.81	3.64	3.51	3.40	3.31	3.24
22	7.95	5.72	4.82	4.31	3.99	3.76	3.59	3.45	3.35	3.26	3.18
23	7.88	5.66	4.76	4.26	3.94	3.71	3.54	3.41	3.30	3.21	3.14
24	7.82	5.61	4.72	4.22	3.90	3.67	3.50	3.36	3.26	3.17	3.09
25	7.77	5.57	4.68	4.18	3.85	3.63	3.46	3.32	3.22	3.13	3.06
30	7.56	5.39	4.51	4.02	3.70	3.47	3.30	3.17	3.07	2.98	2.91
35	7.42	5.27	4.40	3.91	3.59	3.37	3.20	3.07	2.96	2.88	2.80
40	7.31	5.18	4.31	3.83	3.51	3.29	3.12	2.99	2.89	2.80	2.73
45	7.23	5.11	4.25	3.77	3.45	3.23	3.07	2.94	2.83	2.74	2.67
50	7.17	5.06	4.20	3.72	3.41	3.19	3.02	2.89	2.78	2.70	2.63
55	7.12	5.01	4.16	3.68	3.37	3.15	2.98	2.85	2.75	2.66	2.59
60	7.08	4.98	4.13	3.65	3.34	3.12	2.95	2.82	2.72	2.63	2.56
75	6.99	4.90	4.05	3.58	3.27	3.05	2.89	2.76	2.65	2.57	2.49
100	6.90	4.82	3.98	3.51	3.21	2.99	2.82	2.69	2.59	2.50	2.43
INF	6.63	4.61	3.78	3.32	3.02	2.80	2.64	2.51	2.41	2.32	2.25

TABLE A.4C—Continued

Denominator df	Numerator df										
	12	13	14	15	16	20	24	30	45	60	120
1	6106	6126	6143	6157	6170	6209	6235	6261	6296	6313	6339
2	99.4	99.4	99.4	99.4	99.4	99.4	99.5	99.5	99.5	99.5	99.5
3	27.1	27	26.9	26.9	26.8	26.7	26.6	26.5	26.4	26.3	26.2
4	14.4	14.3	14.2	14.2	14.2	14	13.9	13.8	13.7	13.7	13.6
5	9.89	9.82	9.77	9.72	9.68	9.55	9.47	9.38	9.26	9.20	9.11
6	7.72	7.66	7.60	7.56	7.52	7.40	7.31	7.23	7.11	7.06	6.97
7	6.47	6.41	6.36	6.31	6.28	6.16	6.07	5.99	5.88	5.82	5.74
8	5.67	5.61	5.56	5.52	5.48	5.36	5.28	5.20	5.09	5.03	4.95
9	5.11	5.05	5.01	4.96	4.92	4.81	4.73	4.65	4.54	4.48	4.40
10	4.71	4.65	4.60	4.56	4.52	4.41	4.33	4.25	4.14	4.08	4.00
11	4.40	4.34	4.29	4.25	4.21	4.10	4.02	3.94	3.83	3.78	3.69
12	4.16	4.10	4.05	4.01	3.97	3.86	3.78	3.70	3.59	3.54	3.45
13	3.96	3.91	3.86	3.82	3.78	3.66	3.59	3.51	3.40	3.34	3.25
14	3.80	3.75	3.70	3.66	3.62	3.51	3.43	3.35	3.24	3.18	3.09
15	3.67	3.61	3.56	3.52	3.49	3.37	3.29	3.21	3.10	3.05	3.96
16	3.55	3.50	3.45	3.41	3.37	3.26	3.18	3.10	2.99	2.93	2.84
17	3.46	3.40	3.35	3.31	3.27	3.16	3.08	3.00	2.89	2.83	2.75
18	3.37	3.32	3.27	3.23	3.19	3.08	3.00	2.92	2.81	2.75	2.66
19	3.30	3.24	3.19	3.15	3.12	3.00	2.92	2.84	2.73	2.67	2.58
20	3.23	3.18	3.13	3.09	3.05	2.94	2.86	2.78	2.67	2.61	2.52
21	3.17	3.12	3.07	3.03	2.99	2.88	2.80	2.72	2.61	2.55	2.46
22	3.12	3.07	3.02	2.98	2.94	2.83	2.75	2.67	2.55	2.50	2.40
23	3.07	3.02	2.97	2.93	2.89	2.78	2.70	2.62	2.51	2.45	2.35
24	3.03	2.98	2.93	2.89	2.85	2.74	2.66	2.58	2.46	2.40	2.31
25	2.99	2.94	2.89	2.85	2.81	2.70	2.62	2.54	2.42	2.36	2.27
30	2.84	2.79	2.74	2.70	2.66	2.55	2.47	2.39	2.27	2.21	2.11
35	2.74	2.69	2.64	2.60	2.56	2.44	2.36	2.28	2.16	2.10	2.00
40	2.66	2.61	2.56	2.52	2.48	2.37	2.29	2.20	2.08	2.02	1.92
45	2.61	2.55	2.51	2.46	2.43	2.31	2.23	2.14	2.02	1.96	1.85
50	2.56	2.51	2.46	2.42	2.38	2.27	2.18	2.10	1.97	1.91	1.80
55	2.53	2.47	2.42	2.38	2.34	2.23	2.15	2.06	1.94	1.87	1.76
60	2.50	2.44	2.39	2.35	2.31	2.20	2.12	2.03	1.90	1.84	1.73
75	2.43	2.38	2.33	2.29	2.25	2.13	2.05	1.96	1.83	1.76	1.65
100	2.37	2.31	2.27	2.22	2.19	2.07	1.98	1.89	1.76	1.69	1.57
INF	2.18	2.13	2.08	2.04	2.00	1.88	1.79	1.70	1.55	1.47	1.32

TABLE A.4D

THE F DISTRIBUTION p = 0.005

Denominator df	Numerator df										
	1	2	3	4	5	6	7	8	9	10	11
1	6,000	20,000	22,000	22,000	23,000	23,000	24,000	24,000	24,000	24,000	24,000
2	199	199	199	199	199	199	199	199	199	199	199
3	55.6	49.8	47.5	46.2	45.4	44.8	44.4	44.1	43.9	43.7	43.5
4	31.3	26.3	24.3	23.2	22.5	22	21.6	21.4	21.1	21	20.8
5	22.8	18.3	16.5	15.6	14.9	14.5	14.2	14	13.8	13.6	13.5
6	18.6	14.5	12.9	12	11.5	11.1	10.8	10.6	10.4	10.3	10.1
7	16.2	12.4	10.9	10.1	9.52	9.16	8.89	8.68	8.51	8.38	8.27
8	14.7	11	9.60	8.81	8.30	7.95	7.69	7.50	7.34	7.21	7.10
9	13.6	10.1	8.72	7.96	7.47	7.13	6.88	6.69	6.54	6.42	6.31
10	12.8	9.43	8.08	7.34	6.87	6.54	6.30	6.12	5.97	5.85	5.75
11	12.2	8.91	7.60	6.88	6.42	6.10	5.86	5.68	5.54	5.42	5.32
12	11.8	8.51	7.23	6.52	6.07	5.76	5.52	5.35	5.20	5.09	4.99
13	11.4	8.19	6.93	6.23	5.79	5.48	5.25	5.08	4.94	4.82	4.72
14	11.1	7.92	6.68	6.00	5.56	5.26	5.03	4.86	4.72	4.60	4.51
15	10.8	7.70	6.48	5.80	5.37	5.07	4.85	4.67	4.54	4.42	4.33
16	10.6	7.51	6.30	5.64	5.21	4.91	4.69	4.52	4.38	4.27	4.18
17	10.4	7.35	6.16	5.50	5.07	4.78	4.56	4.39	4.25	4.14	4.05
18	10.2	7.21	6.03	5.37	4.96	4.66	4.44	4.28	4.14	4.03	3.94
19	10.1	7.09	5.92	5.27	4.85	4.56	4.34	4.18	4.04	3.93	3.84
20	9.94	6.99	5.82	5.17	4.76	4.47	4.26	4.09	3.96	3.85	3.76
21	9.83	6.89	5.73	5.09	4.68	4.39	4.18	4.01	3.88	3.77	3.68
22	9.73	6.81	5.65	5.02	4.61	4.32	4.11	3.94	3.81	3.70	3.61
23	9.63	6.73	5.58	4.95	4.54	4.26	4.05	3.88	3.75	3.64	3.55
24	9.55	6.66	5.52	4.89	4.49	4.20	3.99	3.83	3.69	3.59	3.50
25	9.48	6.60	5.46	4.84	4.43	4.15	3.94	3.78	3.64	3.54	3.45
30	9.18	6.35	5.24	4.62	4.23	3.95	3.74	3.58	3.45	3.34	3.25
35	8.98	6.19	5.09	4.48	4.09	3.81	3.61	3.45	3.32	3.21	3.12
40	8.83	6.07	4.98	4.37	3.99	3.71	3.51	3.35	3.22	3.12	3.03
45	8.71	5.97	4.89	4.29	3.91	3.64	3.43	3.28	3.15	3.04	2.96
50	8.63	5.90	4.83	4.23	3.85	3.58	3.38	3.22	3.09	2.99	2.90
55	8.55	5.84	4.77	4.18	3.80	3.53	3.33	3.17	3.05	2.94	2.85
60	8.49	5.79	4.73	4.14	3.76	3.49	3.29	3.13	3.01	2.90	2.82
75	8.37	5.69	4.63	4.05	3.67	3.41	3.21	3.05	2.93	2.82	2.74
100	8.24	5.59	4.54	3.96	3.59	3.33	3.13	2.97	2.85	2.74	2.66
INF	7.88	5.30	4.28	3.72	3.35	3.09	2.90	2.74	2.62	2.52	2.43

TABLE A.4D—Continued

Denominator df	Numerator df										
	12	13	14	15	16	20	24	30	45	60	120
1	24,000	25,000	25,000	25,000	25,000	25,000	25,000	25,000	25,000	25,000	25,000
2	199	199	199	199	199	199	199	199	199	199	199
3	43.4	43.3	43.2	43.1	43	42.8	42.6	42.5	42.3	42.1	42
4	20.7	20.6	20.5	20.4	20.4	20.2	20	19.9	19.7	19.6	19.5
5	13.4	13.3	13.2	13.1	13.1	12.9	12.8	12.7	12.5	12.4	12.3
6	10	9.95	9.88	9.81	9.76	9.59	9.47	9.36	9.20	9.12	9.00
7	8.18	8.10	8.03	7.97	7.91	7.75	7.64	7.53	7.38	7.31	7.19
8	7.01	6.94	6.87	6.81	6.76	6.61	6.50	6.40	6.25	6.18	6.06
9	6.23	6.15	6.09	6.03	5.98	5.83	5.73	5.62	5.48	5.41	5.30
10	5.66	5.59	5.53	5.47	5.42	5.27	5.17	5.07	4.93	4.86	4.75
11	5.24	5.16	5.10	5.05	5.00	4.86	4.76	4.65	4.52	4.45	4.34
12	4.91	4.84	4.77	4.72	4.67	4.53	4.43	4.33	4.19	4.12	4.01
13	4.64	4.57	4.51	4.46	4.41	4.27	4.17	4.07	3.94	3.87	3.76
14	4.43	4.36	4.30	4.25	4.20	4.06	3.96	3.86	3.73	3.66	3.55
15	4.25	4.18	4.12	4.07	4.02	3.88	3.79	3.69	3.55	3.48	3.37
16	4.10	4.03	3.97	3.92	3.87	3.73	3.64	3.54	3.40	3.33	3.22
17	3.97	3.90	3.84	3.79	3.75	3.61	3.51	3.41	3.28	3.21	3.10
18	3.86	3.79	3.73	3.68	3.64	3.50	3.40	3.30	3.17	3.10	2.99
19	3.76	3.70	3.64	3.59	3.54	3.40	3.31	3.21	3.07	3.00	2.89
20	3.68	3.61	3.55	3.50	3.46	3.32	3.22	3.12	2.99	2.92	2.81
21	3.60	3.54	3.48	3.43	3.38	3.24	3.15	3.05	2.91	2.84	2.73
22	3.54	3.47	3.41	3.36	3.31	3.18	3.08	2.98	2.84	2.77	2.66
23	3.47	3.41	3.35	3.30	3.25	3.12	3.02	2.92	2.78	2.71	2.60
24	3.42	3.35	3.30	3.25	3.20	3.06	2.97	2.87	2.73	2.66	2.55
25	3.37	3.30	3.25	3.20	3.15	3.01	2.92	2.82	2.68	2.61	2.50
30	3.18	3.11	3.06	3.01	2.96	2.82	2.73	2.63	2.49	2.42	2.30
35	3.05	2.98	2.93	2.88	2.83	2.69	2.60	2.50	2.36	2.28	2.16
40	2.95	2.89	2.83	2.78	2.74	2.60	2.50	2.40	2.26	2.18	2.06
45	2.88	2.82	2.76	2.71	2.66	2.53	2.43	2.33	2.19	2.11	1.99
50	2.82	2.76	2.70	2.65	2.61	2.47	2.37	2.27	2.13	2.05	1.93
55	2.78	2.71	2.66	2.61	2.56	2.42	2.33	2.23	2.08	2.00	1.88
60	2.74	2.68	2.62	2.57	2.53	2.39	2.29	2.19	2.04	1.96	1.83
75	2.66	2.60	2.54	2.49	2.45	2.31	2.21	2.10	1.96	1.88	1.74
100	2.58	2.52	2.46	2.41	2.37	2.23	2.13	2.02	1.87	1.79	1.65
INF	2.36	2.29	2.24	2.19	2.14	2.00	1.90	1.79	1.63	1.53	1.36

TABLE A.5

DURBIN–WATSON TEST BOUNDS[a]

	Level of significance $\alpha = .05$									
	$m = 1$		$m = 2$		$m = 3$		$m = 4$		$m = 5$	
n	D_L	D_U	D_L	D_U	D_L	D_U	D_L	D_U	D_L	D_U
15	1.08	1.36	0.95	1.54	0.82	1.75	0.69	1.97	0.56	2.21
16	1.10	1.37	0.98	1.54	0.86	1.73	0.74	1.93	0.62	2.15
17	1.13	1.38	1.02	1.54	0.90	1.71	0.78	1.90	0.67	2.10
18	1.16	1.39	1.05	1.53	0.93	1.69	0.82	1.87	0.71	2.06
19	1.18	1.40	1.08	1.53	0.97	1.68	0.86	1.85	0.75	2.02
20	1.20	1.41	1.10	1.54	1.00	1.68	0.90	1.83	0.79	1.99
21	1.22	1.42	1.13	1.54	1.03	1.67	0.93	1.81	0.83	1.96
22	1.24	1.43	1.15	1.54	1.05	1.66	0.96	1.80	0.86	1.94
23	1.26	1.44	1.17	1.54	1.08	1.66	0.99	1.79	0.90	1.92
24	1.27	1.45	1.19	1.55	1.10	1.66	1.01	1.78	0.93	1.90
25	1.29	1.45	1.21	1.55	1.12	1.66	1.04	1.77	0.95	1.89
26	1.30	1.46	1.22	1.55	1.14	1.65	1.06	1.76	0.98	1.88
27	1.32	1.47	1.24	1.56	1.16	1.65	1.08	1.76	1.01	1.86
28	1.33	1.48	1.26	1.56	1.18	1.65	1.10	1.75	1.03	1.85
29	1.34	1.48	1.27	1.56	1.20	1.65	1.12	1.74	1.05	1.84
30	1.35	1.49	1.28	1.57	1.21	1.65	1.14	1.74	1.07	1.83
31	1.36	1.50	1.30	1.57	1.23	1.65	1.16	1.74	1.09	1.83
32	1.37	1.50	1.31	1.57	1.24	1.65	1.18	1.73	1.11	1.82
33	1.38	1.51	1.32	1.58	1.26	1.65	1.19	1.73	1.13	1.81
34	1.39	1.51	1.33	1.58	1.27	1.65	1.21	1.73	1.15	1.81
35	1.40	1.52	1.34	1.58	1.28	1.65	1.22	1.73	1.16	1.80
36	1.41	1.52	1.35	1.59	1.29	1.65	1.24	1.73	1.18	1.80
37	1.42	1.53	1.36	1.59	1.31	1.66	1.25	1.72	1.19	1.80
38	1.43	1.54	1.37	1.59	1.32	1.66	1.26	1.72	1.21	1.79
39	1.43	1.54	1.38	1.60	1.33	1.66	1.27	1.72	1.22	1.79
40	1.44	1.54	1.39	1.60	1.34	1.66	1.29	1.72	1.23	1.79
45	1.48	1.57	1.43	1.62	1.38	1.67	1.34	1.72	1.29	1.78
50	1.50	1.59	1.46	1.63	1.42	1.67	1.38	1.72	1.34	1.77
55	1.53	1.60	1.49	1.64	1.45	1.68	1.41	1.72	1.38	1.77
60	1.55	1.62	1.51	1.65	1.48	1.69	1.44	1.73	1.41	1.77
65	1.57	1.63	1.54	1.66	1.50	1.70	1.47	1.73	1.44	1.77
70	1.58	1.64	1.55	1.67	1.52	1.70	1.49	1.74	1.46	1.77
75	1.60	1.65	1.57	1.68	1.54	1.71	1.51	1.74	1.49	1.77
80	1.61	1.66	1.59	1.69	1.56	1.72	1.53	1.74	1.51	1.77
85	1.62	1.67	1.60	1.70	1.57	1.72	1.55	1.75	1.52	1.77
90	1.63	1.68	1.61	1.70	1.59	1.73	1.57	1.75	1.54	1.78
95	1.64	1.69	1.62	1.71	1.60	1.73	1.58	1.75	1.56	1.78
100	1.65	1.69	1.63	1.72	1.61	1.74	1.59	1.76	1.57	1.78

[a]Source: Reprinted, with permission, from J. Durbin and G. S. Watson, "Testing for Serial Correlation in Least Squares Regression. II," *Biometrika* **38** (1951), pp. 159–178.

TABLE A.5—Continued

Level of significance $\alpha = .01$

n	m = 1 D_L	m = 1 D_U	m = 2 D_L	m = 2 D_U	m = 3 D_L	m = 3 D_U	m = 4 D_L	m = 4 D_U	m = 5 D_L	m = 5 D_U
15	0.81	1.07	0.70	1.25	0.59	1.46	0.49	1.70	0.39	1.96
16	0.84	1.09	0.74	1.25	0.63	1.44	0.53	1.66	0.44	1.90
17	0.87	1.10	0.77	1.25	0.67	1.43	0.57	1.63	0.48	1.85
18	0.90	1.12	0.80	1.26	0.71	1.42	0.61	1.60	0.52	1.80
19	0.93	1.13	0.83	1.26	0.74	1.41	0.65	1.58	0.56	1.77
20	0.95	1.15	0.86	1.27	0.77	1.41	0.68	1.57	0.60	1.74
21	0.97	1.16	0.89	1.27	0.80	1.41	0.72	1.55	0.63	1.71
22	1.00	1.17	0.91	1.28	0.83	1.40	0.75	1.54	0.66	1.69
23	1.02	1.19	0.94	1.29	0.86	1.40	0.77	1.53	0.70	1.67
24	1.04	1.20	0.96	1.30	0.88	1.41	0.80	1.53	0.72	1.66
25	1.05	1.21	0.98	1.30	0.90	1.41	0.83	1.52	0.75	1.65
26	1.07	1.22	1.00	1.31	0.93	1.41	0.85	1.52	0.78	1.64
27	1.09	1.23	1.02	1.32	0.95	1.41	0.88	1.51	0.81	1.63
28	1.10	1.24	1.04	1.32	0.97	1.41	0.90	1.51	0.83	1.62
29	1.12	1.25	1.05	1.33	0.99	1.42	0.92	1.51	0.85	1.61
30	1.13	1.26	1.07	1.34	1.01	1.42	0.94	1.51	0.88	1.61
31	1.15	1.27	1.08	1.34	1.02	1.42	0.96	1.51	0.90	1.60
32	1.16	1.28	1.10	1.35	1.04	1.43	0.98	1.51	0.92	1.60
33	1.17	1.29	1.11	1.36	1.05	1.43	1.00	1.51	0.94	1.59
34	1.18	1.30	1.13	1.36	1.07	1.43	1.01	1.51	0.95	1.59
35	1.19	1.31	1.14	1.37	1.08	1.44	1.03	1.51	0.97	1.59
36	1.21	1.32	1.15	1.38	1.10	1.44	1.04	1.51	0.99	1.59
37	1.22	1.32	1.16	1.38	1.11	1.45	1.06	1.51	1.00	1.59
38	1.23	1.33	1.18	1.39	1.12	1.45	1.07	1.52	1.02	1.58
39	1.24	1.34	1.19	1.39	1.14	1.45	1.09	1.52	1.03	1.58
40	1.25	1.34	1.20	1.40	1.15	1.46	1.10	1.52	1.05	1.58
45	1.29	1.38	1.24	1.42	1.20	1.48	1.16	1.53	1.11	1.58
50	1.32	1.40	1.28	1.45	1.24	1.49	1.20	1.54	1.16	1.59
55	1.36	1.43	1.32	1.47	1.28	1.51	1.25	1.55	1.21	1.59
60	1.38	1.45	1.35	1.48	1.32	1.52	1.28	1.56	1.25	1.60
65	1.41	1.47	1.38	1.50	1.35	1.53	1.31	1.57	1.28	1.61
70	1.43	1.49	1.40	1.52	1.37	1.55	1.34	1.58	1.31	1.61
75	1.45	1.50	1.42	1.53	1.39	1.56	1.37	1.59	1.34	1.62
80	1.47	1.52	1.44	1.54	1.42	1.57	1.39	1.60	1.36	1.62
85	1.48	1.53	1.46	1.55	1.43	1.58	1.41	1.60	1.39	1.63
90	1.50	1.54	1.47	1.56	1.45	1.59	1.43	1.61	1.41	1.64
95	1.51	1.55	1.49	1.57	1.47	1.60	1.45	1.62	1.42	1.64
100	1.52	1.56	1.50	1.58	1.48	1.60	1.46	1.63	1.44	1.65

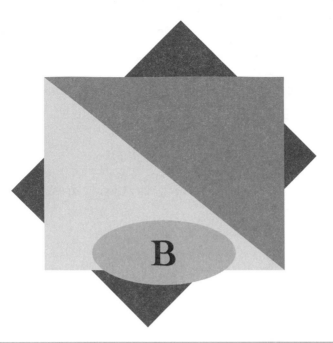

A Brief Introduction to Matrices

Matrix algebra is widely used for mathematical and statistical analysis. The use of the matrix approach is practically a necessity in multiple regression analysis, since it permits extensive systems of equations and large arrays of data to be denoted compactly and operated upon efficiently. This appendix provides a brief introduction to matrix notation and the use of matrices for representing operations involving systems of linear equations. The purpose here is not to provide a manual for performing matrix calculations, but rather to provide for an understanding and appreciation of the various matrix operations as they apply to regression analysis.

DEFINITION   A **matrix** is a rectangular array of elements arranged in rows and columns.

A matrix is much like a table and can be thought of as a multidimensional number. Matrix algebra consists of a set of operations or algebraic rules that allow the manipulation of matrices. In this section we present those operations that will enable the reader to understand the fundamental building blocks of a multiple regression analysis. Additional information is available in a number of texts (such as Graybill, 1983).

The **elements** of a matrix usually consist of numbers or symbols representing numbers. Each element is indexed by its location within the matrix, which is identified by its row and column in that order. For example, the matrix A shown below has 3 rows and 4 columns. The element a_{ij} identifies the element in the ith row and jth column. Thus, the element a_{21} identifies the element in the second row and first column:

$$A = \begin{bmatrix} a_{11} & a_{12} & a_{13} & a_{14} \\ a_{21} & a_{22} & a_{23} & a_{24} \\ a_{31} & a_{32} & a_{33} & a_{34} \end{bmatrix}.$$

The notation for this matrix follows the usual convention of denoting a matrix by a capital letter and its elements by the same lowercase letter with the appropriate row and column subscripts.

An example of a matrix with three rows and columns is

$$B = \begin{bmatrix} 3 & 7 & 9 \\ 1 & 4 & -2 \\ 9 & 15 & 3 \end{bmatrix}.$$

In this matrix, $b_{22} = 4$ and $b_{23} = -2$.

A matrix is characterized by its **order**, which is the number of rows and columns it contains. The matrix B just shown is a 3×3 matrix, since it contains 3 rows and 3 columns. A matrix with equal numbers of rows and columns, such as B, is called a **square matrix**. A 1×1 matrix is known as a **scalar**.

In a matrix, the elements whose row and column indicators are equal, say, a_{ii}, are known as **diagonal elements** and lie on the **main diagonal** of the matrix. For example, in matrix B, the main diagonal consists of the elements $b_{11} = 3$, $b_{22} = 4$, and $b_{33} = 3$.

A matrix that contains nonzero elements only on the main diagonal is a *diagonal matrix*. A diagonal matrix whose nonzero elements are all unity is an *identity* matrix. It has the same function as the scalar 1 in that if a matrix is multiplied by an identity matrix, it is unchanged.

Matrix Algebra

Two matrices A and B are **equal** only if all corresponding elements of A are the same as those of B. Thus, $A = B$ implies $a_{ij} = b_{ij}$ for all i and j. It follows that two equal matrices must be of the same order.

The **transpose** of a matrix A of order $(r \times c)$ is defined as a matrix A' of order $(c \times r)$ such that

$$a'_{ij} = a_{ji}.$$

For example, if

$$A = \begin{bmatrix} 1 & -5 \\ 2 & 2 \\ 4 & 1 \end{bmatrix}, \text{ then } A' = \begin{bmatrix} 1 & 2 & 4 \\ -5 & 2 & 1 \end{bmatrix}.$$

In other words, the rows of A are the columns of A' and vice versa. This is one matrix operation that is not relevant to scalars.

A matrix A for which $A = A'$ is said to be **symmetric**. A symmetric matrix must obviously be square, and each row has the same elements as the corresponding column. For example, the following matrix is symmetric:

$$C = \begin{bmatrix} 5 & 4 & 2 \\ 4 & 6 & 1 \\ 2 & 1 & 8 \end{bmatrix}.$$

The operation of **matrix addition** is defined as follows:

$$A + B = C$$

if $a_{ij} + b_{ij} = c_{ij}$, for all i and j. Thus, addition of matrices is accomplished by the addition of corresponding elements. As an example, let

$$A = \begin{bmatrix} 1 & 2 \\ 4 & 9 \\ -5 & 4 \end{bmatrix} \text{ and } B = \begin{bmatrix} 4 & -2 \\ 1 & 2 \\ 5 & -6 \end{bmatrix}.$$

Then

$$C = A + B = \begin{bmatrix} 5 & 0 \\ 5 & 11 \\ 0 & -2 \end{bmatrix}.$$

In order for two matrices to be added, that is, to be **conformable** for addition, they must have the same order. Subtraction of matrices follows the same rules.

The process of **matrix multiplication** is more complicated. The definition of matrix multiplication is as follows:

$$C = A \cdot B,$$

if

$$c_{ij} = \Sigma_k \, a_{ik} \, b_{kj}.$$

The operation may be better understood when expressed in words:

> The element of the ith row and jth column of the product matrix C (c_{ij}) is the pairwise sum of products of the corresponding elements of the ith row of A and the jth column of B.

In order for A and B to be conformable for multiplication, then, the number of columns of A must be equal to the number of rows of B. The order of the product matrix C will be equal to the number of rows of A by the number of columns of B.

As an example, let

$$A = \begin{bmatrix} 2 & 1 & 6 \\ 4 & 2 & 1 \end{bmatrix} \text{ and } B = \begin{bmatrix} 4 & 1 & -2 \\ 1 & 5 & 4 \\ 1 & 2 & 6 \end{bmatrix}.$$

Note that the matrix A has three columns and that B has three rows; hence, these matrices are conformable for multiplication. Also, since A has two rows and B has three columns, the product matrix C will have two rows and three columns. The elements of $C = AB$ are obtained as follows:

$$c_{11} = a_{11}b_{11} + a_{12}b_{21} + a_{13}b_{31} =$$
$$= (2)(4) + (1)(1) + (6)(1) = 15$$
$$c_{12} = a_{11}b_{12} + a_{12}b_{22} + a_{13}b_{32} =$$
$$= (2)(1) + (1)(5) + (6)(2) = 19$$

$$\cdots$$

$$c_{23} = a_{21}b_{13} + a_{22}b_{23} + a_{23}b_{33} =$$
$$= (4)(-2) + (2)(4) + (1)(6) = 6.$$

The entire matrix C is

$$C = \begin{bmatrix} 15 & 19 & 36 \\ 19 & 16 & 6 \end{bmatrix}.$$

Note that even if A and B are conformable for the multiplication AB, it may not be possible to perform the operation BA. However, even if the matrices are conformable for both operations, usually

$$AB \neq BA,$$

although exceptions occur for special cases.

An interesting corollary of the rules for matrix multiplication is that

$$(AB)' = B'A',$$

that is, the transpose of a product is the product of the individual transposed matrices in reverse order.

There is no matrix division as such. If we require that matrix A is to be "divided" by matrix B, we first obtain the **inverse** (sometimes called reciprocal) of B. Denoting that matrix by C, we then multiply A by C to obtain the desired result.

The inverse of a matrix A, denoted A^{-1}, is defined by the property

$$AA^{-1} = I,$$

where I is the identity matrix which, as defined earlier, has the role of the number "1." Inverses are defined only for square matrices. However, not all square matrices are invertible (see later discussion).

Unfortunately, the definition of the inverse of a matrix does not suggest a procedure for computing it. In fact, the computations required to obtain the inverse of a matrix are quite tedious. Procedures for inverting matrices using hand or desk calculators are available but will not be presented here. Instead, we will always present inverses that have been obtained by a computer.

The following will serve as an illustration of the inverse of a matrix. Consider the two matrices A and B, where $A^{-1} = B$:

$$A = \begin{bmatrix} 9 & 27 & 45 \\ 27 & 93 & 143 \\ 45 & 143 & 245 \end{bmatrix}, B = \begin{bmatrix} 1.47475 & -0.113636 & -0.204545 \\ -0.113636 & 0.113636 & -0.045455 \\ -0.204545 & -0.0454545 & 0.068182 \end{bmatrix}.$$

The fact that B is the inverse of A is verified by multiplying the two matrices. The first element of the product AB is the sum of products of the elements of the first row of A with the elements of the first column of B:

$$(9)(1.47475) + (27)(-0.113636) + (45)(-0.2054545) = 1.000053.$$

This element should be unity; the difference is due to roundoff error, which is a persistent feature of matrix calculations. Most modern computers carry sufficient precision to make roundoff error insignificant, but this is not always guaranteed. The reader is encouraged to verify the correctness of the preceding inverse for at least a few other elements.

Other properties of matrix inverses are as follows:

(1) $AA^{-1} = A^{-1}A$.
(2) If $C = AB$ (all square), then $C^{-1} = B^{-1}A^{-1}$. Note the reversal of the ordering, just as for transposes.
(3) If $B = A^{-1}$, then $B' = (A')^{-1}$.
(4) If A is symmetric, then A^{-1} is also symmetric.
(5) If an inverse exists, it is unique.

Certain matrices do not have inverses; such matrices are called *singular*. For example, the matrix

$$A = \begin{bmatrix} 2 & 1 \\ 4 & 2 \end{bmatrix}$$

cannot be inverted.

Solving Linear Equations

Matrix algebra is of interest in performing regression analyses because it provides a shorthand description for the solution to a set of linear equations. For example, assume we want to solve the following set of equations:

$$5x_1 + 10x_2 + 20x_3 = 40$$

$$14x_1 + 24x_2 + 2x_3 = 12$$

$$5x_1 - 10x_2 = 4.$$

This set of equations can be represented by the matrix equation

$$A \cdot X = B,$$

where

$$A = \begin{bmatrix} 5 & 10 & 20 \\ 14 & 24 & 2 \\ 5 & -10 & 0 \end{bmatrix}, X = \begin{bmatrix} x_1 \\ x_2 \\ x_3 \end{bmatrix}, \text{ and } B = \begin{bmatrix} 40 \\ 12 \\ 4 \end{bmatrix}.$$

The solution to this set of equations can be represented by matrix operations. Premultiply both sides of the matrix equation by A^{-1} as follows:

$$A^{-1} \cdot A \cdot X = A^{-1} \cdot B.$$

Now $A \cdot A^{-1} = I$, the identity matrix; hence, the equation can be written

$$X = A^{-1} \cdot B,$$

which is a matrix equation representing the solution.

We can now see the implications of the singular matrix shown earlier. Using that matrix for the coefficients and adding a right-hand side produces the equations:

$$2x_1 + x_2 = 3$$

$$4x_1 + 2x_2 = 6.$$

Note that these two equations are really equivalent; therefore, any of an infinite number of combination of x_1 and x_2 satisfying the first equation are also a solution to the second equation. On the other hand, changing the right-hand side produces the equations

$$2x_1 + x_2 = 3$$

$$4x_1 + 2x_2 = 10,$$

which are inconsistent and have no solution. In regression applications it is usually not possible to have inconsistent sets of equations.

It must be noted that the matrix operations presented here are but a small subset of the field of knowledge about and uses of matrices. Furthermore, we will not actually be performing many matrix calculations. However, an understanding and appreciation of this material will make more understandable the material in this book.

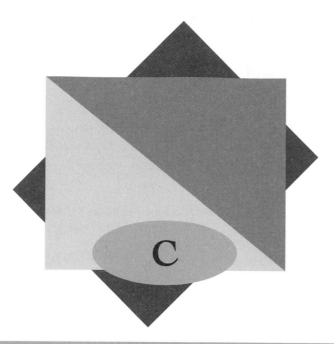

Estimation Procedures

This appendix will discuss two commonly used methods of estimation, the least squares procedure and the maximum likelihood procedure. In many cases the two yield the same estimators and have the same properties. In other cases, they will be different. This appendix is not intended to be a manual for doing these estimation procedures, but rather to provide an understanding and appreciation of the estimation procedures as they apply to regression analysis. Good presentations and discussions of these estimation procedures can be found in many references, including Neter *et al.* (1996), Draper and Smith (1988), and Mendenhall *et al.* (1990).

Least Squares Estimation

Least squares estimation is introduced in Chapter 2 as an alternative method of estimating the mean of a single population and is used throughout the book for estimating parameters in both linear and nonlinear models. In fact, least squares is probably the most often used method of estimating unknown parameters in the general statistical model. The form of the general statistical model is

$$y = f(x_1, \ldots, x_m, \beta_1, \ldots, \beta_p) + \epsilon,$$

where the x_i are independent variables and the β_i are the unknown parameters. The function f constitutes the deterministic portion of the model and the ϵ terms, called random errors, are the stochastic or statistical portion. We are interested in obtaining estimates of the unknown parameters based on a sample of size n $(m+1)$-tuples, $(y_i, x_{1i}, \ldots, x_{mi})$. The procedure minimizes the following sums of squares (hence the name "least squares"):

$$\Sigma \epsilon^2 = \Sigma[(y - f(x_1, \ldots x_m, \beta_1, \ldots, \beta_p)]^2.$$

This quantity is considered a function of the unknown parameters, β_i, and is minimized with respect to them. Depending on the nature of the function, this is often accomplished through calculus.

As an example, let us find the least squares estimate for a single mean, μ, based on a random sample of $y_1, \ldots, y_n$. As in Section 13, we will assume the model:

$$y_i = \mu + e_i, i = 1, \ldots, n.$$

We want to minimize the sums of squares of the errors:

$$\Sigma \epsilon_i^2 = \Sigma(y_i - \mu)^2 = \Sigma(y_i^2 - 2\mu y_i + \mu^2).$$

We will use differential calculus to obtain this minimum. By taking the derivative with respect to μ, we get

$$\frac{d(\Sigma \epsilon_i^2)}{d\mu} = 2\Sigma y_i \quad 2n\mu.$$

Setting equal to zero yields

$$\Sigma y_i = n\hat{\mu}.$$

Note that convention requires the value of the unknown quantity, μ in this case, to be replaced by its estimate, $\hat{\mu}$ in this equation, known as the normal equation.

Solving this equation yields

$$\hat{\mu} = \frac{\Sigma y_i}{n} = \bar{y}.$$

It is easy to show that this estimate results in the minimum sum of squares. This, of course, is the solution given in Section 1.3.

We now find the least squares estimates for the two unknown parameters in the simple linear regression model. We assume a regression model of

$$y_i = \beta_0 + \beta_1 x_i + \epsilon_i, i = 1, \ldots, n.$$

The sums of squares is

$$\Sigma \epsilon_i^2 = \Sigma(y_i - \beta_0 - \beta_1 x_i)^2.$$

To minimize this function, we will use partial derivatives:

$$\frac{\partial(\Sigma \epsilon_i^2)}{\partial \beta_0} = -2\Sigma(y_i - \beta_0 - \beta_1 x_i)$$

$$\frac{\partial(\Sigma \epsilon_i^2)}{\partial \beta_1} = -2\Sigma x_i(y_i - \beta_0 - \beta_1 x_i).$$

Equating these derivatives to zero gives us

$$\Sigma y_i - n\hat{\beta}_0 - \hat{\beta}_1 \Sigma x_i = 0$$
$$\Sigma x_i y_i - \hat{\beta}_0 \Sigma x_i - \hat{\beta}_1 \Sigma x_i^2 = 0.$$

The solutions to these equations are the least squares estimators given in Section 2.3:

$$\hat{\beta}_1 = \frac{\Sigma xy - \dfrac{(\Sigma x)(\Sigma y)}{n}}{\Sigma x^2 - (\Sigma x)^2/n}$$

$$\hat{\beta}_0 = \bar{y} - \hat{\beta}_1 \bar{x}.$$

The general regression model has more than two parameters and is very cumbersome to handle without using matrix notation. Therefore, the least squares estimates can be best obtained using matrix calculus. Since this topic is beyond the scope of this book, we will simply give the results, in matrix form. The general regression model is written in matrix form in Section 3.3 as

$$Y = XB + E,$$

where Y is an $n \times 1$ matrix of observed values, X is an $n \times (m+1)$ matrix of independent variables, B is an $(m+1) \times 1$ matrix of the unknown parameters, and E is an $n \times 1$ matrix of error terms. The sums of squares to be minimized is written in matrix form as

$$E'E = (Y - XB)'(Y - XB) = Y'Y - B'X'Y + B'X'XB.$$

To minimize this function, we take the derivative with respect to the matrix B and get the following:

$$\frac{\partial(E'E)}{\partial B} = -2X'Y + 2X'XB.$$

Equating to zero yields the matrix form of the normal equations given in Section 3.3:

$$(X'X)\hat{B} = X'Y.$$

The solutions to this matrix equation are

$$\hat{B} = (X'X)^{-1}X'Y.$$

Maximum Likelihood Estimation

The maximum likelihood estimation procedure is one of several estimation procedures that use the underlying probability distribution of the random variable. For example, in our earlier illustration from Chapter 2, we considered the variable y as having a normal distribution with mean μ and standard deviation σ. The maximum likelihood procedure maximizes what is called the likelihood function. Suppose we sample from a population with one unknown parameter θ. The probability distribution of that population is denoted by $f(y;\theta)$. If we consider a sample of size n as n independent realizations of the random variable, y, then the likelihood function of the sample is simply the joint distribution of the y_1, y_2, ..., y_n, denoted as

$$L(\theta) = \Pi f(y;\theta).$$

Note that the likelihood can be expressed as a function of the parameter θ.

As an illustration of the logic behind the maximum likelihood method, consider the following example. Suppose we have a box that contains three balls, the colors of which we do not know. We do know that there are either one or two red balls in the box, and we would like to estimate the number of red balls in the box. We sample one ball from the box and observe that it is red. We replace the ball and randomly draw another and observe that it is red also.

Obviously, at least one ball is red. If only one of the balls in the box is red and the others are some other color, then the probability of drawing a red ball on one try is 1/3. The probability of getting two red balls is then (1/3)(1/3) = 1/9. If two of the balls in the box are red and the other is some other color, the probability of drawing a red ball on one try is 2/3. The probability of getting two red balls is then (2/3)(2/3) = 4/9. It should seem reasonable to choose two as our estimate of the number of red balls because that estimate maximizes the probability of the observed sample. Of course, it is possible to have only one red ball in the box, but the observed outcome gives more credence to two.

Returning to our example from Chapter 2, we consider a sample of size n from a normal distribution with mean μ and known standard deviation σ. The form of the probability distribution is

$$f(y;\mu) = \frac{1}{\sigma\sqrt{2\pi}} e^{-(y-\mu)^2/2\sigma^2}.$$

The likelihood function is then

$$L(\mu) = \frac{1}{(\sigma\sqrt{2\pi})^n} e^{-\Sigma(y_i-\mu)^2/2\sigma^2}.$$

Notice that we express the likelihood as a function of the unknown parameter μ only, since we know the value of σ. To maximize the likelihood, we take advantage of the fact that the optimum of this function occurs at the same places as the natural log of the function. So taking the log of the likelihood function gives us

$$\log(L) = -\frac{n}{2}\log(\sigma^2) - \frac{n}{2}\log(2\pi) - \frac{\Sigma(y_i - \mu)^2}{2\sigma^2}.$$

To obtain the maximum likelihood estimate of the unknown parameter μ, we use calculus. Taking the derivative with respect to μ gives us

$$\frac{d\log(L)}{d\mu} = \frac{\Sigma(y_i - \mu)}{\sigma^2}.$$

Equating to zero gives us

$$\frac{\Sigma(y_i - \mu)}{\sigma^2} = 0$$

$$\Sigma y_i - n\hat{\mu} = 0$$

$$\hat{\mu} = \frac{\Sigma y_i}{n} = \bar{y}.$$

This is the same estimate we obtained using least squares.

We use the same procedure to obtain the maximum likelihood estimator for β_0 and β_1 in the simple linear regression model with normal error terms. The likelihood now has three unknown parameters, β_0, β_1, and σ^2, and is given by

$$L(\beta_0, \beta_1, \sigma^2) = \frac{1}{(\sigma\sqrt{2\pi})^n} e^{-\Sigma(y_i - \beta_0 - \beta_1 x_i)^2/2\sigma^2}.$$

We again take advantage of the correspondence between the function and the natural log of the function and maximize the following equation:

$$\log(L) = -\frac{n}{2}\log(\sigma^2) - \frac{n}{2}\log(2\pi) - \frac{\Sigma(y_i - \beta_0 - \beta_1 x_i)^2}{2\sigma^2}.$$

Taking partial derivatives with respect to the parameters gives

$$\frac{\partial \log(L)}{\partial \beta_0} = \frac{1}{\sigma^2}\Sigma(y_i - \beta_0 - \beta_1 x_i)$$

$$\frac{\partial \log(L)}{\partial \beta_1} = \frac{1}{\sigma^2}\Sigma x_i(y_i - \beta_0 - \beta_1 x_i)$$

$$\frac{\partial \log(L)}{\partial \sigma^2} = -\frac{n}{2\sigma^2} + \frac{1}{2\sigma^4}\Sigma(y_i - \beta_0 - \beta_1 x_i)^2.$$

Equating to zero and simplifying yields

$$\Sigma y_i - n\hat{\beta}_0 - \hat{\beta}_1 \Sigma x_i = 0$$

$$\Sigma x_i y_i - \hat{\beta}_0 \Sigma x_i - \hat{\beta}_1 \Sigma x_i^2 = 0$$

$$\hat{\sigma}^2 = \frac{1}{n}\Sigma(y_i - \hat{\beta}_0 - \hat{\beta}_1 x_i)^2.$$

Notice that these are exactly the same estimates for β_0 and β_1 as we obtained using least squares. The result is exactly the same for the multiple regression equation. The maximum likelihood estimates and the least squares estimates for the coefficients are identical for the regression model as long as the assumption of normality holds. Note that MSE $= \frac{n}{n-2}\hat{\sigma}^2$, therefore the maximum likelihood estimate differs from the least squares estimate by only a constant.

REFERENCES

Afifi, A. A., and Azen, S. P. (1979). *Statistical analysis: a computer oriented approach.* 2nd ed. Academic Press, New York.

Agresti, A. (1984). *Analysis of ordinal categorical data.* Wiley, New York.

Agresti, A. (1990). *Categorical data analysis.* Wiley, New York.

Aylward, G. Pl, Harcher, R. P., Leavitt, L. A., Rao, V., Bauer, C. R., Brennan, M. J. and Gustafson, N. F. (1984). Factors affecting neo-behavioral responses of preterm infants at term conceptual age. *Child Development*, 55, 1155–1165.

Barabba, V. P. ed. (1979). *State and metropolitan data book.* U.S. Census Bureau, Department of Commerce, Washington.

Begg, C. B., and Gray, R.(1984). Calculation of polytomous logistic regression parameters using individualized regressions. *Biometrika* 71, 11–18.

Belsley, D. A., Kuh, E., and Welsch, R. E. (1980). *Regression diagnostics.* Wiley. New York.

Bishop, Y. M. M., Fienberg, S.E., and Holland, P. W. (1975). *Discrete multivariate analysis.* MIT Press, Cambridge, Mass.

Box, G. E. P. and Cox, D. R. (1964). An analysis of transformations. *J. Roy. Statist. Soc.,*B-26, 211–243, discussion 244–252.

Central Bank of Barbados (1994). *1994 Annual statistical digest.* Central Bank of Barbados, Bridgetown, Barbados.

Dickens, J. W. and Mason, D. D. (1962). A peanut sheller for grading samples: an application in statistical design. *Transactions of the ASAE:* Volume 5, Number 1, 42–45.

Draper, N. R., and Smith, H. (1988). *Applied regression analysis,* 2nd ed. Wiley, New York.

Drysdale, C. V., and Calef, W. C. (1977). *Energetics of the United States.* Brookhaven National Laboratory, Upton, NY.

Finney, D.J. (1971). *Probit analysis,* 3rd ed. Cambridge University Press, Cambridge.

Fogiel, M. (1978). *The Statistics problem solver.* Research and Education Association, New York.

Freund, R. J. (1980). The case of the missing cell. *The American Statistician,* 34, 94–98.

Freund, R. J.,and Littell, R. C. (1991). *SAS® system for regression,* 2nd ed. SAS Institute Inc, Cary, NC.

Freund, R. J., and Minton, P. D. (1979). *Regression methods.* Marcel Dekker, New York.

Freund, R. J., and Wilson, W. J. (1993). *Statistical methods.* Academic Press, San Diego.

Freund, R. J., and Wilson, W. J. (1997). *Statistical methods,* Revised edition. Academic Press, San Diego.

Fuller, W. A. (1996). *Introduction to statistical time series,* 2nd ed. Wiley, New York.

Gallant, A. R., and Goebel, J .J. (1976). Nonlinear regression with autoregressive errors. *JASA,* 71, 961– 967.

Graybill, F. A. (1983). *Matrices with applications in statistics,* 2nd ed. Wadsworth, Pacific Grove, CA.

Green, J. A. (1988). Loglinear analysis of cross-classified ordinal data: applications in developmental research. *Child Development* 59, 1–25.

Grizzle, J .E., Starmer, C. F., and Koch, G. G. (1969). Analysis of categorical data by linear models. *Biometrics,* 25, 489–504.

Hamilton, T. R. and Rubinoff, I. (1963). Species abundance: natural regulation of insular abundance. *Science* 142(3599), 1575–1577.

Hosmer, D. W., and Lemeshow, S. (1989). *Applied logistic regression.* Wiley, New York.

Johnson, R. A. and Wichern, D. W. (1988). *Applied multivariate statistical analysis,* 2nd ed. Prentice Hall, Englewood Cliffs, NJ.

Kleinbaum, D. G., Kupper, L. L., and Muller, K. E. (1988). *Applied regression analysis and other multivariable methods,* 2nd. ed. PWS-Kent, Boston.

Li, C. C. (1975). *Path analysis: a primer.* Boxwood Press, Pacific Grove, CA.

Littell, R. C., Freund, R. J., and Spector, P. (1991). *SAS® for linear models,* 3rd ed. SAS Institute Inc, Cary, NC.

Long, J. S. (1997). *Regression models for categorical and limited dependent variables.* Sage Publications, Thousand Oaks.

Mallows, C. L. (1973). Some comments on C_p. *Technometrics,* 15, 661–675.

Mendenhall, W., Wackerly, D. D. and Scheaffer, R. L. (1990). *Mathematical statistics with applications,* 4th ed. Duxbury, Belmont, CA.

Miller, R. G. and Halpern, J. W. (1982). Regression with censored data. *Biometrika* 69, 521–531.

Montgomery, D. C. and Peck, E. A. (1982). *Introduction to linear regression analysis.* Wiley, New York.

Montgomery, D. C. (1997). *Design and analysis of experiments,* 4th ed. Wiley, New York.

Myers, R. (1986). *Classical and modern regression with applications.* Duxbury, Boston.

Myers, R. (1990). *Classical and modern regression with applications,* 2nd ed. PWS-Kent, Boston.

Neter, J., Wasserman, W., and Kutner, M. H. (1989). *Applied linear statistical models,* 2nd ed. Irwin, Homewood Ill.

Neter, J., Kutner, M. H., Wasserman, W., and Nachtsheim, W. (1996). *Applied linear statistical models.* Irwin, Chicago.

Ostle B., and Mensing, R. W. (1975). *Statistics in research,* 3rd ed. Iowa State University Press, Ames, IA.

Rawlings, J. (1988). *Applied regression analysis: A research tool.* Wadsworth, Belmont, CA.

Reaven, G. M. and Miller, R. G. (1979). An attempt to define the nature of chemical diabetes using a multidimensional analysis. *Diabetologia* 16, 17–24.

SAS Institute Inc. (1989). SAS/STAT® *User's Guide, Version 6, Fourth Edition, Volume 1.* SAS Institute Inc, Cary, NC.

SAS Institute Inc. (1989). SAS/STAT® *User's Guide, Version 6, Fourth Edition, Volume 2.* SAS Insitute Inc, Cary, NC.

Seber, G. A. F. (1977). *Linear regression analysis.* Wiley, New York.

Smith, P. L. (1979). Splines as a useful and convenient statistical tool. *The American Statistician* 33, 57–62.

Upton, G. J. G. (1978). *The analysis of cross-tabulated data.* Wiley, New York.

U.S. Bureau of the Census (1986). *State and metropolitan area data book: a statistical abstract supplement.* U.S. Department of Commerce, Washington.

U.S. Bureau of the Census (1988). *Statistical abstract of the United States.* U.S. Department of Commerce, Washington.

U.S. Bureau of the Census (1995). *Statistical abstract of the United States.* U.S. Department of Commerce, Washington.

Van der Leeden, Frits (1990). *The Water encyclopedia,* 2nd ed., Lewis Publications, Chelsea, Mich.

The World almanac and book of facts (1980). Press Pub. Co, New York.

Wright, R. L. and Wilson, S. R. (1979). On the analysis of soil variability, with an example from Spain. *Geoderma* 22, 297–313.

INDEX